Combined Pile-Raft Foundations

This book presents the fundamental features of the design and performance of combined pile-raft foundations (CPRFs). Whereas in a traditional foundation the loads are carried either by the raft or by the piles, the capacity of CPRFs is assessed for the foundation as a whole, reducing total and differential settlements economically.

The five chapters provide an overview of the historical development of piled rafts in practice and research, and of the design concepts developed for piled rafts over the last decades. Fundamental aspects of their bearing behaviour are presented, as well as an overview of the framework of the design process for CPRFs, including the safety concept, the design approach summarised in the *ISSMGE Combined Pile-Raft Foundation Guideline* (ISSMGE TC 212 2013) and the interaction between structural and geotechnical engineering. For numerical analysis based on the finite element method, guidance is given on creating the model and performing the calculations before providing basic information on the requirements for the site investigation, supervision of the construction process and monitoring of the foundation performance. Detailed case studies illustrate the design and performance of CPRFs, and a design example for the foundation of a multi-storey office building founded in non-cohesive soil is investigated, carrying out 3D finite element analysis to estimate deformations and design parameters for structural engineering.

Based on the combined experience of the authors obtained in the last decades working in the industry and research, the book particularly suits consulting engineers engaged in foundation engineering, as well as graduate students and researchers interested in the bearing behaviour of piled rafts and pile groups.

Applied Geotechnics series
William Powrie (ed.)

Particulate Discrete Element Modelling
Catherine O'Sullivan

Groundwater Lowering in Construction
Martin Preene et al.

Practical Rock Mechanics
Steve Hencher

Soil Liquefaction, 2nd ed
Mike Jefferies et al.

Drystone Retaining Walls: Design, Construction and Assessment
Paul McCombie et al.

Fundamentals of Shield Tunnelling
Zixin Zhang et al.

Centrifuge Modelling in Geotechnics
Christoph Gaudin et al.

Weak Rock Engineering Geology and Geotechnics
Kevin Stone et al.

Introduction to Tunnel Construction, 2nd ed
David Chapman et al.

Soil Nailing: A Practical Guide
Raymond Cheung et al.

High Resolution Pressuremeters and Geotechnical Engineering
John Hughes et al.

Practical Engineering Geology, 2nd ed
Steve Hencher

Combined Pile-Raft Foundations: Design and Practice
Oliver Reul and Mark Randolph

For more information about this series, please visit: https://www.routledge.com/Applied-Geotechnics/book-series/APPGEOT

Combined Pile-Raft Foundations
Design and Practice

Oliver Reul
Mark Randolph

CRC Press
Taylor & Francis Group
Boca Raton London New York

CRC Press is an imprint of the
Taylor & Francis Group, an **informa** business

Cover image: Oliver Reul and Mark Randolph

First edition published 2025
by CRC Press
4 Park Square, Milton Park, Abingdon, Oxon, OX14 4RN

and by CRC Press
2385 NW Executive Center Drive, Suite 320, Boca Raton FL 33431

© 2025 Oliver Reul and Mark Randolph

CRC Press is an imprint of Informa UK Limited

The right of Oliver Reul and Mark Randolph to be identified as authors of this work has been asserted in accordance with sections 77 and 78 of the Copyright, Designs and Patents Act 1988.

All rights reserved. No part of this book may be reprinted or reproduced or utilised in any form or by any electronic, mechanical, or other means, now known or hereafter invented, including photocopying and recording, or in any information storage or retrieval system, without permission in writing from the publishers.

For permission to photocopy or use material electronically from this work, access www.copyright.com or contact the Copyright Clearance Center, Inc. (CCC), 222 Rosewood Drive, Danvers, MA 01923, 978-750-8400. For works that are not available on CCC please contact mpkbookspermissions@tandf.co.uk

Trademark notice: Product or corporate names may be trademarks or registered trademarks, and are used only for identification and explanation without intent to infringe.

ISBN: 978-1-032-15550-0 (hbk)
ISBN: 978-1-032-15551-7 (pbk)
ISBN: 978-1-003-24464-6 (ebk)

DOI: 10.1201/9781003244646

Typeset in Sabon
by SPi Technologies India Pvt Ltd (Straive)

Contents

About the authors ix

1 Introduction 1

1.1 Combined pile-raft foundation 1
1.2 Definitions 1
    1.2.1 Foundation systems 1
    1.2.2 Parameters for the quantification of the bearing behaviour of piled rafts 2
1.3 Early developments: applications and research 6
1.4 Early developments: design concepts 15
    1.4.1 General remarks 15
    1.4.2 Conventional design 16
    1.4.3 Modified conventional design 16
    1.4.4 "Creep pile" design 16
    1.4.5 Japanese design approach 16
    1.4.6 Compensated piled raft foundations 17
    1.4.7 Raft-enhanced pile groups vs. pile-enhanced rafts 17
References 18

2 Bearing behaviour of piled rafts 22

2.1 General remarks 22
2.2 Piled rafts subjected to monotonic vertical loading 22
    2.2.1 Numerical study of the bearing behaviour of vertically loaded piled rafts 22
    2.2.2 Pile-pile and pile-raft interaction 31
    2.2.3 Stiffness 39
    2.2.4 Settlements 41
    2.2.5 Efficient reduction of settlements 44
    2.2.6 Bending moments 49

           2.2.7   Ultimate capacity under vertical loading 51
           2.2.8   Example design optimisation 57
    2.3 Long-term bearing behaviour of piled rafts considering
         consolidation and creep 61
           2.3.1   Overview 61
           2.3.2   Influence of consolidation on the bearing
                   behaviour of piled rafts 63
           2.3.3   Messeturm, Frankfurt – long-term bearing
                   behaviour of a piled raft in overconsolidated clay 68
    2.4 Other loading conditions 81
           2.4.1   Piled rafts subjected to monotonic vertical and
                   lateral loading 81
           2.4.2   Piled rafts subjected to periodic
                   and dynamic loading 84
    References 93

3 Design, construction and monitoring of CPRF 101

    3.1 Limit state approach and technical regulations for the
         design process 101
           3.1.1   Overall factor of safety 101
           3.1.2   Limit state design 101
           3.1.3   Combined pile-raft foundation (CPRF) according
                   to ISSMGE technical committee TC 212 (2013) 105
           3.1.4   Loads 109
           3.1.5   Design process – interaction between structural
                   and geotechnical engineering 109
    3.2 Analysis methods 110
           3.2.1   General remarks 110
           3.2.2   Analysis methods for investigating the SLS 111
           3.2.3   Analysis methods for investigating the ULS 125
    3.3 Requirements for the site investigation 126
    3.4 Construction 128
           3.4.1   Construction of CPRF according to the CPRF
                   guideline (ISSMGE TC 212 2013) 128
           3.4.2   Pile types 128
           3.4.3   Pile integrity tests 128
    3.5 Monitoring 129
           3.5.1   Monitoring of CPRF according to the CPRF
                   guideline (ISSMGE TC 212 2013) 129
           3.5.2   Measurement devices 130
           3.5.3   Pile load tests 135

3.6 Suggestions for the design of CPRF under mainly vertically loading 137
References 138

4 Case histories   141

4.1 General remarks 141
4.2 Early developments in overconsolidated London Clay and Frankfurt Formation 141
    4.2.1 Overconsolidated London Clay 141
    4.2.2 Overconsolidated Frankfurt Formation 143
4.3 Foundations designed as CPRFs 147
    4.3.1 WestendDuo, Frankfurt am Main 147
    4.3.2 Park Tower, Frankfurt am Main 152
    4.3.3 Omniturm, Frankfurt am Main 163
    4.3.4 Neue Messehalle 3, Frankfurt am Main 171
    4.3.5 Haus der Wirtschaft, Offenbach 178
    4.3.6 Bahntower, Berlin 185
    4.3.7 Treptowers, Berlin 190
    4.3.8 Weser Tower, Bremen 195
    4.3.9 Hegau-Tower, Singen 199
References 203

5 Design example   208

5.1 General remarks 208
5.2 Building 208
5.3 Subsoil conditions 209
5.4 Foundation configurations 210
    5.4.1 General remarks 210
    5.4.2 Raft foundation 211
    5.4.3 Pile foundation (piled raft – conventional design) 211
    5.4.4 CPRF 214
5.5 3D FEA 215
    5.5.1 3D FEA – model 215
    5.5.2 3D FEA – results 218
    5.5.3 Proof of the external bearing capacity for the ULS 221
    5.5.4 Proof of the external serviceability for the SLS 221

5.6 *Structural analysis of raft and piles* 223
    5.6.1 *Structural analysis of raft and piles – model* 223
    5.6.2 *Comparison of the structural analysis model and 3D FEA* 225
    5.6.3 *Structural design* 226
References 229

Appendix   231
Index   251

About the authors

Oliver Reul is Professor at the University of Kassel, Germany and heads the Department of Geotechnical Engineering. As a consulting engineer and accredited proof engineer for geotechnical engineering, he has been involved in the design and construction of numerous combined pile-raft foundations.

Mark Randolph is Emeritus Professor and Honorary Fellow in the Centre for Offshore Foundation Systems at the University of Western Australia and a consultant with Fugro Australia Pty Ltd in Perth. He is widely published, including two previous books: *Piling Engineering* and *Offshore Geotechnical Engineering*, and over 300 journal articles. He was appointed an Officer of the Order of Australia in 2021.

Chapter 1

Introduction

1.1 COMBINED PILE-RAFT FOUNDATION

In the scope of this book a combined pile-raft foundation (CPRF) is considered to be a foundation where the capacity of the foundation is assessed for the foundation as a whole, i.e. piles and raft are not considered separately with respect to capacity and no separate factors of safety are applied for these two components of the foundation. This approach is the underlying basis of the German "KPP-Richtlinie" (DGGT 2001), a guideline for the design, dimensioning and construction of piled rafts (CPRF; in German: Kombinierte Pfahl-Plattengründung/KPP) which was developed taking into account the experiences gained with piled rafts in Frankfurt in the 1980s and 1990s (Section 1.3). The design philosophy of the German "KPP-Richtlinie" has been adopted more or less one-to-one in the ISSMGE Combined Pile-Raft Foundation Guideline (ISSMGE TC 212 2013) which will be discussed in detail in Chapter 3.

1.2 DEFINITIONS

1.2.1 Foundation systems

In the scope of this book relevant foundation systems are defined as follows:

Raft foundation (R): A foundation consisting only of a reinforced concrete slab of significant plan area is termed a raft foundation (Figure 1.1a).
Single pile (SP): A single pile (SP) is defined as a pile for which the load-bearing behaviour is not influenced by any adjacent structures, such as a foundation slab or piles.
Freestanding pile group (FPG): In the case of a freestanding pile group (Figure 1.1b), the entire structural load is transferred by the piles into the ground. There is no contact between the pile cap and the soil. In construction practice, this case is relevant, e.g. for offshore foundations, but also for pile groups with a suspended pile cap or soft ground at the surface.

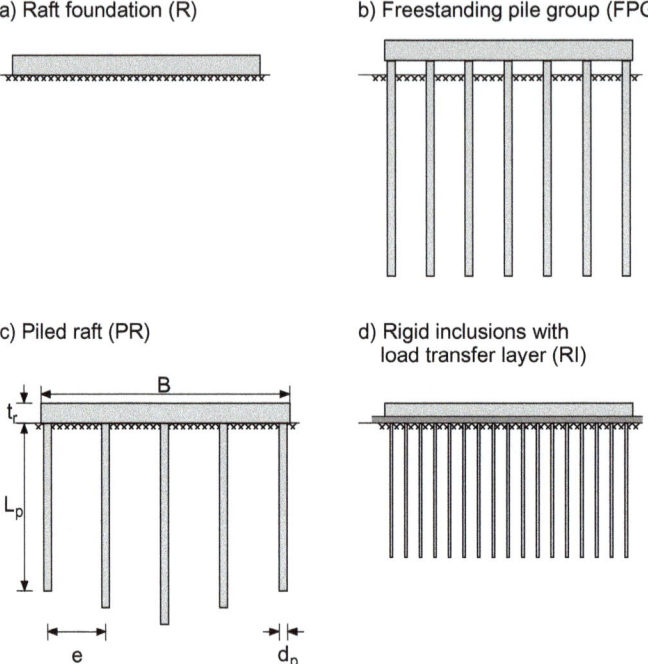

Figure 1.1 Foundation types.

    Piled raft (PR): The raft located on top of the piles is in contact with the soil (Figure 1.1c). The soil has a reasonable stiffness and strength which allows for transfer of load to the soil by means of the raft. However, in the design process this load transfer, except for the self-weight of the raft, is not necessarily considered.

    Rigid inclusions (RI) with load transfer layer (LTL): Rigid inclusions (RI) are concrete columns with – compared to bored piles – relatively small diameters, usually between $d = 0.2$ m and $d = 0.4$ m. Between the RI and the raft a load transfer layer (LTL) is located made of a granular material possibly reinforced with geosynthetic (Figure 1.1d).

1.2.2 Parameters for the quantification of the bearing behaviour of piled rafts

1.2.2.1 Pile load and pile resistance

Assuming equilibrium the resistance of the pile is equal to the load acting on the pile measured for example with a load cell at the pile head (Figure 1.2a). For piles under axial loading without negative shaft friction the axial pile resistance R equals the axial pile load N at the pile head (Figure 1.2b).

Introduction 3

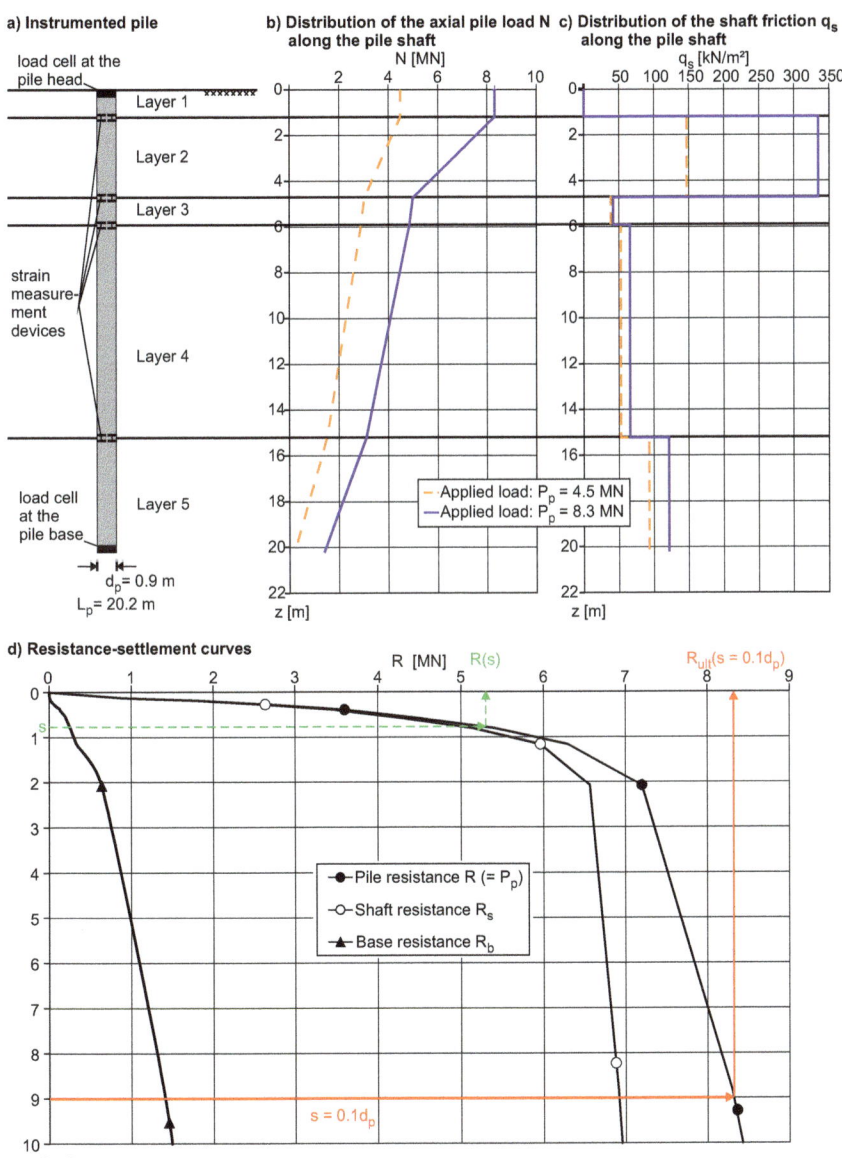

Figure 1.2 Bearing behaviour of a single pile.

From the distribution of the axial pile load N along the pile shaft, which can be derived from strain measurements, the distribution of the shaft friction q_s (Figure 1.2c) can be established:

$$q_{s,i} = \frac{\Delta N_i}{A_{s,i}} \tag{1.1}$$

where $q_{s,i}$ = shaft friction in layer i; ΔN_i = change of axial pile load in layer i; $A_{s,i}$ = shaft area in layer i.

For the sake of simplicity in the scope of this book, a load P_p transferred from the raft, pile cap or directly from the superstructure to the pile head is referred to as pile resistance R. If no other information is available, the ultimate capacity of a single pile is assumed to be the resistance of the pile at a settlement equal to 10% of the pile diameter, i.e. $R_{ult} = R(s = 0.1\ d_p)$ (Figure 1.2d).

1.2.2.2 Equivalent pile spring stiffness

The equivalent pile spring stiffness c_{pile} is used to model the bearing behaviour of a pile for structural analysis of the building and pile cap (or raft) based on the subgrade reaction modulus method (Winkler halfspace).

$$c_{pile} = \frac{R(s)}{s} \tag{1.2}$$

where $R(s)$ = pile resistance for a pile head settlement of s.

The equivalent pile spring stiffness depends on the load level (Figure 1.2d) and the position of the pile within a pile group (Chapter 2).

1.2.2.3 Piled raft coefficient

The piled raft coefficient, α_{pr}, describes the ratio of the sum of all mobilised pile resistances to the total resistance mobilised by the foundation which is equivalent to the total load on the foundation:

$$\alpha_{pr} = \frac{\Sigma R_{pile}}{R_{tot}} = \frac{\Sigma R_{pile}}{P_{tot}} \tag{1.3}$$

where ΣR_{pile} = sum of all mobilised pile resistances; R_{tot} = total mobilised resistance of the foundation and P_{tot} = total load on the foundation. In the scope of this publication all parameters listed above are taken as characteristic values, i.e. there are no partial factors of safety involved. Note also that the resistances R are those mobilised as a result of the applied load; the values would eventually be limited by the geotechnical capacity (shaft plus base) of each pile.

A piled raft coefficient of unity indicates a freestanding pile group whereas a piled raft coefficient of zero describes a raft foundation (unpiled raft). It has to be noted that the piled raft coefficient is not a constant value for a certain foundation configuration but depends on the load level as can be seen in Figure 1.3 for the example of a piled raft in overconsolidated clay

Introduction 5

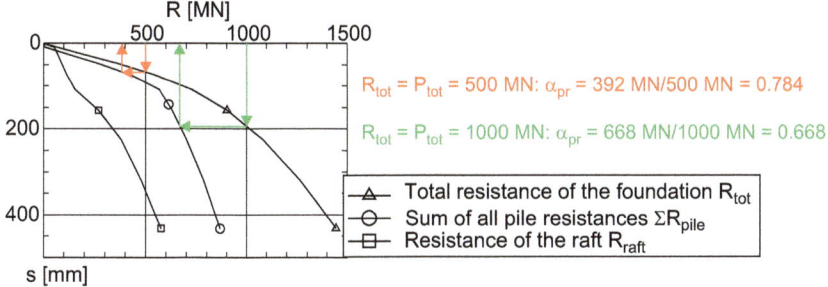

Figure 1.3 Establishing the piled raft coefficient from the resistance settlement curves of a piled raft ($n_p = 49; L_p = 30$ m; $d_p = 1$ m).

with the nonlinear resistance–settlement curves derived from numerical analysis. This phenomenon will be discussed in detail in Chapter 2.

1.2.2.4 Coefficients for average, maximum and differential settlement and for maximum positive bending moment

As will be expanded upon in the further course of this book, in many cases the serviceability is the decisive design criterion for piled rafts. The performance of a raft foundation is usually taken as the reference for evaluating the performance of piled rafts. Constraints, which ensure the satisfactory bearing behaviour of the piled raft, can be described in terms of the behaviour of the raft foundation as follows:

$$s_{pr} \leq \xi_s \cdot s_r \tag{1.4}$$

$$s_{pr,max} \leq \xi_{s,max} \cdot s_{r,max} \tag{1.5}$$

$$\Delta s_{pr} \leq \xi_{\Delta s} \cdot \Delta s_r \tag{1.6}$$

$$m_{pr+} \leq \xi_{m+} \cdot m_{r+} \tag{1.7}$$

where s_r, s_{pr} = average settlement of the raft foundation and the piled raft, respectively; $s_{r,max}, s_{pr,max}$ = maximum settlement of the raft foundation and the piled raft, respectively; $\Delta s_r, \Delta s_{pr}$ = differential settlement of the raft foundation and the piled raft, respectively; m_{r+}, m_{pr+} = maximum positive bending moment of the raft foundation and the piled raft, respectively; and $\xi_s, \xi_{s,max}, \xi_{\Delta s}, \xi_{m+}$ = coefficients for average, maximum and differential settlement and for maximum positive bending moment, respectively.

So for example, for $\xi_{s,max} = 0.5$ the maximum settlement of the raft foundation has been reduced to 50% by the respective piled raft.

6 Combined Pile-Raft Foundations

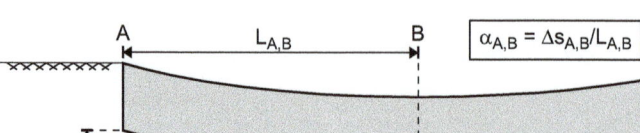

Figure 1.4 Angular distortion.

1.2.2.5 Differential settlements of structures – angular distortions

In the scope of this book the angular distortion α as defined in Figure 1.4 is applied to evaluate the serviceability of the foundation:

$$\alpha = \frac{\Delta s_{A,B}}{L_{A,B}} \tag{1.8}$$

where $\Delta s_{A, B}$ = differential settlement between points A and B and $L_{A, B}$ = distance between points A and B.

In their summary of criteria for settlement and differential settlement of structures, Poulos et al. (2001) suggest limiting $\alpha \leq 1/500$ to prevent cracking in walls.

1.3 EARLY DEVELOPMENTS: APPLICATIONS AND RESEARCH

In the last decades piled rafts were widely used especially for the foundation of high-rise buildings. This section aims to give a concise overview on the historical development from piled rafts to CPRF in practice and research. A summary of early piled raft projects was given by O'Neill et al. (1996). A detailed treatise on case histories of piled rafts and especially CPRF is given in Chapter 4.

Although not equipped with a continuous raft but a grid of beams, probably to save weight, La Azteca building (Figure 1.5) constructed from 1954 to 1955 in Mexico City appears to be the first documented case history where, in the design process, piles and footings were both considered to participate in the load transfer to the subsoil (Zeevaert 1957). Similar foundation beams were used for the "Latino Americana" building, also located in Mexico City (Zeevaert 1956). Poulos (2005) referred to the foundation as an example for a so-called compensated piled raft (Section 1.4.6). A total of 83 concrete piles were driven to a depth of 24 m below ground level in a soft clay (Zeevaert 1957). The piles were assumed to carry only the weight of the building minus the weight of the soil mass excavated. Approximately half a year after the building was finished the settlements amounted to s ≈ 21 cm (Zeevaert 1957).

Starting in 1965 in Gothenburg, Sweden, the Östra Nordstaden building complex was constructed in soft plastic clay (Hansbo et al. 1973). For building

Introduction 7

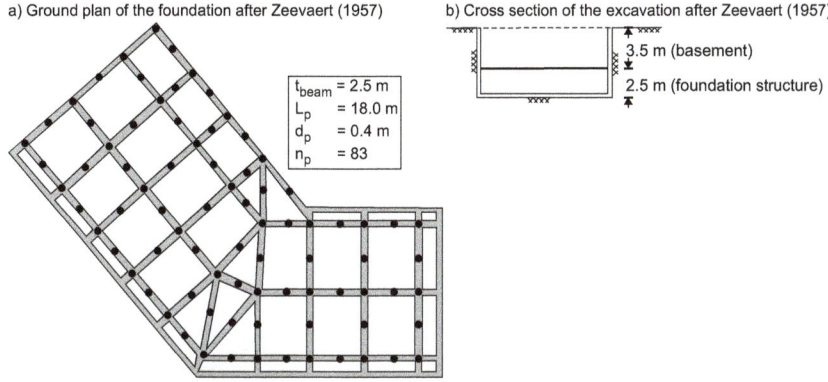

Figure 1.5 La Azteca building, Mexico City.

zones 5 and 6 of this building complex constructed in the early 1970s the foundation comprised a raft (zone 5: $t_r = 1.05$ m, zone 6: $t_r = 1.40$ m) and driven timber piles (zone 5: $L_p = 18$ m, zone 6: $L_p = 20$ m) with a base diameter of $d_b = 0.125$ m and uniform pile spacings of e = 2.4 m (zone 5) and e = 1.9 m (zone 6), respectively. In some areas in zone 5 with pile spacing up to e = 4.8 m, it was assumed that the building load is mainly carried by means of contact pressure between raft and clay. For zone 5 with building loads of up to p = 80 kN/m² in high loaded areas, maximum settlements of approximately s = 5.5 cm were measured (Hansbo et al. 1973). Based on the positive experiences gained in this project the "creep pile" design concept for piled rafts was proposed (Section 1.4.4).

One of the earliest documented case histories of a piled raft (Section 1.4.3) was the high-rise Hyde Park Cavalry Barracks in London built from 1967 to 1970 (Figure 1.6). The 90-m-high tower was founded on an approximately 25 m × 25 m large raft ($t_r = 1.52$ m) and 51 piles ($d_p = 0.91$ m, $L_p = 24.8$ m) in the London Clay (Hooper 1973). The bored piles were enlarged to 2.4 m in diameter at the pile base. Geotechnical measurements, conducted by means of load cells on the pile heads of 3 piles and 3 contact pressure transducers under the raft, showed an increase in pile loads with a simultaneous decrease in contact normal stresses after the end of construction. The resistances (pile head loads) were relatively uniformly distributed within the pile group regardless of the positions of the piles. The high-rise building settled by a maximum of s = 2.2 cm up to about three years after completion of the construction. According to Hooper (1973) the piles carried about 60% of the total building load at the end of construction.

While in London design concepts for piled rafts were developed from nominally freestanding pile foundations, i.e. the piles were considered to transfer 100% of the building load to the subsoil, the development in Frankfurt, Germany, was different. In Frankfurt high-rise buildings have been founded on raft foundations in the Frankfurt Formation comprising

8 Combined Pile-Raft Foundations

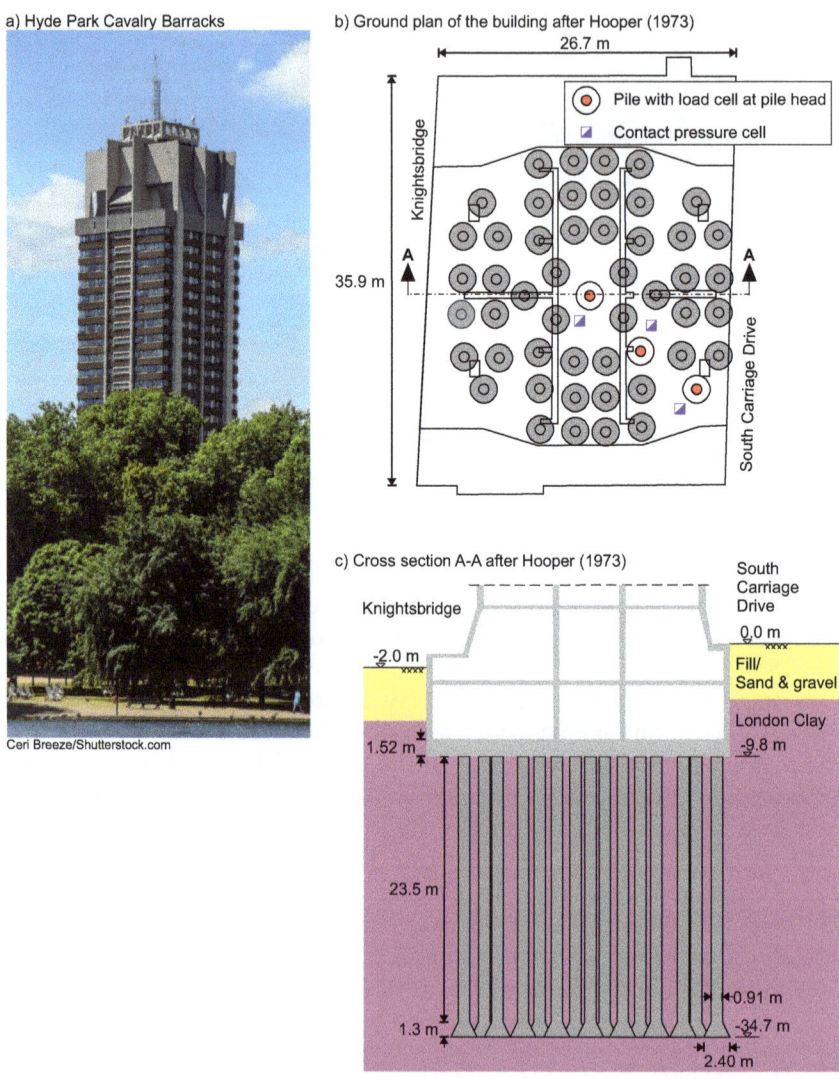

Figure 1.6 Hyde Park Cavalry Barracks, London.

mainly stiff overconsolidated clay since the 1950s. In the older literature (e.g. Sommer 1977, Sommer et al. 1985, 1991, Sommer and Hoffmann 1991, Katzenbach et al. 1994) the Frankfurt Formation was generally termed "Frankfurt Clay". However, in the remainder of this book the term "Frankfurt clay" will be used in a more rational way only for the clayey soils of this layer. A detailed description of the subsoil conditions in Frankfurt is given in Section 4.2.2. With raft thicknesses ranging between t_r = 2 m and t_r = 4 m for the high-rise buildings with maximum heights up to 166 m, settlements s > 30 cm were measured. The maximum settlement was recorded for the 159-m-high Marriott-Hotel constructed from 1972 to 1976 with s_{max} =

33 cm (Katzenbach et al. 1999). As a rule of thumb for the subsoil conditions and typical foundation configurations in Frankfurt, the differential settlements can be estimated to be up to 30% of the average settlements.

For high-rise buildings constructed of reinforced concrete, tilting can be compensated by continuously building along the plumb line which in the most extreme cases, e.g. the historic leaning tower in Pisa (Jamiolkowski and Viggiani 2007), will yield a curved, almost banana-shaped, superstructure. To ensure the serviceability of the raft foundations of the first-generation of Frankfurt high-rise buildings subjected to these large settlements, extensive, in some cases maybe even excessive, measures were taken over the decades, e.g. jacking systems or settlement joints between high-rise and adjacent low buildings.

For the 166-m-high Dresdner Bank (Figure 1.7a), a high-rise building constructed from 1974 to 1978, there was concern that the building load acting eccentrically on the raft (t_r = 4.0 m) would lead to significant

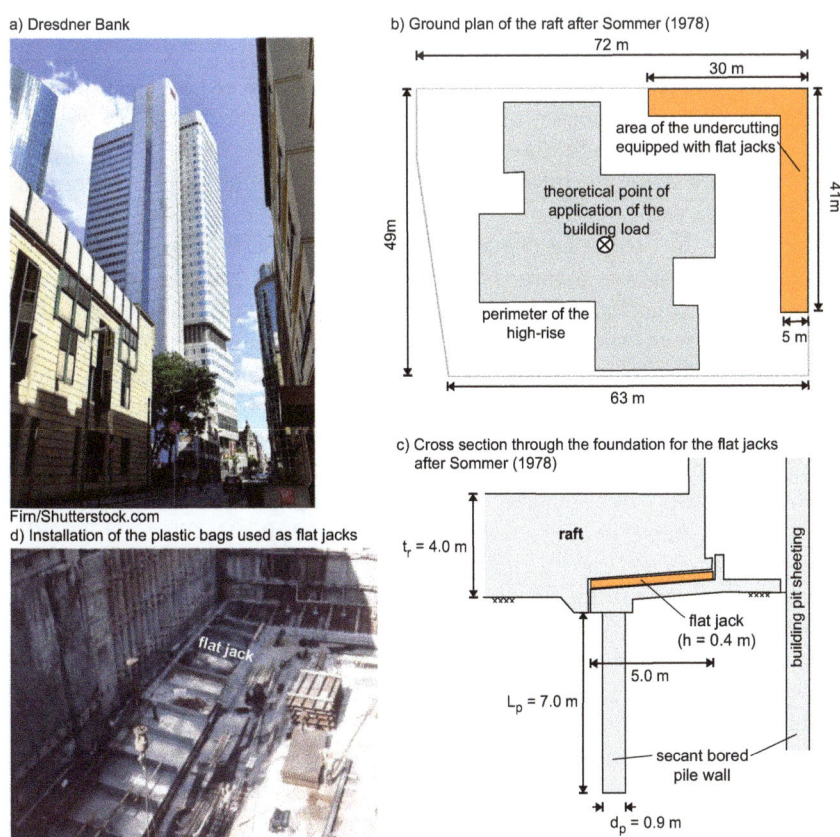

Figure 1.7 Dresdner Bank, Frankfurt (Germany).

differential settlements and tilting (Sommer 1977). Therefore parts of the raft edges were undercut to move the estimated location of the reaction force of the subsoil beneath the theoretical point of application of the building load (Figure 1.7b). In order to cope with varying eccentricities occurring during the construction process, a total of 22 flat jacks were installed in individual compartments under the undercut as a means of correction. The compartments made of reinforced concrete were founded on a secant bored pile wall (Figure 1.7c). The 3 m × 5 m large flat jacks consisted mainly of large plastic bags (Figure 1.7d) which could be filled with water and emptied as required controlling the pressure of the flat jacks during the construction of the high-rise building in such a way that the new building experienced only slight tilting (Sommer 1977). The maximum measured settlement amounted to s_{max} = 20 cm (Ripper and El-Mossallamy (1999).

Finally, the Torhaus Messe (Figure 1.8a) constructed between 1983 and 1986 was the first building in Germany with a foundation designed as a piled raft, considering both bored piles and raft to participate in the load transfer to the subsoil (Sommer et al. 1985). A total number of 84 bored piles with a length of L_p = 20 m and a diameter of d_p = 0.9 m were located under two 17.5 m × 24.5 m large rafts (Figure 1.8b). The distance between the two rafts was 10 m. Since the building had no underground storeys, the bottom of the 2.5-m-thick raft lied just 3 m below ground level (Figure 1.8a). The subsoil comprises quaternary sand and gravel up to 2.5 m below the bottom of the rafts followed by the Frankfurt Formation. The Frankfurt Limestone is outside the influence of the foundation. The groundwater level lies below the rafts.

Similar to the creep pile concept (Section 1.4.4) in the design process of the foundation it was assumed that the piles would fail, i.e. would reach their ultimate capacity, under working load with 25% of the load carried by the raft (Sommer et al. 1985). The geotechnical measurements comprised results from 6 instrumented piles, 11 contact pressure cells and 3 multi-point borehole extensometers. The position of the measurement devices is shown in (Figure 1.8b). From the last documented settlement measurement in 1988 (Sommer 1991), an average centre settlement for the two rafts of 12.4 cm was estimated. Figure 1.8c shows that the pile resistances increased from a centre pile (Pile 1), to the edge piles (Pile 2, Pile 4, Pile 6) and finally to the corner piles (Pile 3, Pile 5), which is typical for a piled raft under working load conditions. The variation of the resistances with the pile position was due to the varying mobilisation of shaft friction. Because of the block deformation of the pile group which can be concluded from the extensometer measurements (Reul and Randolph 2003), there were only small differential displacements between the piles at the centre of the raft and the surrounding soil. Hence, the pile shaft resistances mobilised by the centre piles was substantially smaller than the values mobilised by the edge or corner piles, even though the pile base resistances mobilised were similar. From the last documented pile measurement in February 1986 (Sommer 1991),

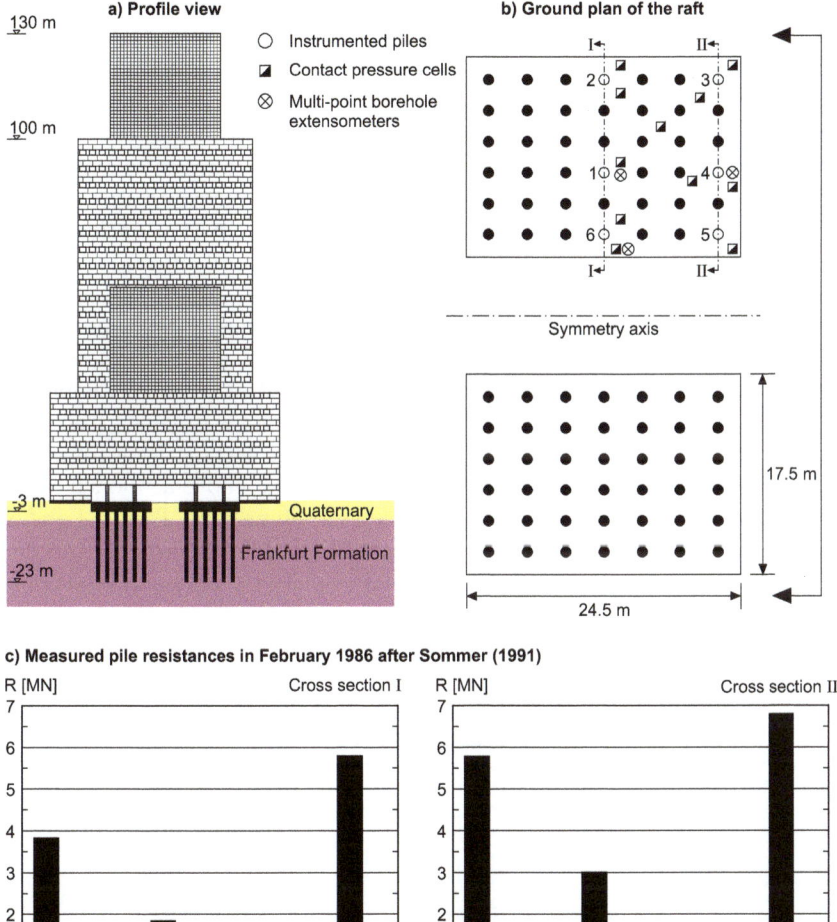

Figure 1.8 Torhaus, Frankfurt (Germany).

a piled raft coefficient of $\alpha_{pr} = 0.67$ was derived. Reul and Randolph (2003) presented the results of a detailed numerical back-analysis of the measurements on the Torhaus with a three-dimensional finite element model where the nonlinear material behaviour of the soil was simulated with an elasto-plastic cap-model.

Further high-rise buildings were constructed in the late 1980s to the mid-1990s following some sort of modified conventional design approach (Section 1.4.3) e.g. the Messeturm (Section 2.3.3) and the Westend 1 (Section 3.2.2) and eventually leading to the development of the design approach of

12 Combined Pile-Raft Foundations

the Combined Pile-Raft Foundation (CPRF) documented in the German "KPP-Richtlinie" (DGGT 2001); this has been adopted by the ISSMGE TC 212 (Section 3.1.3).

While the first piled rafts were mainly used to transfer predominantly vertical, monotonic loads to the subsoil, since the early 2000s, especially in Japan, piled rafts were also designed in increasing number to withstand dynamic lateral loading caused by earthquakes (e.g. Yamashita and Yamada 2009, Yamashita et al. 2011, Section 2.4.2, Appendix B.2). To achieve this task, piled rafts were frequently combined with soil improvement techniques (e.g. Yamashita and Yamada 2009).

The bearing behaviour of piled rafts is strongly influenced by various interactions shown in Figure 1.9, namely between pile and soil, pile and pile, raft and soil and especially raft and pile, which are discussed in detail in Chapter 2. Valuable insight in the bearing behaviour of piled rafts was gained from case histories where foundations instrumented with geotechnical measurement devices were carefully monitored. In addition to the abovementioned Hyde Park Cavalry Barracks and Torhaus, the high-rise Stonebridge Park in London (Cooke et al. 1981, Section 4.2.1, Appendix B.2) and the Messeturm in Frankfurt (e.g. Sommer and Hoffmann 1991, Sommer et al. 1991, Section 2.3.3) should

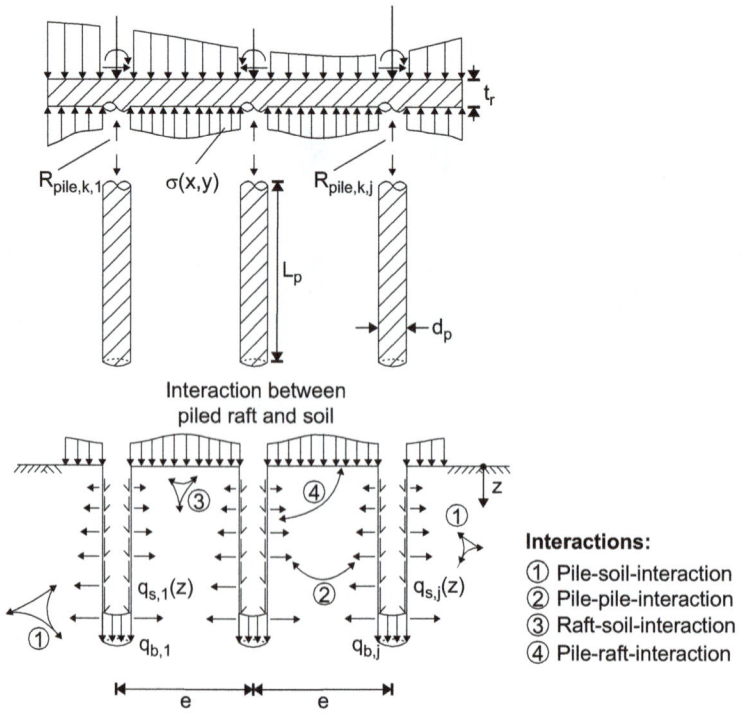

Figure 1.9 Interactions influencing the bearing behaviour of piled rafts. (Adapted from ISSMGE TC 212 2013)

be emphasised, for which extensive measurement results are available that were, and still are, the basis of many further investigations.

Besides in situ measurements on prototypes, model tests and numerical analysis provided the basis for research on the bearing behaviour of piled rafts.

Model tests allow the study of the bearing behaviour of piled rafts under controlled and reproducible boundary conditions for more or less homogeneous subsoil conditions. However, due to possible scale effects and the difficulty of simulating the pile and raft installation realistically, model tests can only be used to a limited extent to interpret the load-bearing behaviour of prototypes. For example, model piles in cohesive soils are usually jacked into the ground, whereas for piled rafts frequently bored piles are installed. Nevertheless, model tests allow at least a qualitative investigation of significant effects of soil-structure interaction. Small-scale model tests are tests with pile lengths in the decimetre range and pile diameters in the millimetre to centimetre range. Large-scale model tests are defined as tests with pile lengths in the meter range and pile diameters in the centimetre to decimetre range.

Pioneering research on piled rafts in cohesive soils was carried out by Whitaker (1957, 1961) by means of small-scale model tests in homogenised, remoulded London clay. Based on these model tests, Cooke (1986) demonstrated the influence of pile spacing on the settlement reduction achievable with piled rafts. The tests also indicated that significantly higher ultimate capacities are achieved with a piled raft compared to a freestanding pile group with the same number of piles.

Small-scale model tests in homogenised, remoulded London clay were also reported by Butterfield and Ghosh (1980). The piled rafts investigated showed a linear load-settlement behaviour in the working load range of the foundation, with the piled raft coefficient lying in a range of $\alpha_{pr} = 0.6$ to $\alpha_{pr} = 0.7$.

Kishida and Meyerhof (1965) showed with small-scale model tests in sand that the bearing capacities of piled rafts and freestanding pile groups are not significantly affected by small load eccentricities, while large load eccentricities cause a significant reduction in the ultimate bearing capacity.

Vesic (1969) investigated the bearing behaviour of piled rafts and freestanding pile groups in sand with large-scale model tests. Whereas no change in the pile base resistance at ultimate limit state could be identified, due to group effects the shaft resistance of a pile within the pile group increased significantly compared to the shaft resistance of a single pile.

Ranjan and Nagrajan (1972) showed, from small-scale model tests in sand, that piled rafts exhibit a stiffer bearing behaviour compared to freestanding pile groups. Garg (1979) came to the same conclusion on the basis of large-scale model tests.

Centrifuge tests, in which a stress state similar to the prototype is generated in a gravitational field caused by centripetal acceleration thus eliminating scaling errors due to stress level, represent a special type of model tests (e.g. Taylor 1995). Some of the earliest research on the bearing behaviour of piled rafts applying centrifuge modelling appears to be the work by Thaher (1991), who presented tests on piled rafts and raft foundations in overconsolidated Kaolin. In addition to a parametric study to determine the influence of the number of piles, the pile length and the pile diameter on the bearing behaviour of a piled raft, Thaher (1991) also modelled the foundation of the Messeturm in Frankfurt, Germany.

Horikoshi and Randolph (1996) used centrifuge model tests in overconsolidated Kaolin clay to investigate the effect of the number of centrally located piles on the differential settlements of a vertically loaded piled raft. During the tests a number of unloading and reloading cycles were carried out. Their objective was to demonstrate that differential settlements could be minimised by centrally located pile groups even with a very flexible raft.

In order to capture the interactions shown in Figure 1.9, numerical methods must be able to model the three-dimensional aspects of the bearing behaviour of piled rafts. For this reason in the 1970s and 1980s the boundary element method (BEM) based on the Mindlin solution (Mindlin 1936) as described by Banerjee and Butterfield (1981), Poulos and Davis (1980), and Poulos (1989) was applied to investigate the bearing behaviour of pile groups and piled rafts. Using this approach Hain and Lee (1978) carried out a back-analysis of the abovementioned La Azteca and the Hyde Park Cavalry Barracks. However, to reduce the computational resources required for the BEM, the approach frequently was simplified, modelling certain aspects of the bearing behaviour by means of load transfer functions (e.g. O'Neill et al. 1977, Chow 1986, Griffiths et al. 1991, Clancy and Randolph 1993). The analysis method of Clancy and Randolph (1993) was used to investigate the performance of the piled raft foundation for the QV1 building in Perth, Western Australia, where the foundations comprised five independent pile-supported rafts (Randolph and Clancy 1994). They demonstrated that the pile support could be optimised, using fewer piles than the actual design, without compromising serviceability limit state (SLS) constraints on total and differential settlements. Ta and Small (1996) presented an approach for the analysis of piled rafts where the raft is assumed to be a thin elastic plate modelled with the finite element method and the finite layer method is used to model the pile-soil response. With this approach Ta and Small (1996) were able to investigate main aspects of the bearing behaviour of piled rafts such as deformations and bending moments of the raft or the load share between piles and raft.

The finite element method has been used since the beginning of the 1970s to investigate the bearing behaviour of piles. Initially, due to the high computational resources required, mainly single piles or piled rafts simplified as axis-symmetric systems were considered, taking into account simplified linear and nonlinear approaches for the material behaviour of the soil. For example, Hooper (1973) back-analyses the measurements on the Hyde Park

Cavalry Barracks by means of axis-symmetric finite element analysis (FEA) applying a linear elastic model for the London Clay. Facilitated by advances in hardware and software, the focus since the 1990s has increasingly been on three-dimensional (3D) investigations of pile groups and piled rafts, also taking into account more realistic constitutive laws for the soil. The first application of three-dimensional finite element analysis on the bearing behaviour of piled rafts appears to date back to Ottaviani (1975). In this pioneering work parametric studies on piled rafts comprising 3 × 3 and 3 × 5 piles (d_p = 1 m, L_p = 20 m and L_p = 40 m, respectively) with a pile spacing of e = 3 m and a 3-m-thick raft are documented. From the analyses, which assumed linear elastic material behaviour for the soil, it was confirmed, for example, that the distribution of pile resistances within the group varies significantly.

Probably one of the earliest applications of 3D FEA in the design process of piled rafts is documented for the Commerzbank Tower in Frankfurt (Katzenbach et al. 1994). The foundation of the Commerzbank Tower is a piled raft designed as a pile foundation with the 111 piles (d_p = 1.8 m/1.5 m, L_p = 37.6 m to L_p = 45.6 m) transferring the load of the superstructure mainly to a limestone layer at a depth of approximately 33 m below the base of the raft. For a total design load of $P_{design} \approx 2100$ MN Katzenbach et al. (1994) predicted a piled raft coefficient of α_{pr} = 0.95 and maximum settlements of s = 5.3 cm based on the 3D FEA with an elasto-plastic soil model.

1.4 EARLY DEVELOPMENTS: DESIGN CONCEPTS

1.4.1 General remarks

A piled raft is a geotechnical composite construction consisting of the three elements: piles, raft and soil (Figure 1.9). The efficient design of piled rafts differs from traditional foundation design, where the loads are assumed to be carried either by the raft or by the piles, considering the safety factors in each case. A rational design of piled rafts will take the load share between the piles and the raft into account and the pile resistance may be mobilised up to a load level that can be of the same order of magnitude as the bearing capacity of a comparable single pile or even greater. Therefore, a properly designed piled raft foundation allows reduction of settlements and differential settlements in a very economical way compared to traditional foundation concepts since the piles are not required to ensure overall stability of the foundation but rather to act as settlement reducers (Burland et al. 1977). The following section provides an overview on the early design concepts developed for piled rafts over the last decades. The design philosophy of the ISSMGE Combined Pile-Raft Foundation Guideline (ISSMGE TC 212 2013) will be discussed in detail in Section 3.1.3.

1.4.2 Conventional design

Traditionally, pile groups that include a pile cap or raft bearing directly on the ground are designed as a pile foundation, ignoring the contribution of the raft to the load transfer. The total structural load is assumed to be transferred from the piles to the soil while maintaining appropriate safety factors with the pile cap usually considered to carry its own weight only. An example of this design principle is the high-rise building at Stonebridge Park in London (Cooke et al. 1981), where safety factors of 2.0 and 3.0 were applied for the pile shaft resistance and the pile base resistance, respectively.

1.4.3 Modified conventional design

The load transfer of the raft is taken into account by assigning a share of the total load to the raft. This load component must be transferred in compliance with safety factors. The part of the load assigned to the piles is also transferred in observance of safety factors. The aforementioned Torhaus falls in this design category. Randolph (1994) and Randolph and Clancy (1994) report on the QV1 building in Perth, Australia, where the raft was assigned a 25% share of the total load in the design.

1.4.4 "Creep pile" design

In the so-called "creep pile" approach proposed by Hansbo and Källström (1983) for relatively soft cohesive soils, the raft has sufficient stability on its own. After estimating the share of the load that can be transferred from the raft while maintaining permissible settlements, the remaining load is transferred by piles that are assumed to be at the ultimate limit state; i.e. the safety factor of the piles is 1.0. The term "creep pile" originates from the fact that the long-term shear strength ("creep strength") of the soil is used to determine the ultimate bearing capacity of the piles. Jendeby (1986) and Hansbo (1993, 1994) report on foundations in Gothenburg, Enköping, and Uppsala, Sweden, that were designed according to the "creep pile" principle.

1.4.5 Japanese design approach

Yamashita et al. (2011) report on a design approach applied for various buildings in Japan where under static loading for the ultimate bearing capacity a factor of safety of $\eta > 3.0$ is required. Ignoring the effect of the piles the ultimate capacity might be established for the raft alone. Additionally, according to a guideline by the Architectural Institute of Japan (AIJ 2001), written in Japanese but now with an informal translation in English (Matsumoto, 2024), it has to be proved that the maximum settlement and the differential settlement or the maximum angular rotation, respectively, are less than project-specific allowable values.

According to Yamashita et al. (2011) under seismic loading conditions for the ultimate bearing capacity of the piled raft a factor of safety of $\eta > 1.5$ is required for combined vertical and lateral loading. Moreover, a factor of safety of $\eta > 1.5$ is required for the ultimate bearing capacity of each of the piles compared to the maximum axial pile load which differs from other concepts presented in this chapter where the individual piles might be loaded up to the failure load of a single pile; the maximum bending moment and the shear force in the pile cross sections due to lateral loading should be less than the design value of the structural strength of the piles.

1.4.6 Compensated piled raft foundations

Based on the pioneering work of Zeevaert (1957) on La Azteca building in Mexico City, Poulos (2005) defined the concept of the so-called compensated piled raft foundation where the raft is located at the base of a deep excavation. According to Sales et al. (2010), the advantage compared to piled rafts located close to the ground surface is that although the raft takes a higher proportion of the load, i.e. the piled raft coefficient is smaller, the overall and differential settlements are lower because the net pressure applied by the building is reduced. However, since the decision whether there is a basement with its respective building costs or not appears to be based rather on demands of the building use, e.g. space required for car parking, storage, building services, etc., than on structural or geotechnical aspects, it is doubtful if the term "design concept" is justified for compensated piled raft foundation.

1.4.7 Raft-enhanced pile groups vs. pile-enhanced rafts

O'Brien et al. (2012) distinguish between raft-enhanced pile groups and pile-enhanced rafts. For raft-enhanced pile groups, O'Brien et al. (2012) assume that both piles and raft will work within a pseudo-elastic range of behaviour, i.e. the pile group capacity will not be fully mobilised at working load level. The proportioning of load between pile group and raft is governed by their relative stiffnesses, so for the raft to resist a significant proportion of the load from the superstructure, the design must balance the number and length of piles (hence the pile group stiffness) relative to the raft stiffness. On the other hand, for pile-enhanced rafts, the piles will be designed to mobilise close to their ultimate capacity, with the piles usually being located beneath heavily loaded superstructure columns (O'Brien et al. 2012). According to O'Brien et al. (2012), the pile resistance must be maintained at relatively large overall settlements between $s = 5$ cm and $s = 10$ cm, suggesting the need for a ductile load-settlement behaviour. In practice, however, since the raft will force settlement of the underlying soil, it is unlikely that slip between pile shafts and soil will occur over the full length of the piles. Fundamentally, raft-enhanced pile groups and pile-enhanced rafts are opposite ends of a continuum, with the piles carrying the major share of the design load for the former and the raft carrying the major share of the design load for the latter.

REFERENCES

Architectural Institute of Japan (AIJ) (2001). Piled raft foundation, Recommendations for Design of Building Foundations, 339–348 (in Japanese).
Banerjee, P. K., Butterfield, R. (1981). *Boundary Element Methods in Engineering Science*. McGraw-Hill.
Butterfield, R., Ghosh, N. (1980). A linear elastic interpretation of model tests on single piles and groups of piles in clay. *Numerical methods in offshore piling*, ICE, London, 109–118.
Burland, J.B., Broms, B.B., De Mello, V.F.B. (1977). Behaviour of foundations and structures. *Proceedings of the 9th International Conference on Soil Mechanics and Foundation Engineering*, Tokyo, 2, 495–546. Rotterdam: Balkema.
Chow, Y.K. (1986). Analysis of vertically-loaded pile groups. *International Journal for Numerical and Analytical Methods in Geomechanics*, 10, 1, 59–72.
Clancy, P., Randolph, M. F. (1993). An approximate analysis procedure for piled raft foundations. *IJNAMG*, 17, 849–869.
Cooke, R. W. (1986). Piled raft foundations on stiff clays - A contribution to design philosophy. *Géotechnique*, 36, 2, 169–203.
Cooke, R. W., Bryden-Smith, D. W., Gooch, M. N., Sillett, D. F. (1981). Some observations of the foundation loading and settlement of a multi-storey building on a piled raft foundation in London Clay. *Proceedings of the ICE*, 70, Part 1, 433–460.
Deutsche Gesellschaft für Geotechnik (DGGT), Arbeitskreis Pfähle (2001). Richtlinie für den Entwurf, die Bemessung und den Bau von Kombinierten Pfahl-Plattengründungen (KPP) – KPP-Richtlinie.
Garg, K. G. (1979). Bored pile groups under vertical load in sand. *Journal of the Geotechnical Engineering Division*, 105, GT8, 939–956.
Griffiths, D.V., Clancy, P., Randolph, M.F. (1991). Piled raft foundation analysis by finite elements. *Proceedings of the 7th International Conference on Computer Methods and Advances in Geomechanics*, Cairns: 2, 1153–1157.
Hain, S.J., Lee, I.K. (1978). The analysis of flexible pile-raft systems. *Géotechnique*, 28, 1, 65–83.
Hansbo, S. (1993). Interaction problems related to the installations of pile groups. *Proceedings of the Conference Deep Foundations on Bored and Auger Piles*, Ghent, 59–66, Rotterdam: Balkema
Hansbo, S. (1994). Foundations on friction creep piles in soft clays. *Proceedings of the 1st International Conference on Case Histories in Geotechnical Engineering*, 913–922.
Hansbo, S., Källström, R. (1983). A case study of two alternative foundation principles. *Väg-och Vattenbyggaren*, 7–8, 23–27.
Hansbo, S., Hofmann, E., Mosesson, J. (1973). Östra Nordstaden, Gothenburg. Experiences concerning a difficult problem and its unorthodox solution. *Proceedings of the 8th ICSMFE*, Moscow, 2, 2, 105–110.

Hooper, J. A. (1973). Observations on the behaviour of a piled-raft foundation on London Clay. *Proceedings of the ICE*, 55, Part 2, 855–877.
Horikoshi, K., Randolph, M. F. (1996). Centrifuge modelling of piled raft foundations on clay. *Géotechnique*, 46, 4, 741–752.
ISSMGE Technical Committee TC 212 (2013). ISSMGE Combined Pile-Raft Foundation Guideline. Report of the ISSMGE Technical Committee TC 212 – Deep Foundations, ed. Katzenbach, R., Choudhury D, Technische Universität Darmstadt.
Jamiolkowski, M., Viggiani, C. (2007). The restoration of the leaning tower of Pisa. Report of the International Committee for the Safeguard of the Leaning Tower of Pisa.
Jendeby, L. (1986). *Friction piled foundations in soft clay - A study of load transfer and settlements*. Gothenburg, Sweden: Chalmers University of Technology.
Katzenbach, R., Arslan, U., Gutwald, J. (1994). A numerical study on pile foundation on the 300 m high Commerzbank Tower in Frankfurt am Main. *Proceedings of the 3rd European Conference on Numerical Methods in Geomechanics*, Manchester, 271–277, Rotterdam: Balkema.
Katzenbach, R., Moormann, C., Reul, O., Hoffmann, H. (1999). Hochhausgründungen als Motor innovativer, kostengünstiger Fundamentierungs- und Gebäudetechniken. Hochhäuser: Darmstädter Statik-Seminar, Hrsg.: R. Pfeiffer, Berichte des Instituts für Statik der Technischen Universität Darmstadt, Nr. 16, XIII.
Kishida, H., Meyerhof, G. G. (1965). Bearing Capacity of Pile Groups under Eccentric Loads in Sand. *Proceedings of the 6th ICSMFE, Montréal*, II, 270–274.
Matsumoto, T. (2024). Private communication.
Mindlin, R. D. (1936). Force at a point in the interior of a semi-infinite solid. *Physics*, 7, 195–202.
O'Brien, A.S., Burland, J.B., Chapman, T. (2012). Rafts and piled rafts. In: *ICE Manual of Geotechnical Engineering: Volume II*, Chapter 56, 853–886.
O'Neill, M.W., Ghazzaly, O.I., Ha, H.B. (1977). Analysis of three dimensional pile groups with non-linear soil response and pile-soil pile interaction. *Proceedings of the 9th Annual Offshore Technology Conference*, Houston, Paper OTC 2838, 245–256.
O'Neill, M. W., Caputo, V., De Cock, F., Hartikainen, J., Mets, M. (1996). Case histories of pile supported rafts. *Report for ISSMFE TC18*, University of Houston, Texas.
Ottaviani, M. (1975). Three-dimensional finite element analysis of vertically loaded pile groups. *Géotechnique*, 25, 2, 159–174.
Poulos, H. G. (1989). Pile behaviour - theory and application. *Géotechnique*, 39, 3, 365–415.
Poulos, H.G. (2005). Piled raft and compensated piled raft foundations for soft soil sites. In *Advances in designing and testing deep foundations*. Geotechnical Special Publication No. 129. American Society of Civil Engineers (ASCE), Reston, VA, 214–234.
Poulos, H.G., Davis, E.H. (1980). *Pile foundation analysis and design*. New York: Wiley.
Poulos, H.G., Carter, J.C., Small, J.C. (2001). Foundations and retaining structures – Research and practice. *Proceedings of the 15th ICSMFE*, Istanbul, Vol. 4, 2527–2606, Rotterdam: Balkema.
Randolph, M.F. (1994). Design Methods for pile groups and piled rafts. *Proceedings of the 13th ICSMFE*, New Delhi, Vol. 5, 61–82, Rotterdam: Balkema.

Randolph, M. F., Clancy, P. (1994). Design and performance of a piled raft foundation. *Proceedings of the Conference on Vertical and Horizontal Deformations of Foundations and Embankments*, Texas, ASCE Geotechnical Special Publication No. 40, 314–324.

Ranjan, G., Nagrajan, K. (1972). Influence of spacing in model piled foundation groups in sand. *Proceedings of the Symposium on Modern Trends in Civil Engineering*, Roorkee, India, Vol. 1.2, 63–66.

Reul, O., Randolph, M.F. (2003). Piled rafts in overconsolidated clay – Comparison of in-situ measurements and numerical analyses. *Géotechnique*, 53, 3, 301–315.

Ripper, P., El-Mossallamy, Y. (1999). Entwicklung der Hochhausgründungen in Frankfurt. Hochhäuser: Darmstädter Statik-Seminar, Hrsg.: R. Pfeiffer, Berichte des Instituts für Statik der Technischen Universität Darmstadt, Nr. 16, XVI.

Sales, M.M., Small, J.C., Poulos, H.G. (2010). Compensated piled rafts in clayey soils: Behaviour, measurements, and predictions. *Canadian Geotechnical Journal*, 47, 327–345.

Sommer, H. (1977). Pressure Cushions to Correct Tilting of High Buildings. *Proceedings of the 9th ICSMFE*, Tokio, Vol. 1, 735–738.

Sommer, H. (1978). Messungen, Berechnungen, und Konstruktives bei der Gründung Frankfurter Hochhäuser. *Bauingenieur*, 53, 205–211.

Sommer, H. (1991). Entwicklung der Hochhausgründungen in Frankfurt/Main. *Festkolloquium 20 Jahre Grundbauinstitut Prof. Dr.-Ing. H. Sommer und Partner*, 47–62.

Sommer, H., Hoffmann, H. (1991). Load-settlement behaviour of the fairtower (Messeturm) in Frankfurt/Main. *Proceedings of the 4th International Conference on Ground Movements and Structures*, 612–627, London: Pentech Press.

Sommer, H., Wittmann, P., Ripper, P. (1985). Piled raft foundation of a tall building in Frankfurt clay. *Proceedings of the 11th ICSMFE*, San Francisco, Vol. 4, 2253–257, Rotterdam: Balkema.

Sommer, H., Tamaro, G., DeBeneditis, C. (1991). Messe Turm, foundations for the tallest building in Europe. *Proceedings of the 4th International Conference on Piling and Deep Foundations*, 139–145, Rotterdam: Balkema.

Ta, L.D., Small, J.C. (1996). Analysis of piled raft systems in layered soils. *International Journal for Numerical and Analytical Methods in Geomechanics*, 20, 57–72.

Taylor, R. N. (ed.) (1995). *Geotechnical centrifuge technology*. Blackie Academic and Professional, London.

Thaher, M. (1991). Tragverhalten von Pfahl-Platten-Gründungen im bindigen Baugrund, Berechnungsmodelle und Zentrifugen-Modellversuche. Schriftenreihe des Instituts für Grundbau, Wasserwesen und Verkehrswesen der Ruhr-Universität Bochum, Serie Grundbau, Heft 15

Vesic, A. S. (1969). Experiments with instrumented pile groups in sand. *Proceedings of the Performance of Deep Foundations*, ASTM, STP 444, 177–222.

Whitaker, T. (1957). Experiments with model piles in groups. *Géotechnique*, VII, 147–167.

Whitaker, T. (1961). Some experiments on model piled foundations in clay. Proceedings of the Symp. Pile Foundations, *Proceedings of the 6th Congress International Association for Bridge and Structural Engineering*, Stockholm, 124–139.

Yamashita, K., Yamada, T. (2009). Settlement and load sharing of a piled raft with ground improvement on soft ground. *Proceedings of the 17th ICSMGE*, Alexandria, 1236–1239.

Yamashita, K., Yamada, T., Hamada, J. (2011). Investigation of settlement and load sharing on piled rafts by monitoring full-scale structures. *Soils Found*, 51, 3, 513–532.

Zeevaert, L. (1956). Heavy and tall building problems in Mexico City. *ASCE Journal of Structural Division*, Vol. 82, ST 2, paper 917.

Zeevaert, L. (1957). Compensated friction-pile foundation to reduce the settlement of buildings on the highly compressible volcanic clay of Mexico City. *Proceedings of the 4th ICSMFE*, Vol. 3, 81–86.

Chapter 2
Bearing behaviour of piled rafts

2.1 GENERAL REMARKS

In the following the term "piled raft" is used since the mechanical bearing behaviour of this foundation system is the main focus. In accordance with the focus of the ISSMGE Combined Pile-Raft Foundation Guideline (ISSMGE TC 212 2013), which will be discussed in detail in Section 3.1.3, this chapter mainly emphasises the bearing behaviour of foundations under monotonic, vertical loading.

2.2 PILED RAFTS SUBJECTED TO MONOTONIC VERTICAL LOADING

2.2.1 Numerical study of the bearing behaviour of vertically loaded piled rafts

An extensive parametric study on piled rafts subjected to monotonic vertical loading, based on 3D FEA with the program ABAQUS, was carried out by Reul (2000, 2001) with parts subsequently published by Reul and Randolph (2004) and Reul (2004). Unless indicated otherwise, all results presented in Section 2.2 are taken from those studies.

For the presentation of the results of the parametric study a number of parameters are defined in the following. The diameter of the equivalent pier d_{eq} is approximated after Randolph (1994) as:

$$d_{eq} = \sqrt{\frac{4}{\pi} A_g} = 1.13\sqrt{A_g} \qquad (2.1)$$

For pile configuration 3 (Figure 2.3), the plan area of the pile group as a block, A_g, is taken to be the mean value of the pile group at the centre of the raft and a pile group with a diameter given by the piles placed at the edge of the raft.

The pile group-raft area ratio a_{gr} is given by:

$$a_{gr} = \frac{A_g}{A_r} \qquad (2.2)$$

where A_r = raft area.
The relative pile length is defined as:

$$\frac{L_p}{a_{eq}} = \frac{L_p}{\sqrt{\frac{B \cdot L}{\pi}}} \qquad (2.3)$$

where a_{eq} = equivalent circular raft radius; B = breadth of the raft; and L = length of the raft.

After Horikoshi and Randolph (1997) a rational definition of the raft-soil stiffness ratio for rectangular rafts K_{rs} may be expressed as:

$$K_{rs} = 5.57 \frac{E_r}{E_s} \frac{1-v_s^2}{1-v_r^2} \left(\frac{B}{L}\right)^{0.5} \left(\frac{t_r}{L}\right)^3 \qquad (2.4)$$

where E_r, E_s = Young's modulus of raft and soil, respectively; v_r, v_s = Poisson's ratio of raft and soil, respectively; and t_r = thickness of the raft.

For the purpose of this study, an equivalent Young's modulus for the soil $E_{s,eq}$ is defined to take the nonlinear soil behaviour in to account. The equivalent Young's modulus is calculated as the modulus that gives, for a uniformly loaded raft foundation on an elastic layer of finite thickness under the same total load, the same average settlement as the elasto-plastic FEA. The raft-soil stiffness ratios given in this study are the mean values for the investigated load range. The average settlement under a uniform load s is assumed to be the same as the settlement of a rigid foundation and is calculated, using the approximation of Davis and Taylor (1962), as:

$$s \approx \frac{1}{3}\left(2 \cdot s_{centre} + s_{corner}\right) \qquad (2.5)$$

The settlements at the centre and at the corner of the foundation s_{centre} and s_{corner} are calculated from the solutions presented by Ueshita and Meyerhof (1968) for a uniform rectangular load on an elastic layer of finite thickness.

Based on the formulation given by Horikoshi and Randolph (1999) the normalised overall stiffness of a piled raft K^* is calculated as:

$$K^* = \frac{2 \cdot k_{pr}}{G_{s,eq,0} \cdot d_{eq}} = \frac{4 \cdot k_{pr} \cdot (1+v_s)}{E_{s,eq,0} \cdot d_{eq}} \qquad (2.6)$$

where k_{pr} = overall stiffness of the piled raft; and $G_{s,\ eq,\ 0}$, $E_{s,\ eq,\ 0}$ = initial equivalent shear modulus and Young's modulus of the soil, respectively. The initial equivalent Young's modulus of the soil is established as described above under consideration of an initial, small load increment (for approximately elastic conditions).

The normalised differential settlement Δs^* is defined as:

$$\Delta s^* = \frac{\Delta s}{s_r} \tag{2.7}$$

where Δs = differential settlement; and s_r = average settlement of the raft foundation.

The normalised bending moment per unit length m^* is defined as:

$$m^* = \frac{m}{P_s} \tag{2.8}$$

where m = bending moment per unit length; and P_s = load caused by the superstructure.

The ratio of the effective applied load P_{eff} to the combined ultimate capacity R_{ult} of n_p single piles is calculated as:

$$P_t^* = \frac{P_{eff}}{(n_p \cdot R_{ult})} \tag{2.9}$$

A number of other performance measures, comparing the piled raft with a purely raft foundation, have been identified in Section 1.2.2 (see equations 1.4 to 1.7).

The parametric study was carried out modelling a subsoil condition based on that in Frankfurt am Main, Germany, which is characterised mainly by tertiary soils and rock (Section 4.2.2). These comprise the Frankfurt Formation at the top underlain by the rocky Frankfurt limestone. The Frankfurt Formation, in the older literature generally termed "Frankfurt Clay" (e.g. Sommer and Hoffmann, 1991a, Sommer et al., 1991, Reul and Randolph, 2003), consists of an irregular sequence of clays and clay marls, silt and sand layers of varying thickness, as well as limestone and dolomite banks. For simplification, the influence of the Frankfurt limestone can be neglected due to its significantly larger stiffness.

In the FEA the soil was simplified as a single-phase drained medium. The long-term behaviour of this single-phase medium was idealised using the effective strength parameters c' and φ'. The main equations of the elastoplastic cap model used to simulate the soil are given by Reul and Randolph (2004). The contact between structure and soil was described as perfectly rough. This means that no relative motion takes place between the nodes of the finite elements that represent the structure and those of the finite elements that represent the soil. The material behaviour in the contact area was simulated by the material behaviour of the soil.

Verification and calibration of the model was based on the back-analysis of static load tests and the measured bearing behaviour of foundations. Reul and Randolph (2003) presented the results of detailed back-analyses of three piled raft foundations in Frankfurt. The material parameters used in the FEA of the parametric study are summarised in Table 2.1.

Figure 2.1 shows the model conditions and finite element mesh applied in the parametric study. The soil and the piles are represented by first-order solid finite elements of hexahedra (brick) and triangular prism (wedge) shape. First-order shell elements of square and triangular shape with reduced integration were used to model the raft. Circular piles were replaced by square piles with the same shaft circumference. A discussion of the influence of the mesh refinement on the results of the FEA can be found in Section 3.2.2. The foundation level was set 14 m below ground level. Only the soil below the foundation level is modelled with finite elements, with the soil above the foundation level being considered as applying a surcharge only. The base of the compressible soil layer, i.e. the Frankfurt Formation, was set 83 m below ground level.

As shown in Figure 2.2 four different load configurations were studied. For load type I (core-edge loading) half of the load was applied in a core area at the centre of the raft and the other half at the edge of the raft. The core area is 25% of the total area of the raft. This is a typical configuration for tall buildings with stiff structural elements such as elevator shafts and stairways at the centre of the raft and columns at the edge to support the facade. Load type II and III have loads applied only in the core area or the edge area, respectively. Load type IV describes a uniform load over the whole raft area, a load configuration frequently applied in parametric studies on the behaviour of (piled) rafts (e.g. Fraser and Wardle 1976, Horikoshi and Randolph 1998, Sheil 2017, Modak and Singh 2023). The total applied load in the four different load configurations is the same.

The main aim of the application of the four different load configurations is to investigate the global response of the piled raft under uniform and nonuniform loading. Investigation of the need to install piles beneath a raft where column loadings are present, as discussed by Poulos (2001), is outside the scope of what is considered here.

In the parametric study square raft foundations (R), freestanding pile groups (FPG) (only load type IV) and piled rafts (PR) with an edge length of $B = 14$ m (only load type IV), $B = 38$ m and $B = 62$ m (only load type IV) and single piles were considered. Three basic pile configurations were investigated (Figure 2.3). Pile configuration 1 has the piles uniformly distributed under the whole raft area. In pile configuration 2 the piles are placed only in the central area of the raft (beneath the core loading as detailed below).

Table 2.1 Material parameters used in the FEA

Parameter			Frankfurt Formation	Raft	Piles
Young's modulus	E	MPa	$45 + \left(\tanh\left(\dfrac{z-30}{15}\right) + 1\right) \cdot 0.7 \cdot z$	34000	30000
Poisson's ratio	ν	-	0.15	0.2	0.2
Total unit weight of moist soil	γ	kN/m³	19	25	25
Buoyant unit weight	γ'	kN/m³	9	15	15
Coefficient of earth pressure at rest	K_0	-	0.72 (0 ≤ z < 25) 0.57 (z ≥ 25)	-	-
Angle of internal friction	φ'	°	20	-	-
Cohesion	c'	kPa	20	-	-
Shape parameter of the transition surface between cone and cap	α	-	0	-	-
Shape parameter of the cone	K	-	0.795	-	-
Shape parameter of the cap	R	-	0.1	-	-

z: depth below surface of tertiary layers in [m]

Bearing behaviour of piled rafts 27

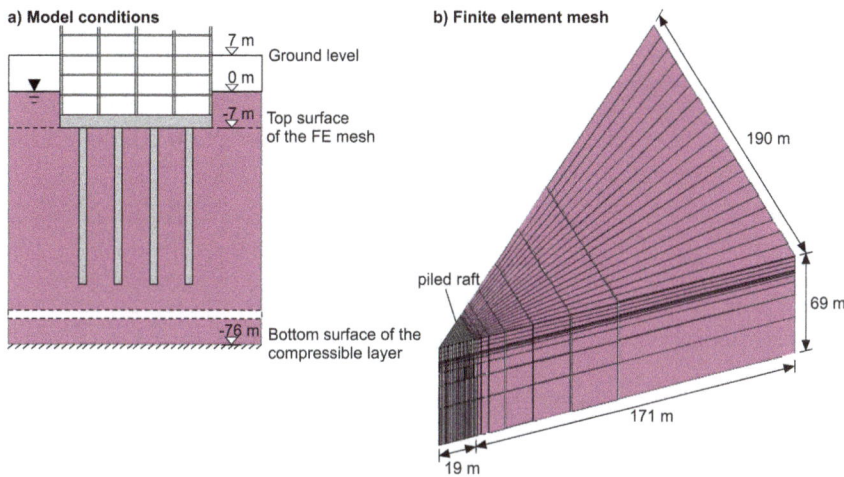

Figure 2.1 Model conditions in the parametric study.

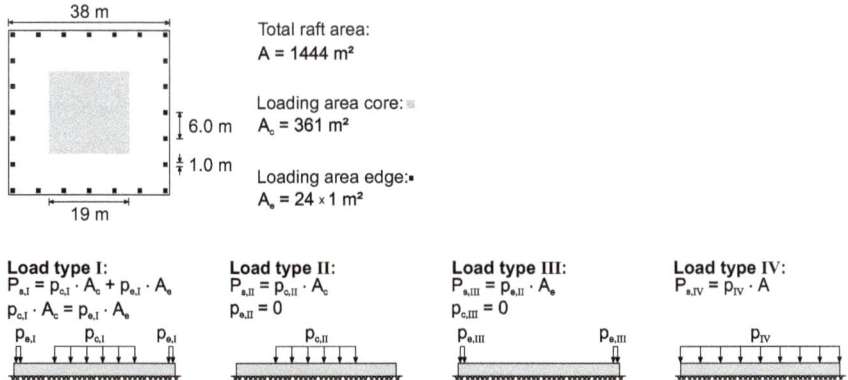

Figure 2.2 Load configurations in the parametric study.

Pile configuration 3 has piles under the central area of the raft as well as under the edges of the raft. Approximately 450 system configurations were analysed in total for this study (Reul 2000, 2001).

Table 2.2 outlines the step-by-step analysis of the construction process in the FEA. For the four different load configurations, the weight of the superstructure was applied on the raft according to Figure 2.2, whereas the weight of the raft was applied to the soil before the stiffness of the raft had been included in the model. The maximum load, including the weight of the raft and the uplift due to the base of the raft being located 7 m below groundwater, amounts to P_{eff} = 721.7 MN for load types I, II and III, which is approximately 20% of the ultimate capacity of an equivalent raft foundation under drained conditions. The load is within the range of values known

28 Combined Pile-Raft Foundations

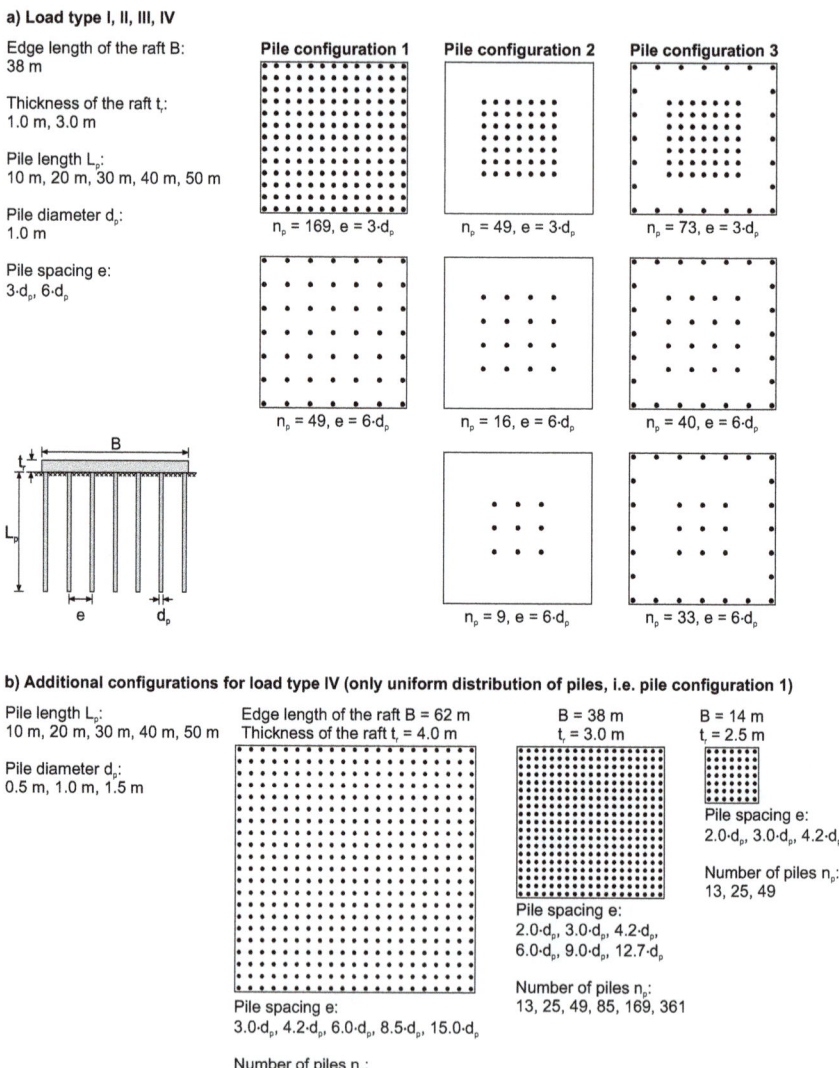

Figure 2.3 Pile configurations in the parametric study.

for piled rafts in Germany (Reul and Randolph 2004, Chapter 4). For load type IV maximum loads of P_{eff} = 166.3 MN (B = 14 m), P_{eff} = 1443.4 MN (B = 38 m) and P_{eff} = 4423.7 MN (B = 62 m) were applied.

For the majority of foundations studied in the scope of the parametric study, raft-soil stiffness ratios between $K_{rs} \approx 0.06$ (B = 38 m, t_r = 1 m) and $K_{rs} \approx 1.5$ (B = 38 m, t_r = 3 m), respectively, were applied. This bandwidth seems to be representative for a wide range of piled rafts for tall buildings (Table 2.3).

Table 2.2 Step-by-step-analysis of the construction process in the FE calculations for the parametric study

Step	Load type I, II, III		Load type IV					
	B = 38 m		B = 14 m		B = 38 m		B = 62 m	
	P_{eff} [MN]	σ'_v [kPa]	P_{eff} [MN]	σ'_v [kPa]	P_{eff} [MN]	σ'_v [kPa]	P_{eff} [MN]	σ'_v [kPa]
I In situ stress state	–	126	–	126	–	126	–	126
II Excavation to a depth of 7 m below ground level	–	63	–	63	–	63	–	63
III Installation of the piles	–	63	–	63	–	63	–	63
IV Excavation to a depth of 14 m below ground level	–	0	–	0	–	0	–	0
V Application of the weight of the raft minus 50% of the uplift as uniform load on the subsoil (zero stiffness of the raft)	57.8	40	5.4	28	57.8	40	249.9	65
VI Installation of the raft	57.8	40	5.4	28	57.8	40	249.9	65
VII Loading of the raft and application of 100% uplift	360.9	250	41.6	212	360.9	250	1105.9	288
IX Further loading of the raft	721.7	500	166.3	848	1443.4	1000	4423.7	1151

P_{eff}: applied load; σ'_v: mean vertical effective stress at foundation level

30 Combined Pile-Raft Foundations

Table 2.3 Raft-soil stiffness ratios of piled rafts for tall buildings

Building	H [m]	P_{eff} [MN]	A [m²]	t_r [m]	n_p [-]	L_p [m]	d_p [m]	K_{rs} [-]
Japan Center, Frankfurt (Appendix)	115	630	1920	3.5	25	22.0	1.3	0.29
Messeturm, Frankfurt (Section 2.3.3)	256	1570	3457	6.0	64	26.9×34.9	1.3	1.41
Torhaus, Frankfurt (Sections 1.3, 2.2.5)	130	2×200	2×429	2.5	2×42	20.0	0.9	2.18
Westend 1, Frankfurt (Section 3.2.2)	208	950	2940	4.7	40	30.0	1.3	0.62
Haus der Wirtschaft, Offenbach (Section 4.3.5)	68	605	5120	2.0	47 6	25.0 37.5×41.0	1.2	0.04
Treptowers, Berlin (Section 4.3.7)	122	632	1376	3.0	54	12.5×16.0	0.9	0.51
Weser Tower, Bremen (Sections 4.3.8)	82	269	836	1.8	31	16.0×22.0	0.9	0.08

H = height of the building; P_{eff} = effective load; A_r = area of the raft; t_r = max. thickness of the raft; n_p = number of piles; L_p = pile length; d_p = pile diameter; K_{rs} = raft-soil stiffness ratio

2.2.2 Pile-pile and pile-raft interaction

Figure 2.4 shows the resistance-settlement curves of a raft foundation (R), a freestanding pile group (FPG) with no contact between raft and soil and a piled raft (PR). The raft has a thickness of 3 m and an edge length of 38 m for each of the three foundation types corresponding to a raft-soil stiffness ratio of $K_{rs} \approx 1.5$. The piles have a length of 30 m ($L_p/a_{eq} \approx 1,4$) and a diameter of 1.0 m. For the freestanding pile group and the piled raft, one analysis was carried out with 49 piles and one with 169 piles, equivalent to a pile spacing of $e/d_p = 6$ and $e/d_p = 3$, respectively.

The raft resistance and the pile resistance of the piled raft are compared separately with the resistance of the raft foundation and the resistance of the freestanding pile group in Figure 2.4. Additionally, the resistance of n_p single piles (SP) is plotted together with the freestanding pile group and the piles of the piled raft. The resistance-settlement curves show nonlinearities for all the foundations investigated but only for the freestanding pile group with 49 piles does the load approach the ultimate resistance. Comparing the resistance settlement curves of the freestanding pile groups and the hypothetical foundations of n_p single piles, the influence of pile-pile interaction is clear. Within the range of loading investigated, the resistance-settlement behaviour of the foundation with 169 single piles (169·SP) is significantly stiffer than for the freestanding pile group with 169 piles. For settlements smaller than approximately 90 mm the 49 single piles (49·SP) exhibit stiffer behaviour than the freestanding pile group with 49 piles, while the freestanding pile group was able to carry higher loads than the 49 single piles for larger settlements. The analyses show, that even for a pile spacing of $e/d_p = 6$, interaction between the piles is evident.

For equal settlement, the pile resistances of the piled rafts are marginally smaller than the resistances of the freestanding pile groups with the same number of piles. Because of pile-pile interaction as well as pile-raft interaction, this behaviour reverses for the piled raft and the freestanding pile group with 49 piles once settlements exceed about 200 mm. The raft resistances of the piled rafts are significantly smaller than the resistances of the raft foundations for the same settlement.

Figure 2.4 clearly indicates the influence of pile-pile and pile-raft interaction on the bearing behaviour of the whole foundation. In the following, these effects are investigated with respect to the pile load and shaft friction distribution along the pile shaft for different pile positions. In Figure 2.5 the pile load distributions from the FEA of a single pile (SP), freestanding pile groups (FPG) and piled rafts (PR) are plotted for settlements of $s = 0.02 \cdot d_p = 20$ mm, $s = 0.1 \cdot d_p = 100$ mm and $s = 0.2 \cdot d_p = 200$ mm.

As observed from measurements, e.g. for the Torhaus (Section 1.3), Figure 2.5 shows the axial pile loads at the pile head, i.e. the pile resistances, increasing from the centre piles to the corner piles for all configurations investigated. For a settlement of 20 mm, the pile resistances of the freestanding pile group are

32 Combined Pile-Raft Foundations

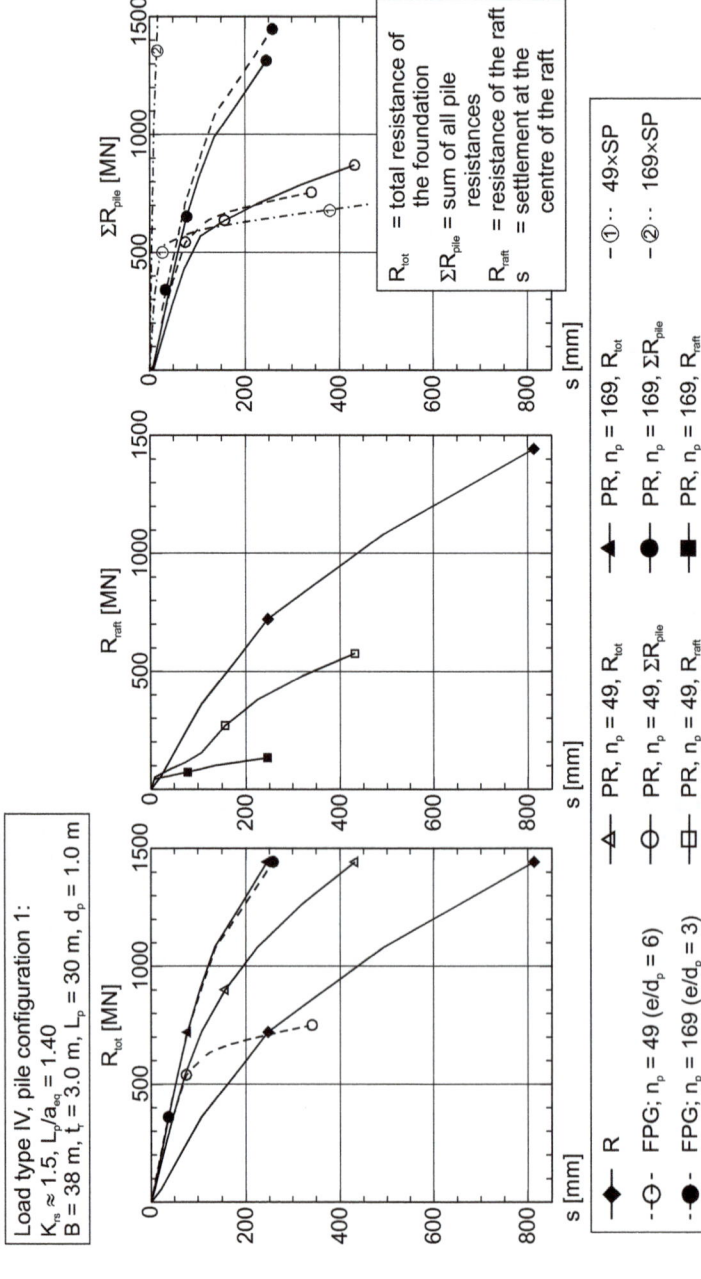

Figure 2.4 Raft foundation (R), freestanding pile group (FPG) and piled raft (PR): Resistance settlement curves.

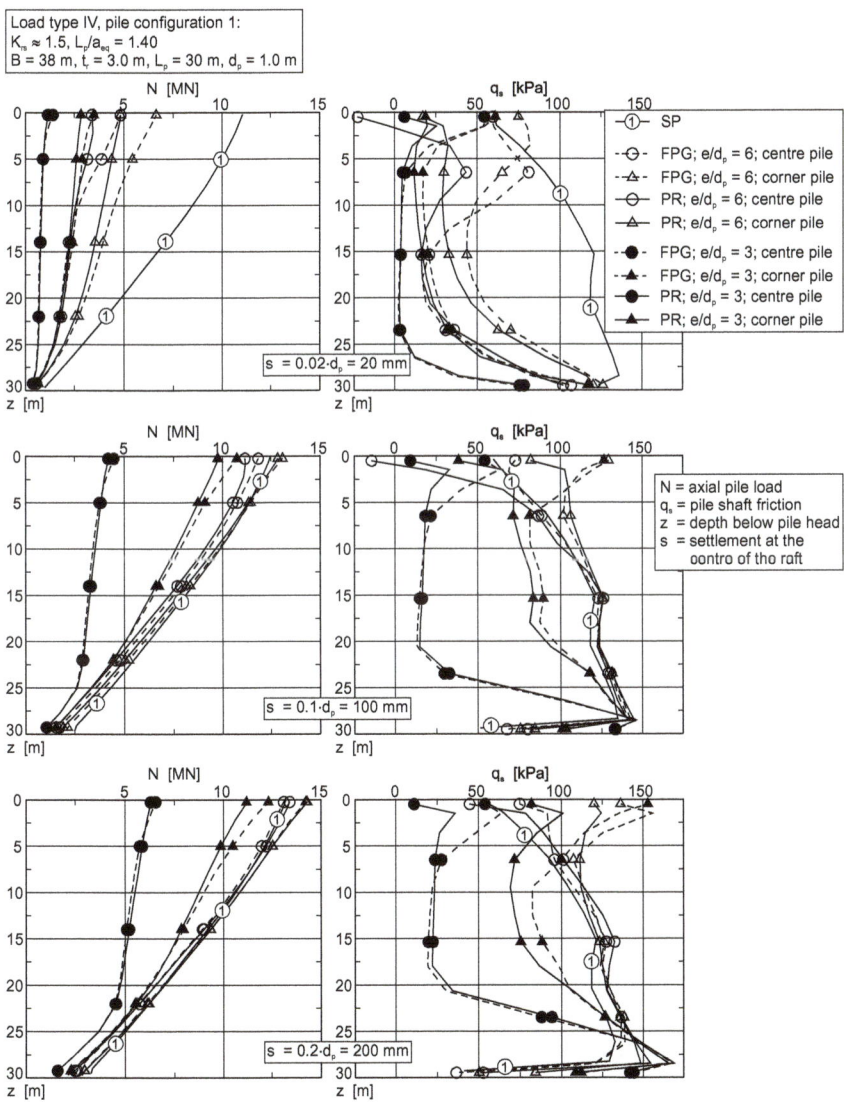

Figure 2.5 Single pile (SP), freestanding pile group (FPG) and piled raft (PR): Axial pile load and pile shaft friction distribution along the pile shaft.

considerably smaller than the resistance of the single pile. This is due to the fact that, in the central area of the freestanding pile group, the piles carry much less load than the average, so only small amounts of shaft friction are mobilised. Except for the centre pile ($e/d_p = 3$; $n_p = 169$), the pile load and shaft friction distribution of all the other investigated piles is comparable for a settlement of 100 mm. It may be seen that within the first 5 m, the corner piles of the freestanding pile group develop shaft friction of more than double the value of the single pile. The differences in pile base resistances are small and independent of the settlement level and the pile position. Further increase of settlement to 200 mm yields no qualitative change in the characteristics of the curves. The pile loads and the shaft friction have increased, but not by the same magnitude as observed from $s = 20$ mm to $s = 100$ mm. The shaft friction of the single pile obviously has reached its ultimate value and could not increase any further.

For settlements of 20 mm the piles of the piled rafts show smaller shaft friction than the piles in the freestanding pile groups, especially within the upper third of the pile shaft. For settlements greater than 100 mm, the shaft friction distributions of piles at the same position in the freestanding pile group and in the piled raft become similar. Only at the upper end of the pile shaft do the piles of the piled raft show a significant decrease of shaft friction. This is due to the prevention of relative movement between pile shaft and soil close to the raft. Similar observations were reported by Matsumoto et al. (2010) from small-scale model tests in dry sand.

The centre pile of the piled raft ($e/d_p = 6$; $n_p = 49$) shows negative shaft friction at the upper pile shaft. This phenomenon is caused by the simulation of the installation of the raft and is in accordance with observations from field measurements e.g. for the Messeturm (Section 2.3.3) and Treptowers (Section 4.3.7). An increase of the settlements to 200 mm leads to an increase in the shaft friction over the upper shaft length of the piles of the piled raft. The negative shaft friction has fully vanished. Katzenbach et al. (1998) show that the piles of the piled raft exhibit increasing shaft friction if the settlements increase further, while for the freestanding pile group the shaft friction remains almost the same. This is due to the increase in the stress level in the soil caused by the load transfer from the raft to the soil (Reul 2000). As shown before for the freestanding pile groups, the differences in the pile base resistances of the piled rafts are small, independent of the settlement level and the pile position.

Figures 2.6 and 2.7 show the contours of the settlements related to the final excavation and the changes of the mean stress (equivalent pressure stress) in the vicinities of a freestanding pile group and a piled raft comprising 49 and 169 piles, respectively, for a load of $P_{eff} = 721.7$ MN ($V_{ult}/P_{eff} \approx 5$). The change in the mean stress, Δp, is defined as:

$$\Delta p = p_P - p_0 \tag{2.10}$$

with $p = \dfrac{1}{3}(\sigma_1 + \sigma_2 + \sigma_3)$

where p_p = mean stress (equivalent pressure stress) after the load P was applied; p_0 = mean stress after the final excavation; and $\sigma_1, \sigma_2, \sigma_3$ = principal stresses. Since a drained analysis was carried out, the stresses correspond to effective stresses.

The piles of the freestanding pile group ($n_p = 49$) punch in the subsoil (Figure 2.6). Within a radius of approximately 1 m around the pile shaft the

Bearing behaviour of piled rafts 35

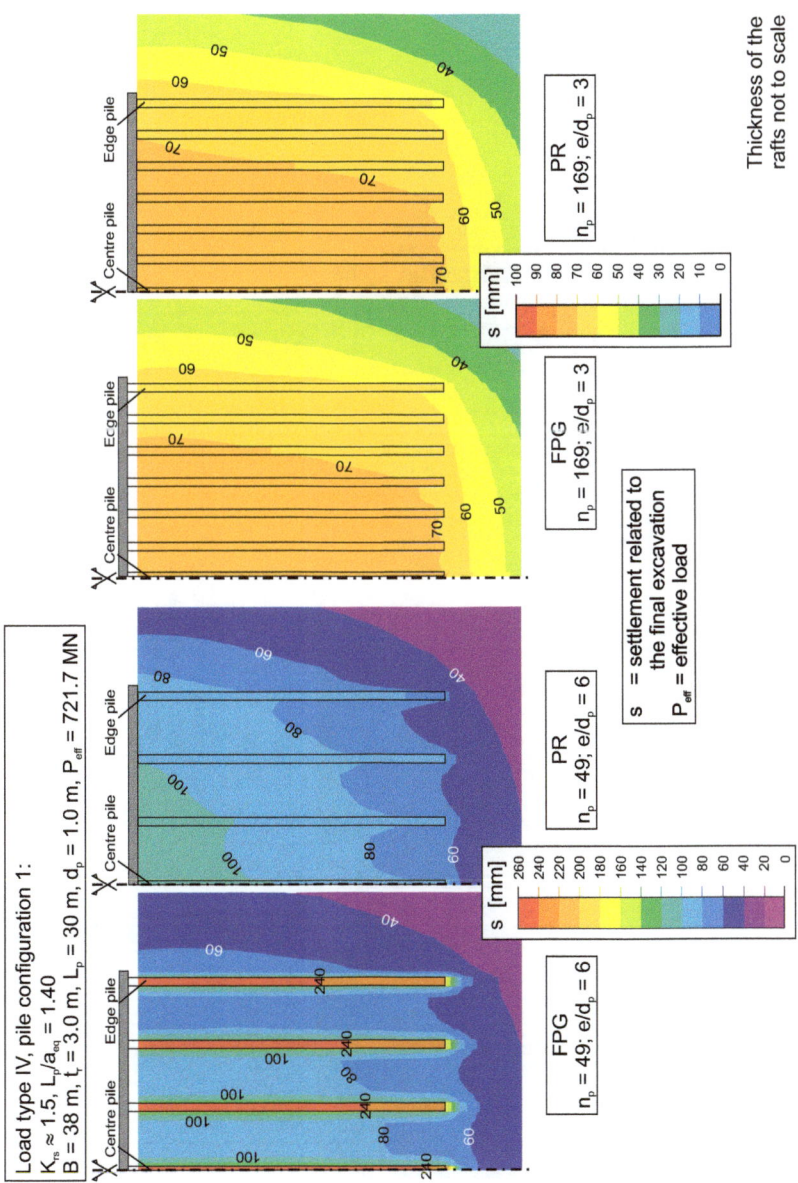

Figure 2.6 Freestanding pile group (FPG) and piled raft (PR): Contour of settlements.

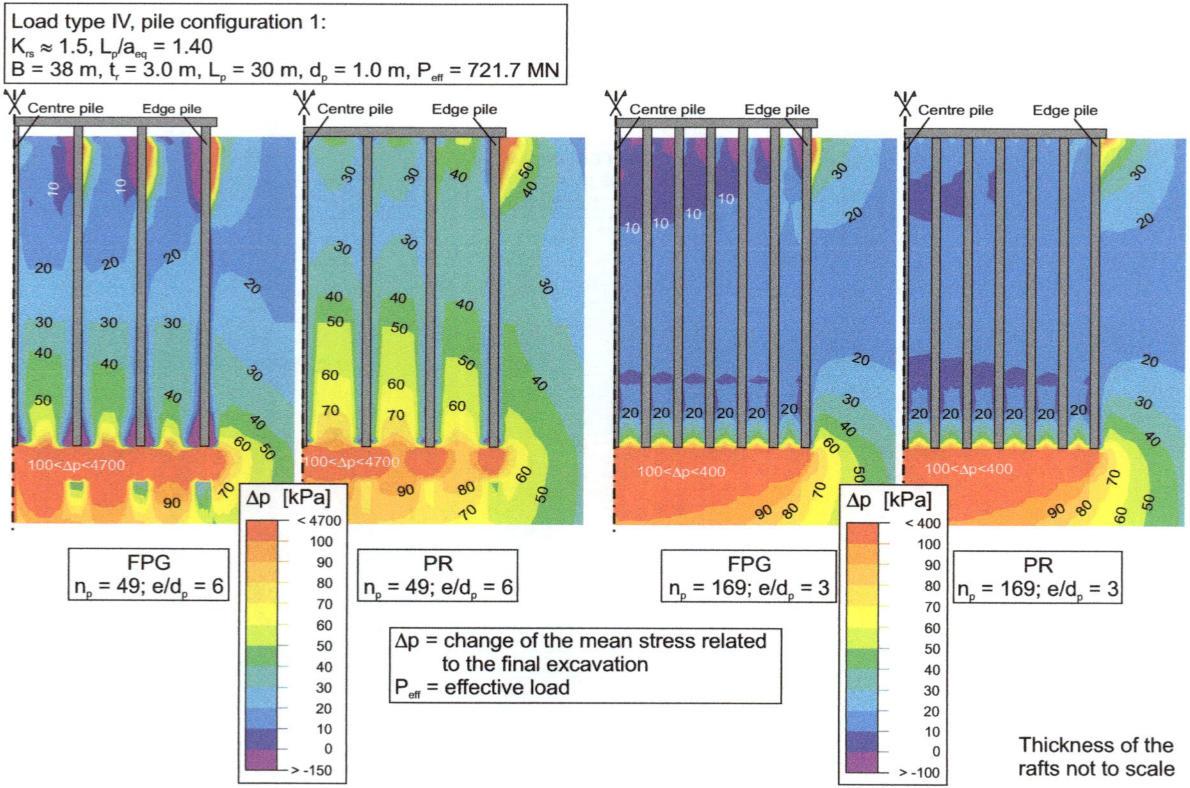

Figure 2.7 Freestanding pile group (FPG) and piled raft (PR): Contour of changes of mean stress.

differential settlements between the soil and pile reached up to 160 mm while the settlement of the pile is approximately 260 mm. The piled raft ($n_p = 49$) shows much smaller settlements of approximately 120 mm but also the differential settlements between foundation and subsoil are significantly smaller. In contrast to the two foundations with 49 piles, the freestanding pile group and the piled raft with 169 piles show block deformation of the piles and subsoil. This means that the piles inside the pile group and the soil have almost the same settlements. The small relative displacements between the piles and the soil explain the low shaft friction mobilised by the centre pile in the upper 25 m, as shown in Figure 2.5.

The scale for the change of mean stress (equivalent pressure stress) in Figure 2.7 was chosen to illustrate the changes in stress between the piles and underneath the raft. The maximum values of $\Delta p = 4700$ kPa ($n_p = 49$) and $\Delta p = 400$ kPa ($n_p = 169$) are located in each case under the pile bases. The largest changes of mean stresses between the piles and underneath the raft can be observed for the piled raft with 49 piles. This is due to the considerable contribution of the raft in the load transfer as shown in Figure 2.4.

Because of the influence of the pile-pile and the pile-raft interaction, the bearing behaviour of piles in a pile group differs significantly from the bearing behaviour of a single pile. This can be noticed from the resistance-settlement curves presented in Figure 2.8 for a piled raft with 49 piles ($e/d_p = 6$) and 169 piles ($e/d_p = 3$), respectively. As discussed before, this is mainly due to the variation in mobilised pile shaft resistance. In contrast to the single pile, no pile of the piled rafts reaches an ultimate shaft resistance although the settlements are fairly high ($s = 0.2 \cdot d_p = 200$ mm). Based on small-scale

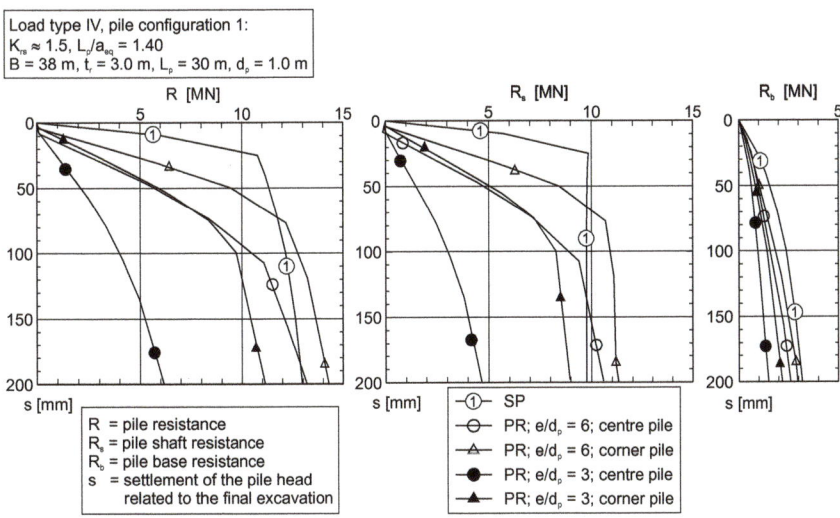

Figure 2.8 Single pile (SP) and piled raft (PR): Pile resistance-settlement curves of different piles.

model tests of piled rafts and FPG in sand, Ranganatham and Kaniraj (1978) pointed out the interaction between raft and piles. They attribute the increased shaft friction of the piles to the increased stress state in the soil due to the load transfer from the raft. Fioravante et al. (2008) also concluded that the increased shaft friction of the piles of a piled raft observed in centrifuge tests on piled rafts and single piles in water saturated fine sand is caused by the load transmitted to the soil by the raft. This is consistent with the study by Hansbo (1993) and Phung (1993) on the "creep-pile" approach (Section 1.4) for piled rafts in non-cohesive soils that the effective stress increase due to load transmitted from the raft leads to an increase in the capacity of the supporting piles. Essentially, the piles will operate below their (enhanced) creep load (Randolph 1994). It may be concluded that under practically relevant loads, at least under drained conditions, the piles of a piled raft do not reach their ultimate bearing capacity (Section 2.2.7).

So far, the results of the numerical study indicate the significance of the pile spacing on the pile-pile and the pile-raft interaction. For a settlement of 50 mm for example, the centre pile ($e/d_p = 3$) shows a pile resistance of 2 MN whereas for a pile spacing of $e/d_p = 6$ the pile resistance increases to almost 6 MN (Figure 2.8).

The bearing behaviour of piled rafts is not only strongly dependent on the geometrical configuration of the foundation but also on the load level and the corresponding settlements, respectively. Figure 2.9 shows the variation of the piled raft coefficient varying with the load level P_{eff}/V_{ult} where the evaluation of the piled raft coefficient is related to the installation of the raft

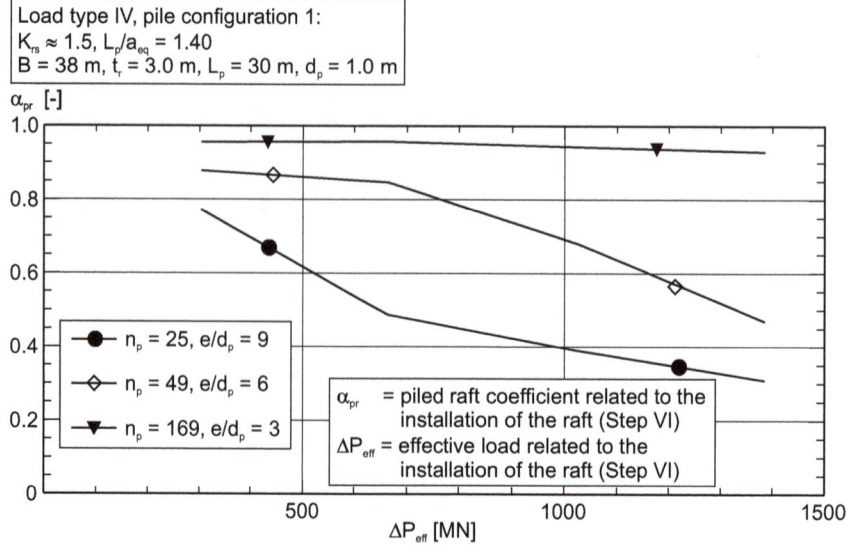

Figure 2.9 Piled raft coefficient depending on the effective load.

(Table 2.2, Step VI), i.e. neither the small resistances mobilised by the piles due to the application of the raft weight (zero stiffness of the raft) nor the weight of the raft were considered. For the foundations with $n_p = 25$ and $n_p = 49$ piles the piled raft coefficient decreases to $\alpha_{pr} = 0.31$ and $\alpha_{pr} = 0.47$ after reaching maximum values of $\alpha_{pr} = 0.77$ and $\alpha_{pr} = 0.88$, respectively. In contrast, the piled raft coefficient of the foundation with $n_p = 169$ piles changed only marginally between $\alpha_{pr} = 0.96$ and $\alpha_{pr} = 0.93$ for the investigated load levels. Based on large-scale model tests on piled rafts and free-standing pile groups in sand, Phung (1993) pointed out that the load distribution between piles and raft depends on the load level or the magnitude of the settlements. While for small settlements the piles transfer the main portion of the load into the soil, for larger settlements there is a significant participation of the raft in the load transfer according to Phung (1993). The dependency of the piled raft coefficient on the load level for relatively large pile spacings also was shown by Horikoshi and Randolph (1996) with centrifuge tests in overconsolidated Kaolin. Horikoshi and Randolph (1996) reported a decrease of the piled raft coefficient between 10% and 40% during the loading of piled rafts with a normalised pile spacing of $e/d_p = 7.94$ and concluded that this decrease of the piled raft coefficient is caused by the nonlinear pile resistance-settlement behaviour. A decrease of the piled raft coefficient with increasing load was also observed by Alnuaim et al. (2015) with centrifuge tests on a piled raft in cohesive soil supported by micropiles. Based on a numerical parametric study for piled rafts in stiff clay, Modak and Singh (2023) suggested an empirical correlation between the piled raft coefficient and the average settlement of the piled raft.

2.2.3 Stiffness

Further evidence for the conclusion that the decrease of the piled raft coefficient is caused by the nonlinear pile resistance-settlement behaviour can be found comparing the pile resistance-settlement curves of the piled rafts with $n_p = 49$ and $n_p = 169$ piles in Figure 2.4. While for the piled raft with 169 piles the ratio of the stiffness of the pile group to the total stiffness of the foundation remains almost constant, this ratio decreases significantly for the piled raft with 49 piles. The piled raft coefficient may therefore be interpreted as the ratio of the stiffness of the pile group to the total stiffness of the foundation:

$$\frac{c_{\Sigma pile}}{c_{tot}} = \frac{\frac{\Sigma R_{pile}}{s}}{\frac{R_{tot}}{s}} = \frac{\Sigma R_{pile}}{R_{tot}} = \alpha_{pr} \tag{2.11}$$

where $c_{\Sigma pile}$ = stiffness of the pile group consisting of the n_p piles of the piled raft; c_{tot} = total stiffness of the foundation; s = settlement; α_{pr} = piled raft coefficient; ΣR_{pile} = sum of all pile resistances and R_{tot} = total resistance of the foundation.

The differences between the pile resistances observed in Figure 2.8 yield substantially different equivalent spring stiffness $c_{pile,i}$ amongst the piles of the pile group. The equivalent spring stiffness is used to model the bearing behaviour of a pile when the foundation is designed based on the subgrade reaction modulus method (Winkler halfspace) (see Sections 3.1.5 and 5.6). Realistic estimation of this quantity is therefore fundamental for an economic and safe foundation design. Figure 2.10 shows the dependency of the equivalent spring stiffness on the position of the pile amongst a pile group and the load level for the two configurations. It should be noted that the absolute value of the equivalent spring stiffness and, particularly for the case of larger pile spacing, the distribution of the stiffness amongst the pile group changes significantly with increasing load level. For comparison it should be mentioned that the stiffness of a single pile subjected to comparable

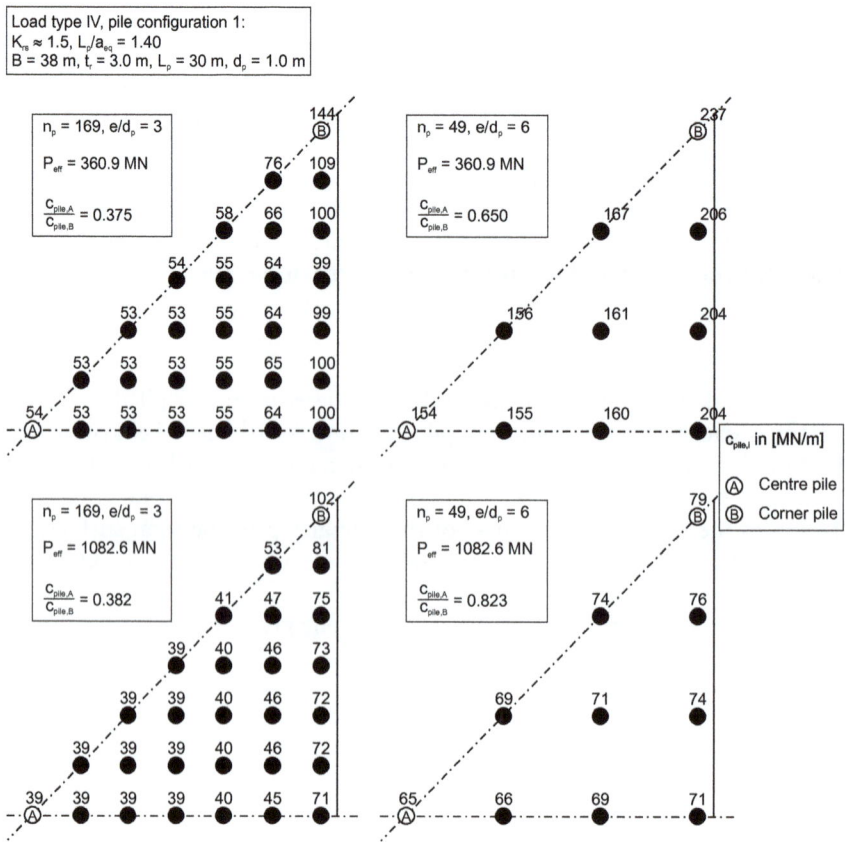

Figure 2.10 Equivalent spring stiffness for the pile depending on the pile position and load level.

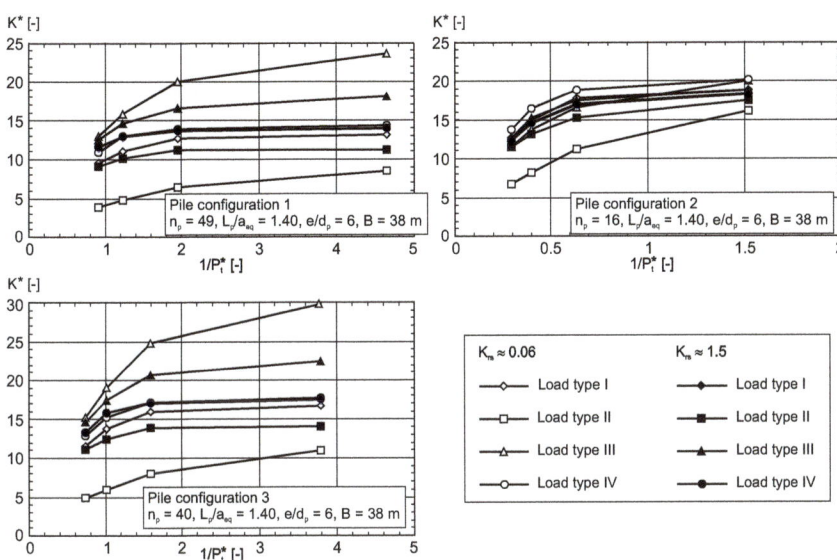

Figure 2.11 Variation of the normalised overall stiffness of the piled raft with the ratio of the combined capacity of n_p single piles to the effective applied load.

settlements as the maximum settlements of the piled rafts lies between c_{pile} = 227 MN/m (s = 50 mm) and c_{pile} = 53 MN/m (s = 250 mm).

The variation of the normalised overall stiffness of the piled raft with the ratio of the combined capacity of n_p single piles to the effective applied load is plotted in Figure 2.11 for the four different load configurations and for various pile configurations (e/d_p = 6·, L_p/a_{eq} = 1.40). While for core-edge loading (load type I) and uniform loading (load type IV), the normalised overall stiffness of the piled raft is determined mainly by the load level, i.e. P_t^*, and the pile configurations, for load type II and load type III the raft-soil stiffness ratio also plays a significant role. Generally, for all load configurations, the normalised overall stiffness decreases with increasing load level due to the nonlinear pile resistance-settlement behaviour.

2.2.4 Settlements

The bearing behaviour of raft foundations frequently is taken as reference for the bearing behaviour of piled rafts. Therefore, first some aspects of the soil-structure interaction of raft foundations subjected to different loading conditions and varying raft-soil stiffness ratios are investigated in the following. Regardless of the varying raft-soil stiffness ratios, which were achieved by varying the raft thickness, the weight of the raft was the same in all the analyses. All results are related to the situation after the installation of the raft; i.e. deformations due to the weight of the raft are not considered.

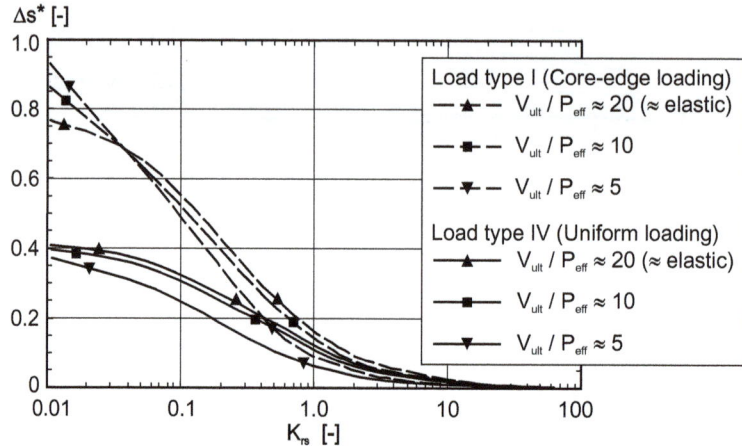

Figure 2.12 Raft foundation: Variation of normalised differential settlements with raft-soil stiffness ratio.

Figure 2.12 shows the variation of the normalised differential settlement with the raft-soil stiffness ratio for uniform loading (load type IV) and core-edge loading (load type I) for different load levels. As expected, the normalised differential settlements decrease with increasing raft-soil stiffness for both load types and become zero for a rigid raft.

For the raft foundation under uniform loading a decrease of the normalised differential settlements with increasing load level can be observed. This is due to the plastic deformations under the edge being larger than those under the centre of the raft, leading to a decrease in the differential settlements. Furthermore, the average settlements increase due to the plastic deformations.

For the core-edge loading the behaviour is more complex since for small raft-soil stiffness ratios ($K_{rs} \leq 0.04$), the normalised differential settlements increase with increasing load level. While the plastic deformations under the edge of the raft are not influenced by the raft-soil stiffness ratio, the plastic deformations under the centre increase significantly for flexible rafts under this type of nonuniform loading. Hence, the differential settlements increase, and, since the increase is larger than the increase in average settlements due to plastic deformations, the normalised differential settlements increase. Furthermore, the normalised differential settlements are significantly higher for this load case than for the uniform loading.

Figure 2.13 shows the variation of the normalised differential settlement of the piled raft with the ratio of the combined capacity of n_p single piles to the effective applied load. The main differences in the normalised differential settlements arise due to the different raft-soil stiffness ratios and load configurations while the load level is of lesser importance. The absolute value of the normalised differential settlement increases with decreasing

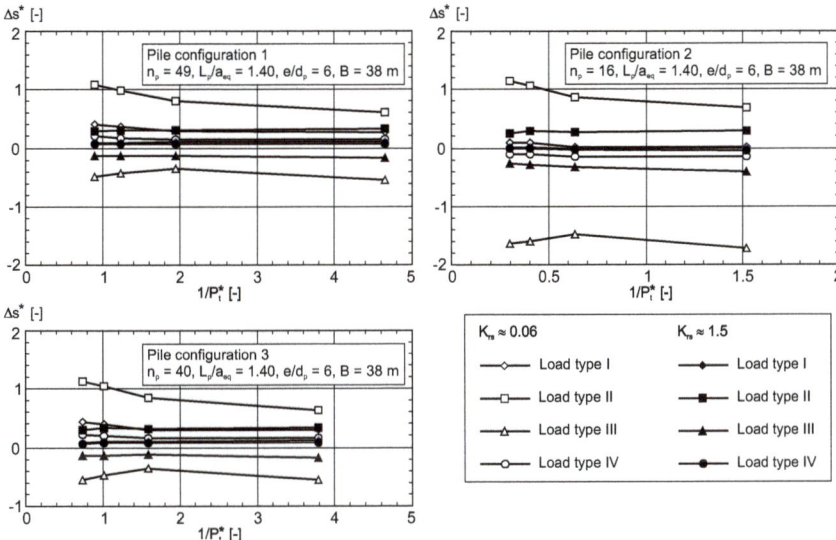

Figure 2.13 Variation of the normalised differential settlement of the piled raft with the ratio of the combined capacity of n_p single piles to the effective applied load.

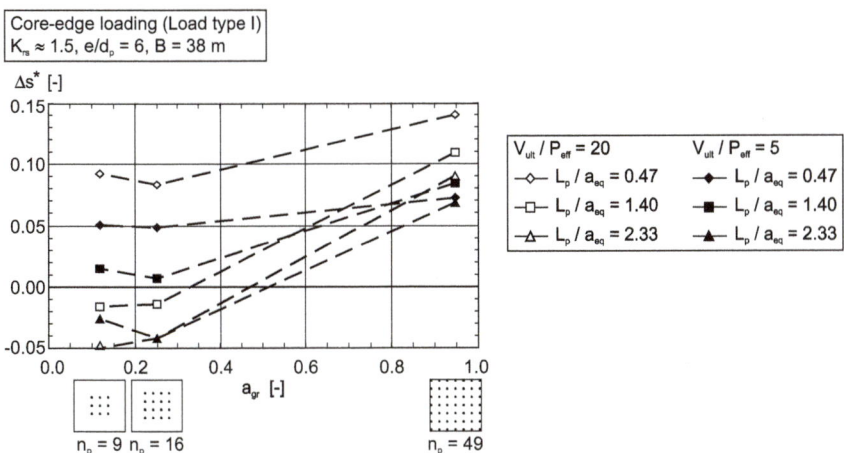

Figure 2.14 Variation of the normalised differential settlement with the pile group-raft area ratio.

raft-soil stiffness ratio. Localised loading of the raft such as load type II and load type III also tends to increase the absolute value of the normalised differential settlement.

Figure 2.14 shows the variation of the normalised differential settlement with the pile group-raft area ratio for piled rafts ($K_{rs} \approx 1.5$, $e/d_p = 6$) subjected to core-edge loading. The normalised differential settlements can be minimised or even reduced to zero with the right pile length if the piles are installed only under the central area of the raft. This supports the design concept for the minimisation of the normalised differential settlements proposed by Horikoshi and Randolph (1998) on the base of the analyses of piled rafts under uniform loading. A minimum pile length for this to be effective appears to be $L_p/a_{eq} \approx 1.4$. Similar results were reported by Prakoso and Kulhawy (2001). They identified the pile group-raft area ratio and the pile length as the most influential elements of the system geometry on the reduction of differential settlements.

The variation of the coefficient for the average settlement with (a) the relative pile length and (b) the total pile length is plotted in Figure 2.15 for six different foundations under core-edge loading (V_{ult}/P_{eff} = 5). For all investigated foundations the coefficient for the average settlement decreases with increasing relative pile length. An efficient reduction of the average settlement can be achieved with a smaller number of longer piles rather than with a large number of shorter piles. A coefficient for the average settlement of ξ_s = 0.6, for example, can be achieved with a total pile length of $n_p \cdot L_p$ = 652 m (pile configuration 2, n_p = 16, L_p/a_{eq} = 1.9) or with $n_p \cdot L_p$ = 2174 m (Pile configuration 1, n_p = 169, L_p/a_{eq} = 0.6). The potential savings of an optimised design process are obvious.

For load type IV (uniform loading) and pile configuration 1, Figure 2.16 shows the range of feasible values of the coefficient for the maximum settlement $\xi_{s,max}$ depending on the piled raft coefficient. For a given piled raft coefficient the coefficient for the maximum settlement appears to be larger than a certain threshold value and, depending on the system configuration and the load level, varies in a certain range.

2.2.5 Efficient reduction of settlements

For a piled raft with B = 38 m, d_p = 1 m and P_{eff} = 721.7 MN ($K_{rs} \approx 1.5$, $V_{ult}/P_{eff} \approx 5$), Figure 2.17 shows contours of the coefficient for the maximum settlement depending on the pile length and the number of piles. For a constant number of piles the coefficient for the maximum settlement decreases with increasing pile length, while for a constant pile length, an increase of the number of piles above $n_p \approx 75$ yields no significant reduction of the coefficient for the maximum settlement. It can be concluded that for certain configurations the total pile length, $n_p \cdot L_p$, can be reduced with only negligible influence on the maximum settlement of the foundation. This was demonstrated by Russo (1998) for the foundation of the multi-storey building Stonebridge Park (Cooke et al. 1981) by means of reducing the number of piles.

For the Torhaus in Frankfurt briefly introduced in Section 1.3, Reul and Randolph (2003) presented the results of a numerical study with a 3D finite element model applying the afore mentioned elasto-plastic cap-model for the Frankfurt Formation. Figure 2.18 shows the coefficients for maximum

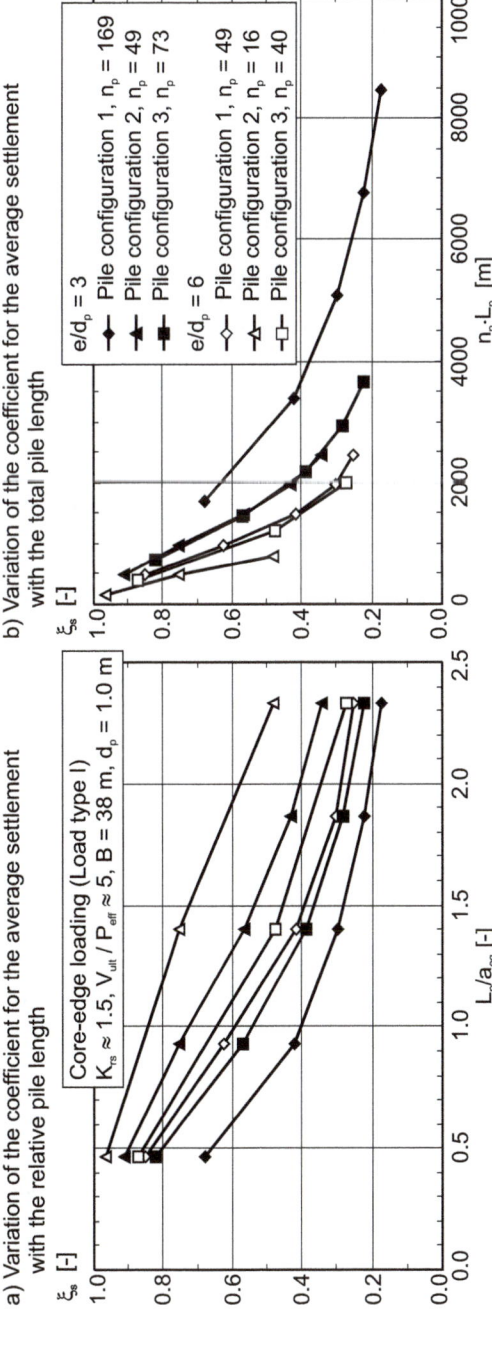

Figure 2.15 Coefficient for the average settlement.

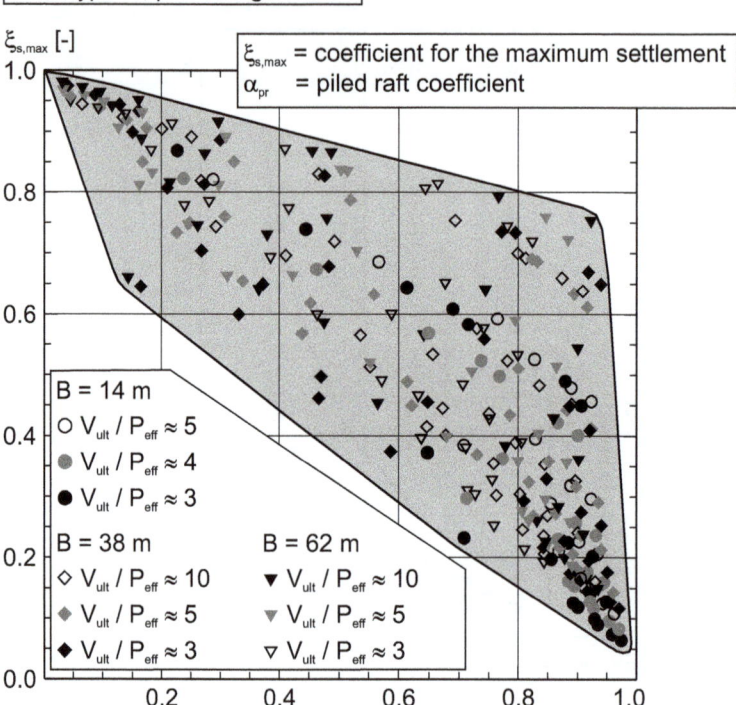

Figure 2.16 Variation of coefficient for maximum settlement depending with piled raft coefficient.

and differential settlement, $\xi_{s,max}$ and $\xi_{\Delta s}$, and the piled raft coefficient, α_{pr}, depending on the total pile length, $n_p \cdot L_p$, for the real pile configuration and two modified pile configurations. For the modified pile configurations A and B, the number of piles was reduced to $n_p = 60$ and $n_p = 40$, respectively. The pile length was varied between $L_p = 20$ m and $L_p = 27.5$ m. Since the raft is loaded uniformly, the necessity to install piles (in view of column loadings as discussed by Poulos (2001)) is not addressed by the modified pile configurations.

For the real pile configuration with a total pile length of $n_p \cdot L_p = 1680$ m, a coefficient for the maximum settlement of $\xi_{s,max} = 0.51$ is achieved. The same value can be attained with a significantly smaller total pile length for a modified pile configuration with longer piles (Figure 2.18c). For pile configuration A the total pile length amounts to $n_p \cdot L_p = 1389$ m ($L_p = 23.1$ m) and for pile configuration B, the total pile length amounts to $n_p \cdot L_p = 1100$ m ($L_p = 27.5$ m).

Moreover, for pile configuration A ($L_p = 23.1$ m) the coefficient for the differential settlement yields only $\xi_{\Delta s} = 0.16$ compared to $\xi_{\Delta s} = 0.50$ for the real pile configuration (Figure 2.18d). Pile configuration B causes hogging of

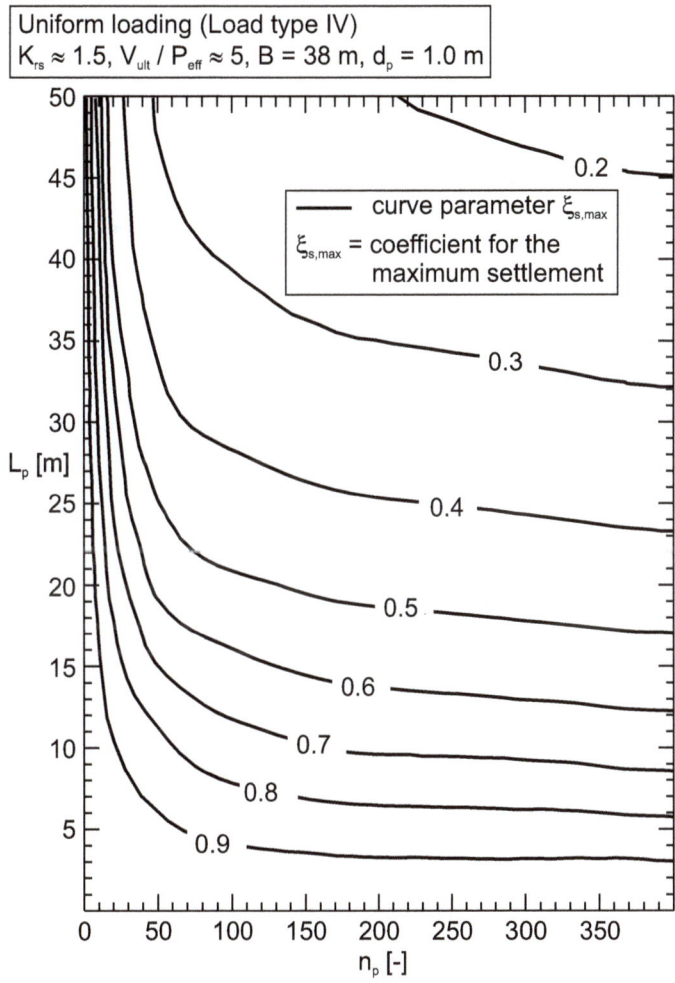

Figure 2.17 Coefficient for maximum settlement depending on the pile length and the number of piles.

the raft ($\xi_{\Delta s} < 0$) for all investigated pile length. These results are in good agreement with the centrifuge model test and numerical studies presented by Horikoshi and Randolph (1998) where the differential settlements of a uniformly loaded raft were minimised by the installation of piles under the central area of the raft.

For the investigated pile configurations, the piled raft coefficient increases with increasing total pile length from $\alpha_{pr} = 0.52$ (Pile configuration B, $L_p = 20$ m) to $\alpha_{pr} = 0.76$ for the real pile configuration (Figure 2.18e). Comparison with the coefficient for the maximum settlement in Figure 2.18c shows that the same maximum settlement can be achieved for different contributions of

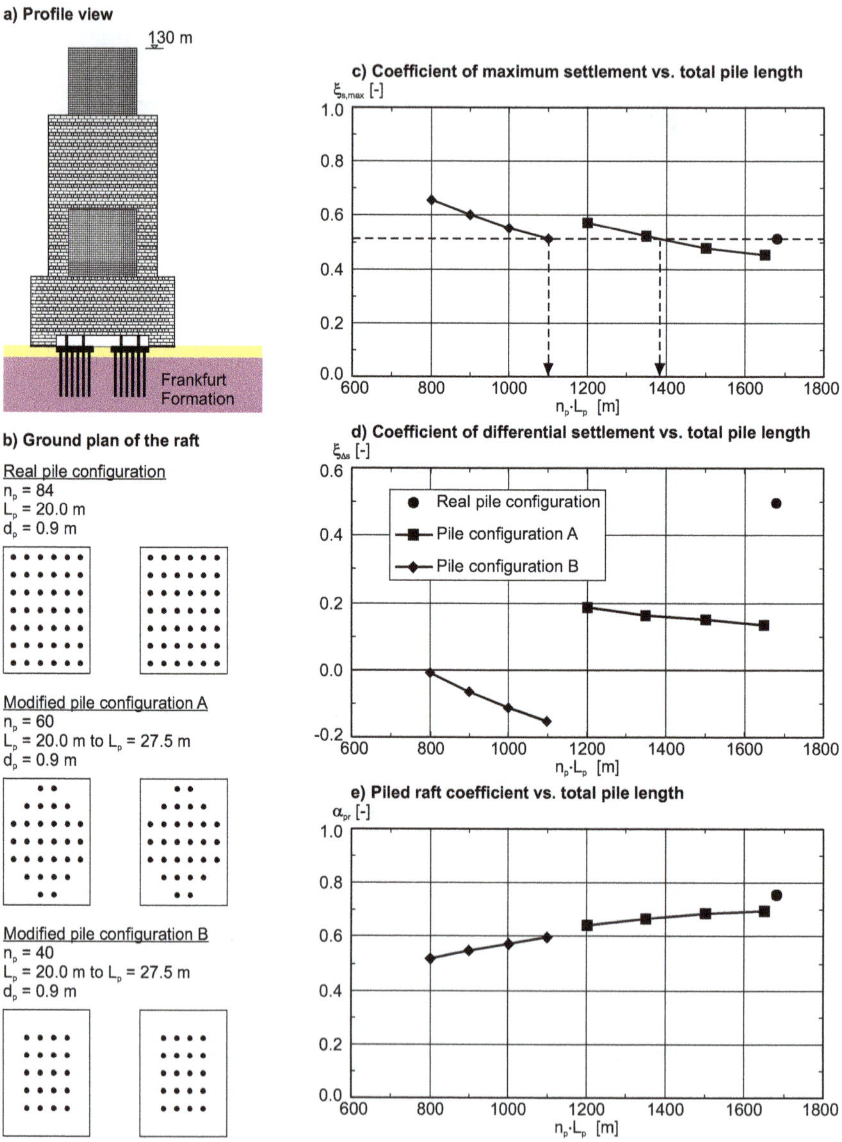

Figure 2.18 Torhaus: Variation of coefficient for maximum and differential settlement and piled raft coefficient with total pile length.

the piles to the load transfer, as was also found in the parametric study (Figure 2.16).

The results presented in this section indicate that it is possible to achieve a required maximum or differential settlement with largely different total pile lengths and contributions of the piles to the load transfer. This becomes even more significant for small settlement reduction coefficients. An economic optimum design would therefore be a design where one or more constraints, such as a maximum settlement, are achieved with the minimum total pile length (for the simplified assumption that the cost of the foundation is proportional to the total pile length). This will be explored further in Section 2.2.8.

2.2.6 Bending moments

After considering the settlement behaviour, this section examines the bending behaviour of a raft foundation. Figure 2.19 shows the variation of the normalised maximum positive bending moment per unit length with the raft-soil stiffness ratio for uniform loading (load type IV) and core-edge loading (load type I) for different load levels. For comparison, the results obtained by Fraser and Wardle (1976) for a uniformly loaded square raft in frictionless contact with a homogeneous isotropic halfspace are also plotted. It should be mentioned that the influence of the finite depth of the compressible soil layer on the solution given by Fraser and Wardle (1976) can be neglected. The difference between the finite element analysis for a very low load level ($V_{ult}/P_{eff} \approx 20$, approximately elastic solution) and the solution by Fraser and Wardle (1976) is therefore due to the nonhomogeneous subsoil conditions, i.e. the increasing Young's modulus with depth.

The normalised maximum positive bending moment per unit length decreases with increasing load level. This observation is in good agreement with the results presented by Hemsley (1998, p. 399) for a circular raft under uniform loading where nonlinearities result from limiting contact pressures.

From the results shown, it can be concluded that the representation of the soil behaviour with an equivalent elastic modulus, which decreases with increasing load level, clearly leads to an overestimation of the maximum positive bending moments. In an analysis that takes account of the nonlinear

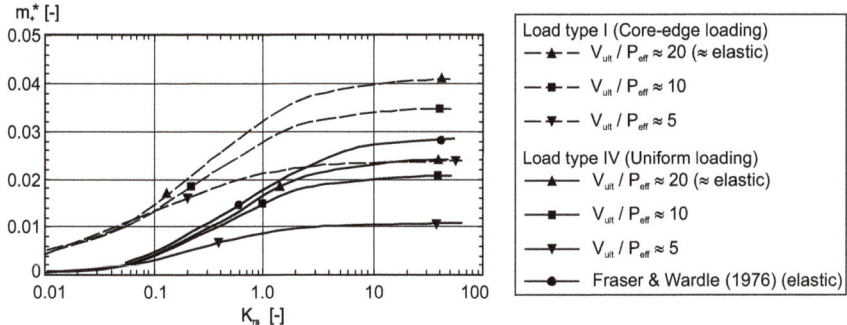

Figure 2.19 Raft foundation: Variation of normalised maximum bending moments per unit length with raft-soil stiffness ratio.

50 Combined Pile-Raft Foundations

soil behaviour and the limiting contact pressure, higher load levels lead to smaller maximum positive bending moments because of the more uniform contact pressure stress distribution below the raft.

The normalised bending moment per unit length acting on a cross section perpendicular to the x-axis is plotted in Figure 2.20a for the cross sections A-A for both uniform and core-edge loading. While for uniform loading the elastic solution (very low load level, $V_{ult}/P_{eff} = 20$) can be assumed to give a

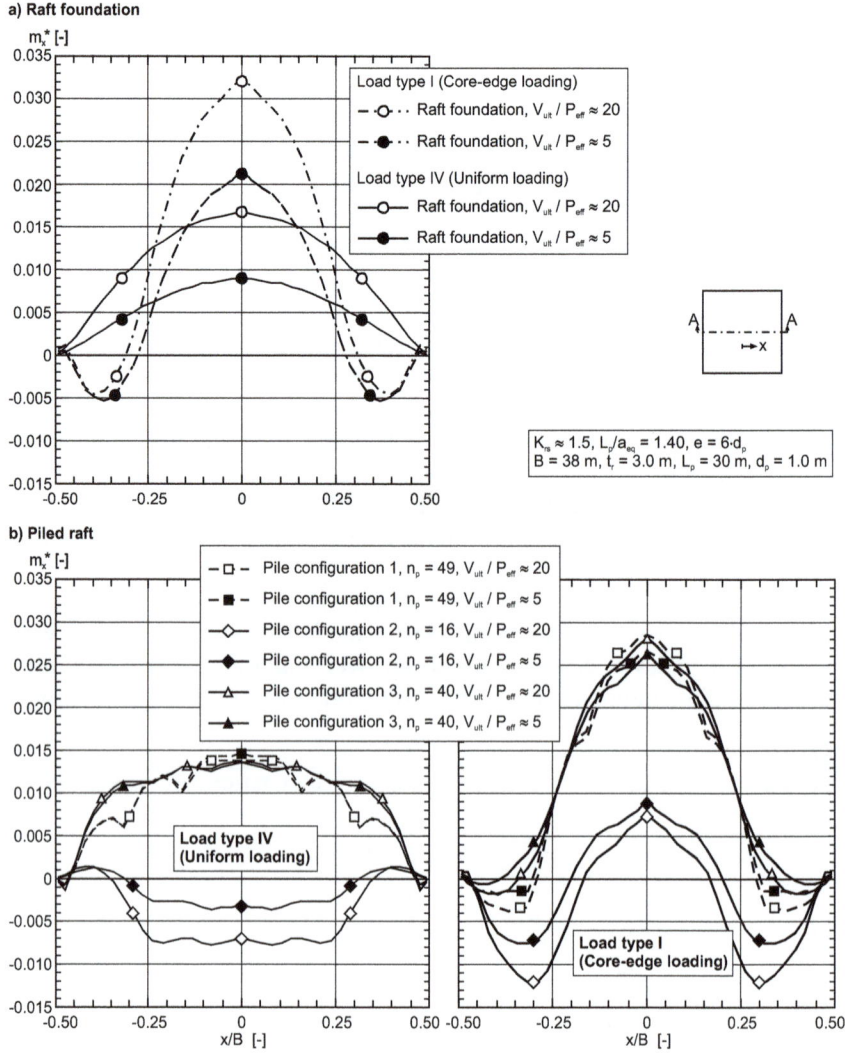

Figure 2.20 Raft foundation and piled raft: Normalised bending moments per unit length in cross section A-A.

conservative upper limit for the bending moments, the situation is not so obvious for more complex loadings such as the core-edge loading. First of all, there are areas of the raft where the absolute value of the (negative) bending moment increases with increasing load level. Moreover, in some areas an elastic analysis predicts positive bending moments while for higher load level the bending moments become negative. However, it should be noted that an analysis that accounts for the nonlinear, elasto-plastic material behaviour of the raft concrete may result in more uniform bending moment distributions. If the yield capacity of the concrete is attained and exceeded in some areas of the raft, moment redistribution will take place, depending on the layout and characteristics of the reinforcement.

In Figure 2.20b three different pile configurations ($K_{rs} \approx 1.5$, $L_p/a_{eq} = 1.40$, $e/d_p = 6$) under uniform and core-edge loading are compared to the raft foundations in Figure 2.20a. For the piled rafts, as well as for the raft foundations, the normalised bending moment distribution depends on the type and level of load.

From Figure 2.20b it can be concluded that, for the configurations investigated, the installation of piles yields no benefits in terms of reduction of the bending moments compared to the raft foundation in the case of uniform loading. Pile configuration 2 with piles placed only under the central area of the raft changes the deformation mode of the raft from sagging to the structurally less favourable hogging.

For core-edge loading, pile configuration 3 reduces the negative bending moments most efficiently at the expanse of slightly increased maximum moments in the centre of the raft for $V_{ult}/P_{eff} \approx 5$, while pile configuration 2 shows increased negative bending moments compared to the raft foundation (see Figure 2.20a).

2.2.7 Ultimate capacity under vertical loading

Although rarely pertinent in practice because of the ensuing large settlements, it is instructive to consider the bearing behaviour of a piled raft under high loads. This is investigated in the following for pile configuration 1 (B = 38 m, $K_{rs} = 1.5$, $n_p = 49$, $L_p = 30$ m, $d_p = 1.0$ m) subjected to load type IV (uniform load). The maximum applied load of $P_{eff} = 2165.1$ MN ($V_{ult}/P_{eff} \approx 2$) causes settlements of s = 99 cm at the centre of the raft. Failure of the foundation, in the sense of loss of external stability characterised by a significant increase of the settlement increment for a given load increment, is not observed (Figure 2.21). This was to be expected as the load applied is still much less than the ultimate capacity of the raft foundation.

The steady increase in pile resistance evident in Figure 2.21 is due to changes of the stress state in the soil due to the load transfer from the raft, as shown in Figure 2.22 by means of the change in mean stress Δp and the degree of utilisation of the deviatoric stress μ_d for loads of $P_{eff} = 1443.4$ MN

52 Combined Pile-Raft Foundations

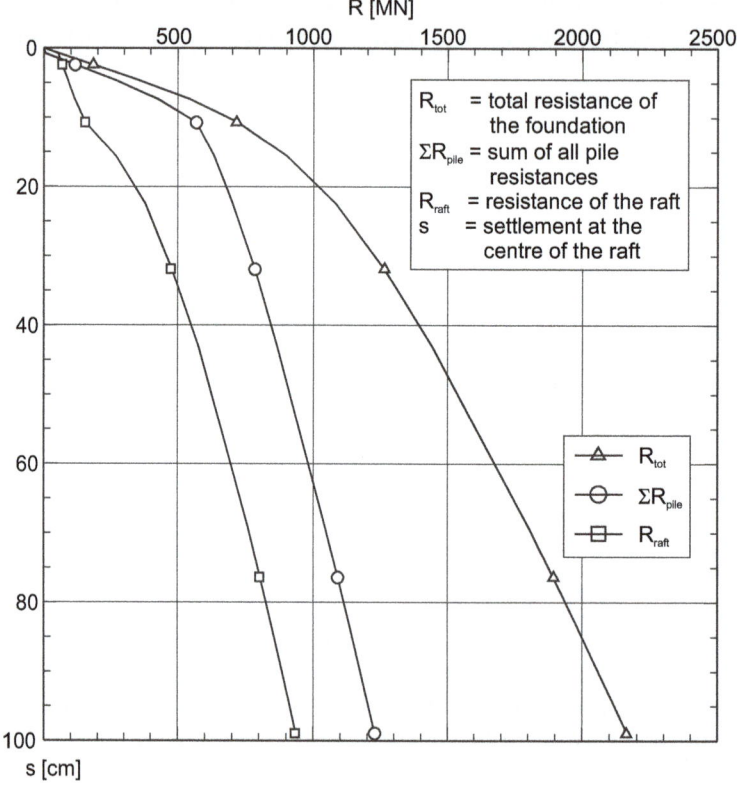

Figure 2.21 Piled raft (PR): Resistance settlement curves.

and P_{eff} = 2165.1 MN. The degree of utilisation of the deviatoric stress μ_d is defined as

$$\mu_d = \frac{q_{current}}{q_{maximum}} \cdot 100\,[\%] \qquad (2.12)$$

where $q_{current}$ = is the current deviatoric stress; and $q_{maximum}$ = maximum possible deviatoric stress for the current mean stress. (Note, the discontinuity in the contour plot in Figure 2.22c is due to the change of the earth pressure coefficient at rest K_0 at a depth of $z = 25$ (Table 2.1).)

The scale chosen for the change of the mean stress in Figure 2.22b is primarily intended to illustrate the stress states below the raft and between the piles. The maximum values below the pile base are therefore only captured qualitatively. In the area between the piles, a change of mean stress from

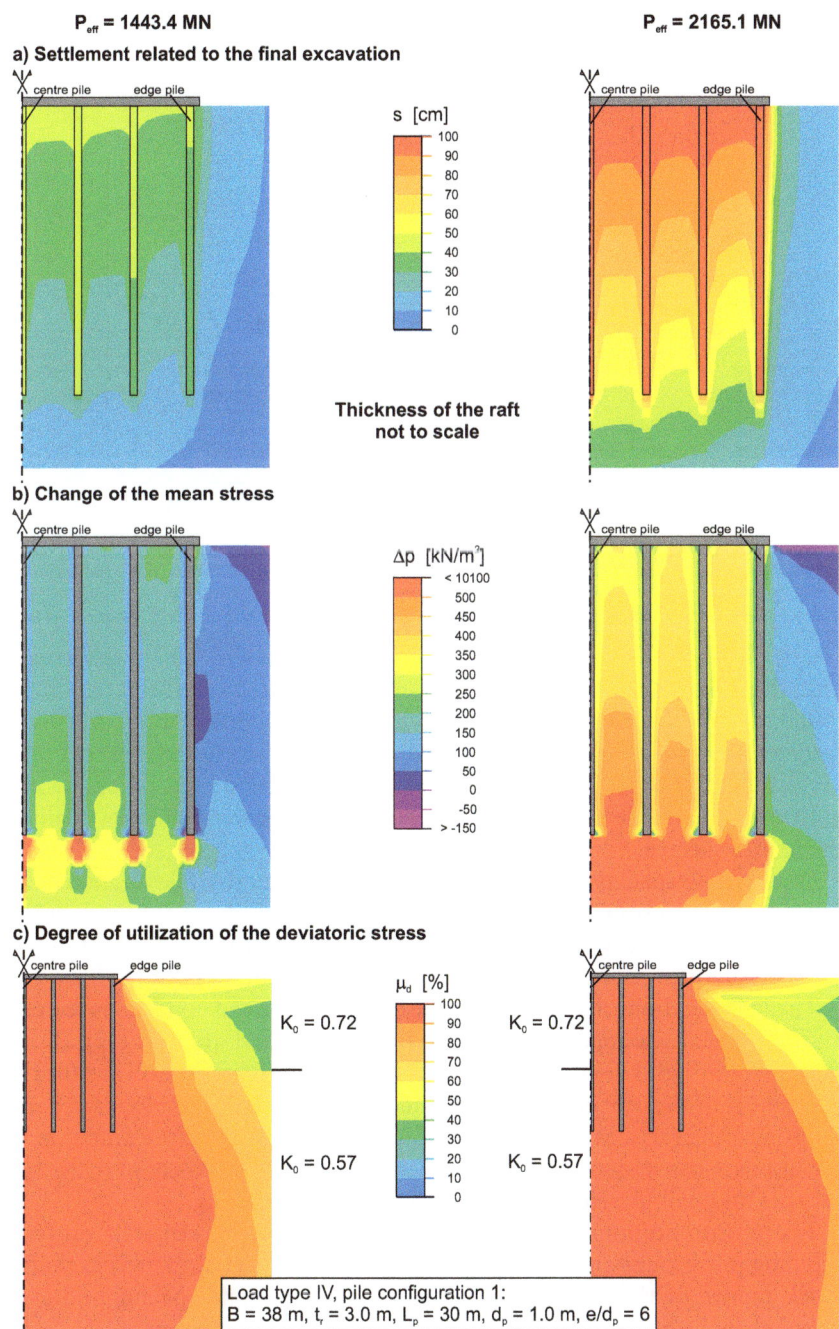

Figure 2.22 Piled raft (PR): Contours of settlements, changes of mean stress and degree of utilisation of the deviatoric stress.

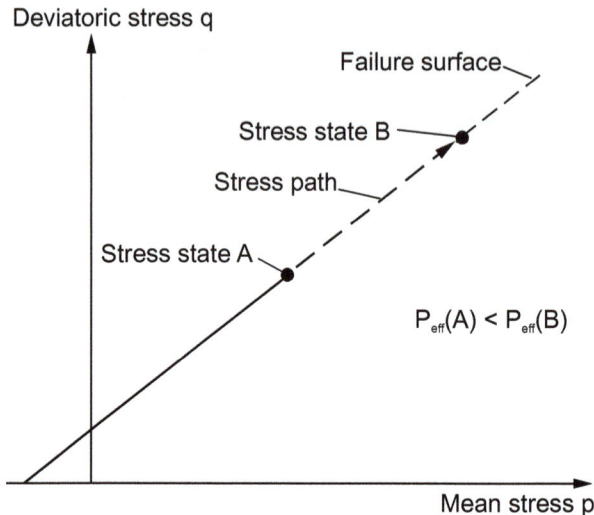

Figure 2.23 Qualitative representation of the stress state in the soil close to the pile shaft.

$\Delta p \approx 175$ kN/m² (P_{eff} = 1443.4 MN) to $\Delta p \approx 375$ kN/m² (P_{eff} = 2165.1 MN) can be observed. This increase of stress is the reason why a further increase of load is possible despite an utilisation of 100% in the entire area of the foundation. For a given load, the maximum possible deviatoric stress was reached; i.e. the stress point reached the failure surface (Figure 2.23). However, any further load also results in an increase in mean stress, which in turn allows the deviatoric stress to increase. The stress state moves along the failure surface.

For the pile spacing of $e/d_p = 6$ investigated here, the piles significantly punch into the ground, as shown in the contour plots of the settlements Figure 2.22a. At the level of the foundation base, there is a significant differential settlement between the piled raft and the surrounding soil.

The investigations indicate that, at least under drained conditions for primarily vertically loaded piled rafts with large ground area, as typical, for example, for high-rise buildings, the ultimate capacity is not decisive for the assessment of the bearing behaviour under practically relevant loads in comparison to the serviceability. Nevertheless, approaches for the estimation of the ultimate capacity of piled rafts have been discussed widely over the last decades and will therefore be summarised in the following.

Taking into account a large number of large-scale model tests on freestanding pile groups and piled rafts in sandy soils, Liu et al. (1985) excluded block failure of the foundation body regardless of pile spacing and suggested that the bearing capacity could be estimated as follows:

$$R_{PR,ult} = R_{FPG,ult} + R_{R,ult} \qquad (2.13)$$

where $R_{PR,ult}$ = ultimate capacity of the piled raft; $R_{FPG,ult}$ = ultimate capacity of a pile group; $R_{R,ult}$ = ultimate capacity of a raft foundation.

Poulos (2000) extended the approach by Liu et al. (1985) in the form:

$$R_{PR,ult} = \min \begin{Bmatrix} R_{FPG,ult} + R_{R,ult} \\ R_{B,ult} + R_{\Delta R,ult} \end{Bmatrix} \quad (2.14)$$

where $R_{B,ult}$ = ultimate capacity of the block containing the piles and $R_{\Delta R,ult}$ = ultimate capacity of the raft outside the periphery of the pile group.

Liu et al. (1994) introduced two coefficients α_{FPG} and α_R to be applied to the ultimate capacity of the pile group and the raft in the piled raft, respectively, and suggested the following approach:

$$R_{PR,ult} = \alpha_{FPG} \cdot R_{FPG,ult} + \alpha_R \cdot R_{R,ult} \quad (2.15)$$

De Sanctis and Mandolini (2006) defined the following two coefficients as:

$$\beta_{pr} = \frac{R_{pr,ult}}{R_{FPG,ult}} \quad (2.16)$$

$$\xi_{pr} = \frac{R_{pr,ult}}{R_{R,ult} + R_{FPG,ult}} \quad (2.17)$$

Evaluating the experimental data by Cooke (1986), Conte et al. (2003), Brand et al. (1972), Liu et al. (1994), Sales (2000) and Borel (2001), De Sanctis and Mandolini (2006) gave ranges for these parameters of β_{pr} = 1.06 to β_{pr} = 9.57 and ξ_{pr} = 0.80 to ξ_{pr} = 1.04.

De Sanctis and Mandolini (2006) presented the results of a numerical parametric study with the FEM for undrained conditions, applying the Tresca model for the soil. In that study, the ultimate capacity of the raft foundation and the piled raft were evaluated for a settlement of $s = 0.1 \cdot B$. As can be expected when the shaft resistance is not influenced by the stress state in the soil, i.e. when the strength of the soil-pile interface has no frictional component, the ultimate capacity of the pile within the piled raft is approximately the same as the ultimate capacity of a pile group, i.e. $\alpha_{FPG} \approx 1$. Based on that, De Sanctis and Mandolini (2006) suggested that the ultimate capacity of a piled raft under vertical loading might be calculated as:

$$R_{PR,ult} = R_{FPG,ult} + \alpha_R \cdot R_{R,ult} \quad (2.18)$$

$$\alpha_R = 1 - 3 \frac{A_{FPG}}{A} \cdot \frac{d_p}{e} \quad (2.19)$$

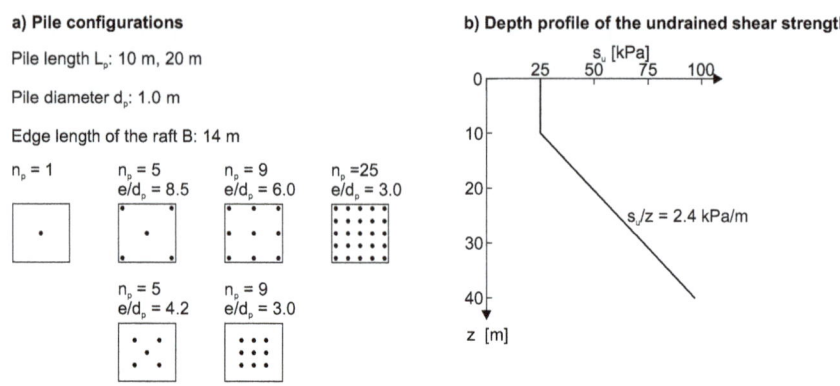

Figure 2.24 Finite element limit analysis (FELA): System configuration.

where A_{FPG} = area occupied by the pile group; A = total area of the raft; d_p = pile diameter and e = pile spacing.

To calculate upper and lower bounds of the failure load the finite element limit analysis (FELA) is an option. Sloan (2013) provides a detailed overview of the FELA including its application on geotechnical boundary value problems. Figure 2.24a shows the system configuration for a FELA carried out with the program OptumG3 (Optum CE 2022). For the study mixed elements of tetrahedral shape were used to give a failure load between the upper and lower bound. The rigid raft (B = 14 m) is placed on top of the soil, i.e. no embedment is considered. The bottom surface of the model is located in a depth of $2 \cdot L_{p,max}$ = 40 m. In horizontal direction the model extends $3 \cdot B$ = 42 m from the centre of the raft, i.e. 35 m behind the edge of the raft. The finite element mesh of the system comprises 1/4 of the complete 3D problem considering the two symmetry planes under vertical, central loading. In the finite element model, the circular piles (d_p = 1 m) were replaced with square piles with an equivalent shaft circumference. For the soil a Tresca model was employed with the undrained shear strength profile plotted in Figure 2.24b taken from the study by De Sanctis and Mandolini (2006).

For one configuration (n_p = 25; L_p = 10 m; e_p/d_p = 3) Figure 2.25 shows the failure mechanism of the piled raft, the raft foundation and the freestanding pile group based on the evaluation of the shear dissipation. For all the investigated configurations in Figure 2.24a, the adaptive mesh refinement resulted in 87200 to 99400 elements for the piled rafts, 93700 elements for the raft foundation and to 63200 to 89800 for the freestanding pile groups.

Figure 2.26a compares the coefficient α_R required for the evaluation of Eq. (2.18) with Eq. (2.19) proposed by De Sanctis and Mandolini (2006) which appears to be a conservative approach for the configurations studied here. Figure 2.26b shows the coefficient defined by Eq. (2.17) derived from

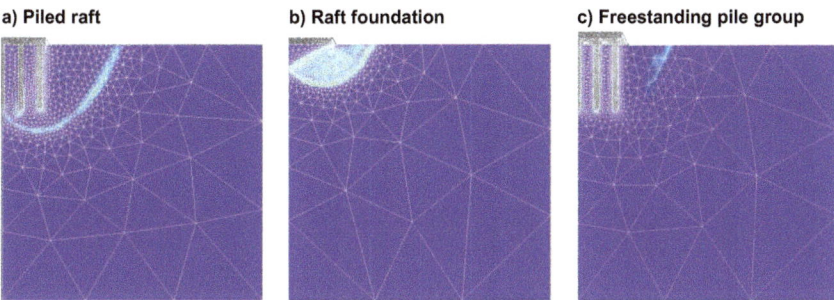

Figure 2.25 Finite element limit analysis (FELA): Failure mechanism based on the evaluation of the shear dissipation ($n_p = 25; L_p = 10$ m; $e_p/d_p = 3$).

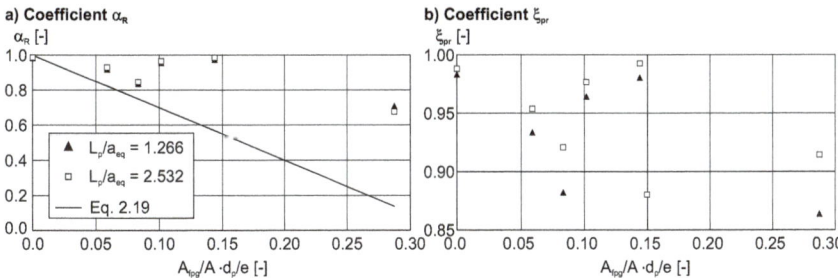

Figure 2.26 Finite element limit analysis (FELA): α_R and ξ_{pr} vs. $\dfrac{A_{FPG}}{A} \cdot \dfrac{d_p}{e}$.

FELA varying between $\xi_{pr} = 0.86$ to $\xi_{pr} = 0.99$ which is within the bandwidth for this parameter summarised by De Sanctis and Mandolini (2006) from experimental data mentioned above.

2.2.8 Example design optimisation

Based on the parametric study reported by Reul (2000, 2001), a set of charts was developed for the load configurations shown in Figure 2.2, where the coefficients for average settlement, differential settlement and maximum positive bending moment are plotted as a function of the total pile length (Reul 2001). Figure 2.27 shows as an example the charts for a foundation under core-edge loading with $n_p = 33$ piles. It is worth mentioning that there are, in fact, system configurations that yield increased differential settlements and maximum positive bending moments of the piled raft compared to the raft foundation ($\xi_{\Delta s} > 1, \xi_{m+} > 1$).

In the following example, an optimised design process is investigated for a foundation ($B = 38$ m, $P_{eff} = 721.7$ MN, $V_{ult}/P_{eff} = 5$) subjected to core-edge loading (load type I). Two different raft thicknesses were considered: $t_r = 1$ m ($K_{rs} \approx 0.06$) and $t_r = 3$ m ($K_{rs} \approx 1.5$). The constraints that ensure satisfactory bearing behaviour of the piled raft are expressed in terms of the behaviour of

58 Combined Pile-Raft Foundations

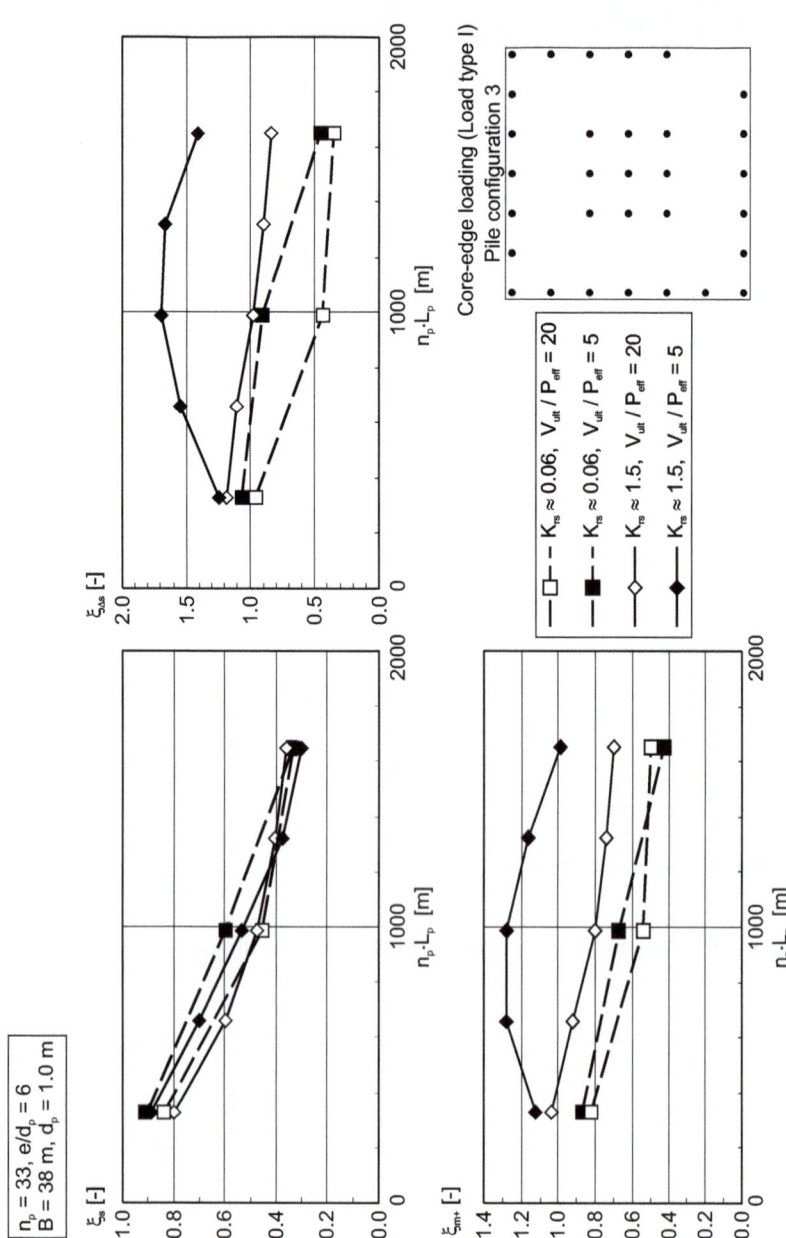

Figure 2.27 Example for the variation of the coefficients for the average settlement, for the differential settlements and the maximum positive bending moments with the total pile length.

Table 2.4 Example for an optimised design process: Results for the raft foundations

t_r [m]	$s_{average}$ [mm]	Δs [mm]	$\Delta s/B$ [-]	m_+ [MNm/m]
1	255	148	1/257	7.8
3	221	13	1/2923	14.6

the corresponding raft foundation. In the example the constraints are assumed to be the average settlement, the differential settlement and the maximum positive bending moment per unit length, as summarised in Table 2.4 for the raft foundation.

The aim of the optimised design process is to find a foundation configuration that gives the minimum total pile length $n_p \cdot L_p$ under the assumption of the following constraints:

$$s_{average} \leq 100 \text{ mm} \tag{2.20}$$

$$\xi_s \leq 0.39 \quad (t_r = 1m)$$
$$\Rightarrow \xi_s \leq 0.45 \quad (t_r = 3m) \tag{2.21}$$

$$\Delta s / B \leq 1/1000 \tag{2.22}$$

$$\xi_{\Delta s} \leq 0.26 \quad (t_r = 1m)$$
$$\Rightarrow \xi_{\Delta s} \leq 2.92 \quad (t_r = 3m) \tag{2.23}$$

No hogging of the raft

$$\Rightarrow \xi_{\Delta s} \geq 0 \quad (t_r = 1m \text{ and } t_r = 3m) \tag{2.24}$$

$$m_{pr+} \leq m_{r+} \tag{2.25}$$

$$\Rightarrow \xi_{m+} \leq 1 \quad (t_r = 1m \text{ and } t_r = 3m) \tag{2.26}$$

The pile diameter is fixed at $d_p = 1$ m and the pile length is limited to $L_p \leq 50$ m.

Application of the charts developed by Reul (2001) for core-edge loading leads to a number of feasible foundation configurations, which are summarised in Table 2.5. The feasible total pile lengths vary significantly for the different pile configurations investigated. The minimum feasible total pile lengths are $n_p \cdot L_p = 2050$ m ($t_r = 1$ m) and $n_p \cdot L_p = 1626$ m ($t_r = 3$ m). For the given load configuration and constraints, pile configuration 3 with piles

60 Combined Pile-Raft Foundations

Table 2.5 Example for an optimised design process: Feasible foundation configurations

t_r [m]	Pile configuration	n_p [-]	ξ_s-constraint [m]	$n \cdot L_p$ for $\xi_{\Delta s}$-constraint [m]	ξ_{m+}-constraint [m]	$n_p \cdot L_p$ for feasible configurations [m]
1	1	49	1840	X1	490	X3
	1	169	4182	4823	1690	4823
	2	9	X1	260	90	X3
	2	16	X1	426	160	X3
	2	49	2232 (X2)	632	490	X3
	3	33	1510	X1	330	X3
	3	40	1614	X1	400	X3
	3	73	2050	1916	730	2050
3	1	49	1387	490	2166	2166
	1	169	3183	1690	4945	4945
	2	9	X1	90	90	X3
	2	16	X1	160	160	X3
	2	49	1893 (X2)	490	490	X3
	3	33	1160	330	1626	1626
	3	40	1297	400	X1	X3
	3	73	1936	730	730	1936

X1: Constraint not fulfilled; X2: Hogging of the raft; X3: No feasible configuration

installed under the core and the edge of the raft is the favourable solution for both raft thicknesses investigated.

Further reduction of the total pile length can be achieved with nonuniform pile lengths. Based on the pile configurations 3 with $n_p = 73$ ($t_r = 1$ m) and $n_p = 33$ ($t_r = 3$ m), the length of the piles under the core and under the edge of the raft was varied. Figure 2.28 and Table 2.6 summarise the optimised design, which was achieved with two additional analyses for $n_p = 73$ ($t_r = 1$ m) and three additional analyses for $n_p = 33$ ($t_r = 3$ m). For both pile configurations, the average settlement turns out to be the critical parameter

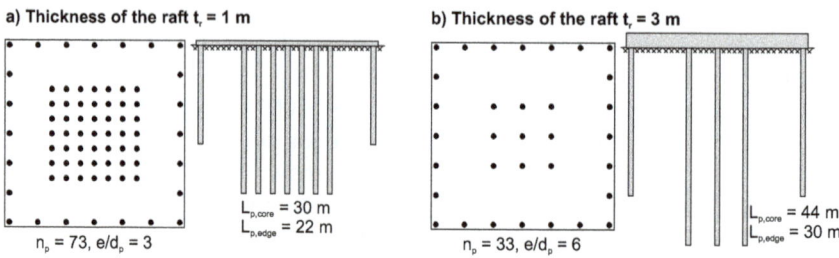

Figure 2.28 Example for an optimised design process: Optimised system configurations.

Table 2.6 Example for an optimised design process: Results achieved with nonuniform pile length

t_r [m]	Pile configuration	n_p [-]	$L_{p,core}$ [m]	$L_{p,edge}$ [m]	ξ_s [-]	$\xi_{\Delta s}$ [-]	ξ_{m+} [-]	$n_p \cdot L_p$ [m]
1	3	73	30	22	0.39	0.08	0.18	1998
3	3	33	44	30	0.45	1.12	0.84	1116

in the design process. Even with this limited number of additional analyses, the total pile length can be reduced further by 3% (n_p = 73, t_r = 1 m) and 31% (n_p = 33, t_r = 3 m).

2.3 LONG-TERM BEARING BEHAVIOUR OF PILED RAFTS CONSIDERING CONSOLIDATION AND CREEP

2.3.1 Overview

The long-term bearing behaviour of foundations in cohesive soil is influenced significantly by consolidation and viscous material behaviour, i.e. creep. Moreover, the overconsolidation ratio OCR is of importance, since applied loads can act completely or at least partially as reloading. The following overview focuses on research on piled rafts in cohesive soils considering time-dependent behaviour and the influence of unloading/reloading cycles including groundwater drawdowns.

The studies by Thaher (1991) and Thaher and Jessberger (1991) on foundations in overconsolidated Kaolin clay were among the earliest research on the bearing behaviour of piled rafts by means of centrifuge tests. In addition to a parametric study to determine the influence of the number of piles, pile length and pile diameter on the bearing behaviour of a piled raft, Thaher (1991) investigated the time-dependent behaviour of the piled raft of the Messeturm in Frankfurt.

Horikoshi and Randolph (1996) carried out centrifuge tests on piled rafts in overconsolidated Kaolin clay to investigate the efficiency of the pile layout in reducing differential settlements. During the tests a number of unloading and reloading cycles were carried out and the development of differential settlements with time was documented.

Cui et al. (2009) carried out coupled 2D FEA with an elasto-viscoplastic constitutive model for the soil. Due to the plane strain conditions considered, the foundation systems modelled corresponded to strip foundations supported by secant bored pile walls or diaphragm walls, rather than piled rafts with individual piles. Cui et al. (2009) reported the influence of negative pore pressures on the shaft friction of the pile rows (walls). In addition, significant differences were found for the development of pile row (wall) resistances with time for the different locations under the raft.

Fattah et al. (2014) studied the influence of pile length, pile spacing and pile-soil stiffness ratios on the development of pore pressures and on the piled raft coefficient by means of coupled 3D FEA with an elasto-plastic soil model. Moreover, effects such as the reduction of settlements and bending moments of the raft as well as of the piled raft coefficient due to a load transfer layer made of granular material placed beneath the raft on the bearing behaviour of the foundation were demonstrated.

Tran et al. (2012) reported results from centrifuge tests on raft foundations and piled rafts in soft clay investigating the influence of groundwater drawdown caused by extraction of groundwater from a deep soil layer. For both foundation types, groundwater drawdown resulted in a significant increase in settlements, with the raft foundations being more affected. Moreover, it was observed that the piled raft coefficient decreased by approximately 15% as a result of the groundwater drawdown.

Rincón et al. (2020) also studied the effect of consolidation processes due to loading and groundwater drawdown on raft foundations and different piled raft configurations with centrifuge tests. Similar to the observations by Tran et al. (2012), a significant increase of settlement was observed when the groundwater was lowered. However, due to the soil beneath the raft settling more than the raft and the resulting separation between raft and soil, the piled raft coefficient actually increased during the drawdown.

Sheil (2017) presented the results of 3D FEA for piled rafts and raft foundations using an elasto-viscoplastic constitutive model to consider time-dependent material behaviour but without considering consolidation. The analyses were performed for a normally consolidated soft clay and for an overconsolidated stiff clay, varying lengths and spacing of the piles. For both soil conditions, preloading resulted in an increase in stiffness for both the piled rafts and the raft foundations, although this effect was more pronounced for the latter.

Hoang and Matsumoto (2020) reported results from small-scale model tests to investigate the long-term bearing behaviour of vertically loaded piled rafts in clay. In the model tests the piled raft coefficient decreased with increasing load level, but increased within the respective load level during the consolidation phase. During the consolidation phase also the pile shaft resistance increased, especially at the upper pile section beneath the raft. During the creep phase at the end of the respective test, settlements increased while the piled raft coefficient remained constant.

Bhartiya et al. (2022) investigated the time-dependent behaviour of rectangular piled rafts in four different saturated, normally consolidated and moderately overconsolidated clayey soils by means of coupled 3D FEA. The modified Cam Clay (MCC) model was adopted for the soil, without considering creep. In the study, it was found that the variation of the degree of consolidation U with the time factor T_v is unaffected by the piled raft geometry but depends on the overconsolidation ratio in the early stages of the consolidation process. Bhartiya et al. (2022) suggested an approach to estimate instant and consolidation settlements of piled rafts and determined a ratio of 1.2 to 2.5 between immediate settlement and total settlement at the end of consolidation. No significant influence of the duration of the construction period on the settlement behaviour was observed, which was probably due to the fact that no rate dependency, i.e. viscosity, was considered in the constitutive model of the soil. However, they found that the piled raft coefficient decreased with increasing OCR and increasing load level. Moreover, a decrease of the piled raft coefficient during the consolidation process was observed which is in contradiction to other findings from numerical and experimental investigations (see Section 2.3.2).

Extensive research on the long-term bearing behaviour of raft foundations and piled rafts in overconsolidated clay was report by Ganal (2024) based on in situ measurements, centrifuge test and numerical simulations. The numerical simulations, based on coupled displacement-pore pressure 3D FEA, implemented the visco-hypoplastic AVISA model (Tafili and Triantafyllidis 2020), which allows consideration of creep effects. The investigations showed, for example, that parameters such as the viscosity index, the OCR and the permeability influence the distribution of pile resistances within the pile group. As a result of the consolidation process, loads are redistributed from the raft to the piles; i.e. the piled raft coefficient increases. Some results of this research are presented in Section 2.3.3.

2.3.2 Influence of consolidation on the bearing behaviour of piled rafts

One of the earliest investigations on the influence of consolidation on the bearing behaviour of piled rafts was carried out by Reul (2000) with parts subsequently published by Reul (2002). Unless indicated otherwise, all results presented in Section 2.3.2 are taken from that study. For analysis of foundations, it is common practice to model the short-term and the long-term behaviour of the system with two separate analyses, i.e. with undrained and drained soil parameters. By contrast, the behaviour of a piled raft throughout the whole consolidation process was studied with a coupled displacement-pore pressure FEA. Additionally, it was investigated whether, assuming an elasto-plastic material model for the soil, the long-term response of a system from coupled displacement-pore pressure FEA would differ from the results of a drained analysis.

For the coupled displacement-pore pressure FEA carried out with the program ABAQUS, it is assumed that the pore fluid flow is governed by Darcy's law, that the soil pores are fully saturated (S_r = 1) and that the pore fluid (water) is incompressible. Figure 2.29 shows the system configuration and finite element mesh used in this study. The modelled piled raft comprises five piles (length L_p = 10 m, diameter d_p = 1 m) and a square rigid raft (breadth B = 8 m). The piles and the raft are assumed to be impermeable. Drainage of the homogeneous soil takes place only at the top surface of the model. In the

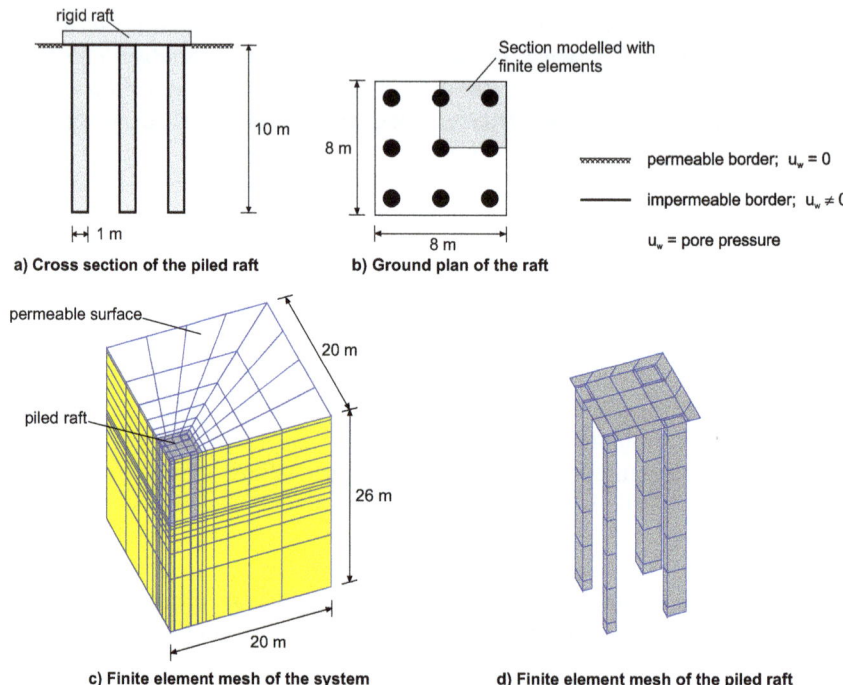

Figure 2.29 System configuration and finite element mesh.

finite element model, the circular piles were replaced with square piles with an equivalent shaft circumference.

The material behaviour of the soil was defined by means of the aforementioned cap model. The soil is overconsolidated, assuming a maximum previous vertical stress of 600 kPa at the current top surface. The material parameters used in the finite element analyses are summarised in Table 2.7. As in the numerical study on the bearing behaviour of vertically loaded piled rafts (Section 2.2.1), the contact between the structure and the soil was described as perfectly rough.

In the FEA the piled raft was loaded in six steps of 2.84 MN over five days, which is equivalent to approximately 114 m^3 concrete per day (Figure 2.30a). During the loading process the piled raft exhibits significant time-dependent settlement (Figure 2.30b). For example, in the last loading step an initial settlement of $s = 5$ mm is observed, giving a total settlement of $s = 25$ mm. On day 15 (10 days after the load increment was applied), the settlement is $s = 31$ mm. Equilibrium conditions are reached after approximately 34 days with a final settlement of $s = 32$ mm.

Figure 2.30c shows the variation of the piled raft coefficient with time. The results of the analysis show that while the piled raft coefficient decreases with increasing load level, during each load step the piled raft coefficient

Table 2.7 Material parameters used in the FEA

Parameter		Unit	Soil	Raft	Piles
Young's modulus	E	MPa	14.5+1.75·z	30,000	30,000
Poisson's ratio	ν	-	0.25	0.2	0.2
Buoyant unit weight	γ'	kN/m³	9	-	15
Coefficient of earth pressure at rest	K_0	-	0.658	-	-
Angle of internal friction	φ'	°	20	-	-
Cohesion	c'	kPa	20	-	-
Shape parameter of the conus	K	-	0.795	-	-
Shape parameter of the cap	R	-	0.1	-	-
Coefficient of permeability	k	m/s	$1 \cdot 10^{-7}$	-	-
Saturation ratio	S_r	m/s	1	-	-

z depth below ground level in [m]

increases with time. For example, at the end of the first load step ($P = 2.84$ MN, $t - 1$ d) $\alpha_{pr} - 0.910$ while at the end of the penultimate load step ($P = 14.19$ MN, $t = 5$ d) the piled raft coefficient has reduced to $\alpha_{pr} = 0.877$. However, within the penultimate load step the piled raft coefficient increases from $\alpha_{pr} = 0.845$ ($t = 4$ d) immediately after application of the load to the final value of $\alpha_{pr} = 0.877$ at the end of the step ($t = 5$ d). The reduction of the piled raft coefficient with increasing load is caused by nonlinear pile resistance-settlement behaviour (Section 2.2.2). The increase in the piled raft coefficient during a load step is caused as consolidation essentially reduces the raft-soil stiffness, resulting in load redistribution between the piles, which must carry an increasing proportion of the total load, and the raft. This tendency was observed in in situ measurements on piled rafts in London Clay (Cooke et al. 1981) and the Frankfurt Formation (Franke and Lutz 1994). More recently the effect was also observed in centrifuge tests by Ganal et al. (2022) and Ganal (2024).

The variation of the normalised pile resistance, $\alpha_{p,i}$, with time is plotted in Figure 2.30d. The normalised pile resistance is defined as

$$\alpha_{p,i} = \frac{\frac{1}{n_{p,i}} \Sigma R_{pile,i}}{\frac{1}{n_p} \Sigma R_{pile}} \tag{2.27}$$

where $\Sigma R_{pile,i}$ = sum of all pile resistances at position i; $n_{p,i}$ = number of piles at position i; ΣR_{pile} = sum of all pile resistances and n_p = number of piles in the pile group. With increasing load level the location of the pile within the pile group becomes less significant. At the start of loading ($P = 2.84$ MN), the resistance of the corner pile is approximately three times the resistance of the centre pile; for $P = 17.02$ MN the difference is reduced to approximately 20%.

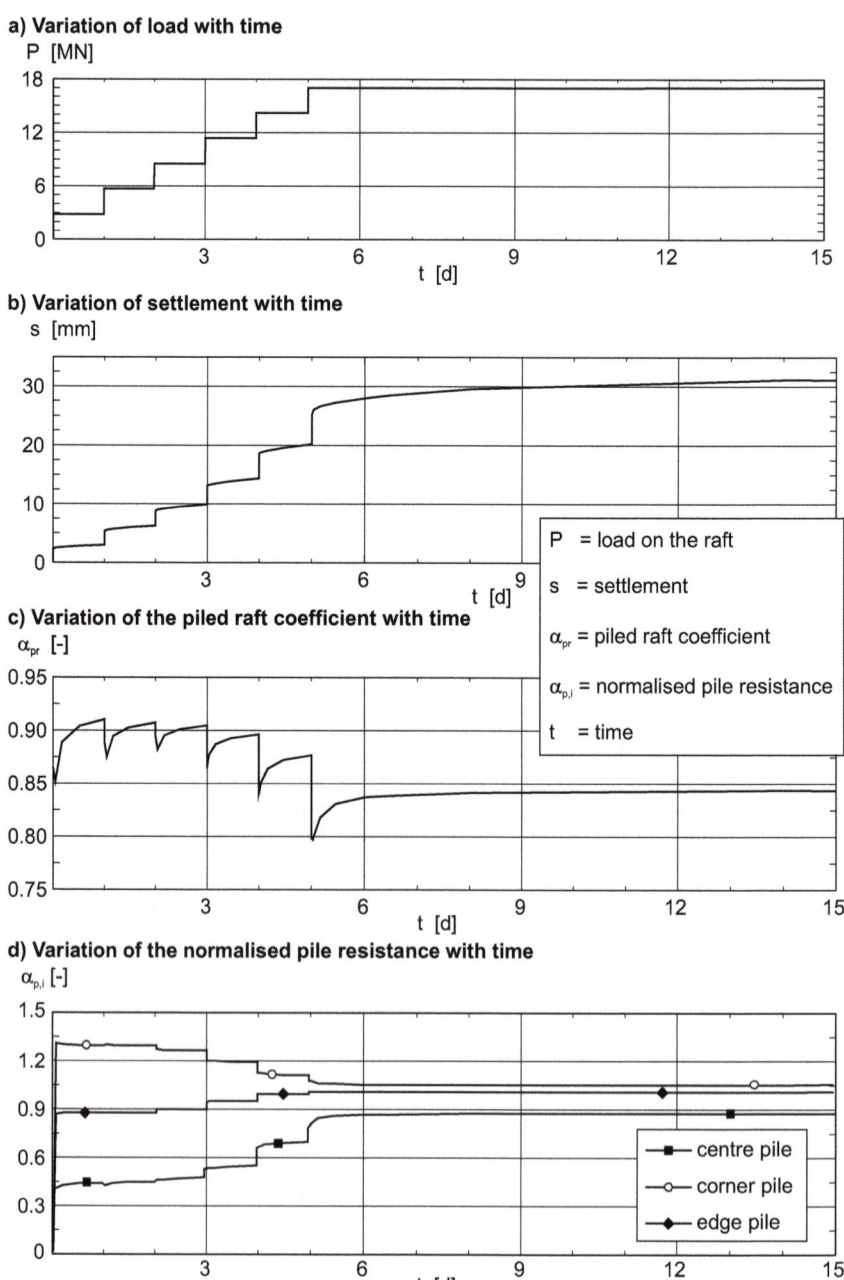

Figure 2.30 Variation of load, settlement, piled raft coefficient and normalised pile resistance with time.

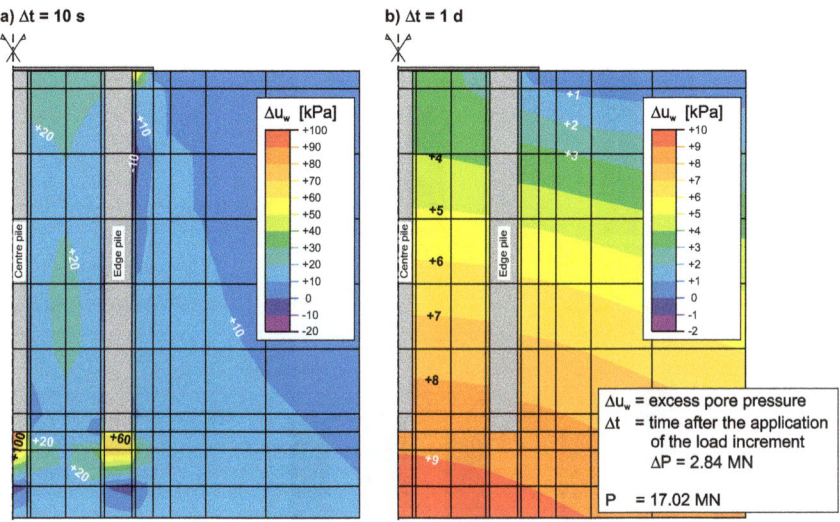

Figure 2.31 Excess pore pressure caused by the loading of the system.

The distribution of excess pore pressures caused by the loading of the system is shown in Figure 2.31 for the maximum applied load of $P = 17.02$ MN. Immediately ($\Delta t = 10$ s) after the application of the load increment $\Delta P = 2.84$ MN, the largest excess pore pressures (up to $\Delta u_w = 100$ kPa) are observed just below the pile base. High excess pore pressures (up to $\Delta u_w = 60$ kPa) are also created in the area between the edge pile and the edge of the raft. Negative excess pore pressures, possibly caused by dilatant soil behaviour, are visible at the shaft of the edge pile. Within one day of the loading, the excess pore pressures have diminished to 10% of their initial values in the vicinity of the piled raft. The increase in excess pore pressures with depth is due to the boundary conditions, which allow drainage only at the top surface of the model (Figure 2.29).

Figure 2.32 compares the normalised pile resistance, the piled raft coefficient and settlement between the coupled displacement-pore pressure finite element analysis (two-phase model) and an analysis without consideration of pore pressures (one-phase model) for two different load levels. The same soil parameters were prescribed for both models (Table 2.7). Therefore, the one-phase model analysis can be described as drained. As would be expected, the short-term settlement is higher in the one-phase model compared to the two-phase model. For example, after a load of $P = 8.51$ MN ($t = 3$ d) is applied, a 20% difference is observed. As steady state conditions are approached, the results converge and the final settlement at $P = 17.02$ MN ($t = 370$ d) is only slightly higher for the coupled analysis with the two-phase model compared to the analysis with the one-phase model. The piled raft coefficients fall

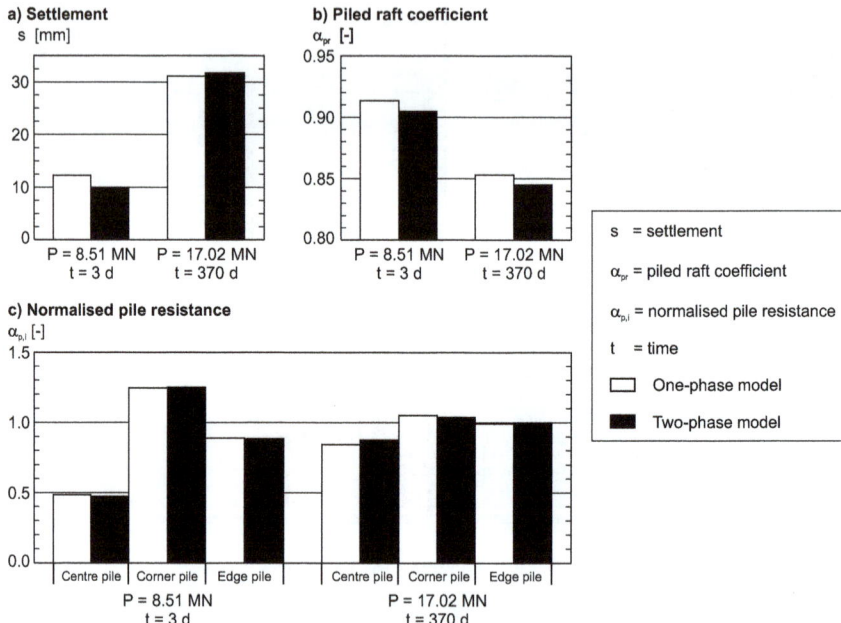

Figure 2.32 Settlement, piled raft coefficient and normalised pile resistance depending on the load.

within 1% of each other where the load share of the piles is smaller in the coupled analysis with the two-phase model.

From the study it can be concluded that coupled displacement-pore pressure FEA are capable of modelling time-dependent behaviour of piled rafts, such as the load share between piles and raft. For the modelled system configuration, subsoil conditions and loading history, the coupled elasto-plastic FEA (two-phase model) yields qualitatively and quantitatively similar results for the long-term behaviour of the foundation as the drained finite analysis with the one-phase model.

2.3.3 Messeturm, Frankfurt – long-term bearing behaviour of a piled raft in overconsolidated clay

This chapter concludes with a case study of the long-term bearing behaviour of a piled raft. In situ measurements and numerical simulations are presented for the well-documented case history of the Messeturm in Frankfurt (e.g., Sommer and Hoffmann 1991a, Sommer et al. 1991). Reul and Randolph (2003) presented the results of a detailed numerical back-analysis of the measurements on the Messeturm with a 3D finite element model where the nonlinear material behaviour of the soil was simulated with the above-mentioned elasto-plastic cap-model, not considering time-dependent effects such as consolidation and creep. Garcia et al. (2006) applied the visco-hypoplastic material model by Niemunis (2003) in their back-analysis of the Messeturm without taking consolidation processes into account.

Unless indicated otherwise, all measurements presented in this section were adapted from Reul (2000), who re-evaluated the data published by Sommer et al. (1990, 1991), Sommer and Hoffmann (1991a, 1991b) and Sommer (1993) for the construction process and carried out long-term measurements. The results of the FEA that were partially published by Tafili et al. (2023) are taken from Ganal (2024).

In the vicinity of the Messeturm the subsoil consists of fill and quaternary sand and gravel up to a depth of 10 m below ground level, which is followed by the Frankfurt Formation (Section 4.2.2) up to a depth of at least 70 m below ground level. The groundwater level is situated 4.5 m to 5.0 m below ground level. From the evaluation of borehole logs, Ganal (2024) determined a total share of 14.6% sand and limestone bands (sand: 5.4%, limestone: 9.2%) in the Frankfurt Formation at the site.

The piled raft of the 256 m high Messeturm comprises 64 bored piles and a square raft with an edge length of 58.8 m. The length of the piles (diameter d_p = 1.3 m) varies from L_p = 26.9 m (outer ring), L_p = 30.9 m (middle ring) to L_p = 34.9 m (inner ring). The foundation level of the 3-m- to 6-m-thick raft lies 11-m- to 14-m-deep below ground level (Figure 2.33a). The construction of the building started in 1988 and was finished in 1991. Figure 2.33c shows the variation of the total settlement inducing building load with time, which amounts to $P = G_{raft} + G + Q/3$ = 1860 MN.

The foundation design followed a modified conventional design (Section 1.4). According to Sommer et al. (1991), in the design process two scenarios were considered, namely the piles carrying 30% and 55% of the total load, respectively. While the ultimate structural capacity of the piles was sufficient to carry the total building load, it was assumed that the piles would reach their ultimate shaft and base resistance well before reaching their structural capacity.

The behaviour of the foundation was monitored from the construction period until more than seven years after the building was finished by means of geodetic and geotechnical measurements with 12 instrumented piles, 13 contact pressure cells, 1 pore pressure cell and 3 multi-point borehole extensometers. The positions of the measurement devices are plotted in the ground plan of the raft (Figure 2.33b).

For all 12 instrumented piles, the axial pile strains were measured at up to 11 depths by means of strain gauges. The pile head loads were measured on three of these piles with load cells at the pile head. For the other 9 piles the pile loads were derived from the strain measurements at the strain gauges placed immediately below the pile head and the pile axial stiffness. The mean Young's modulus of the piles of E_{pile} = 25,000 MPa was derived from the in situ measurements (Reul 2000). Since the pile loads are calculated from strain measurements, the pile loads presented in this section as well as

70 Combined Pile-Raft Foundations

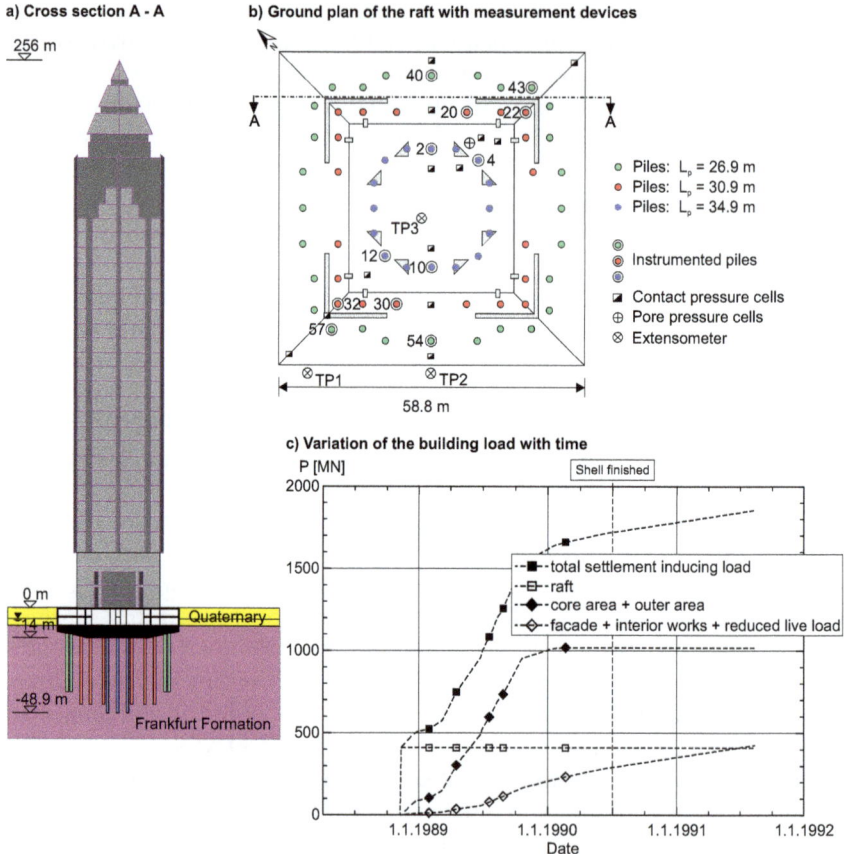

Figure 2.33 Messeturm, Frankfurt.

by Reul and Randolph (2003) are smaller than the values previously published (Sommer et al. 1990, 1991, Sommer & Hoffmann 1991a, 1991b, Sommer 1993) where a Young's modulus of the pile concrete alone of $E_{concrete}$ = 30,000 MPa was assumed (Sommer and Hoffmann 1991a).

For the construction of a subway tunnel with a station 47 m east of the Messeturm, groundwater had to be drawn down more than 12 m at the tunnel (Sommer et al. 1991). As a result, the groundwater level in the vicinity of the Messeturm decreased about 10 m, which led to changes of the uplift on the raft of 287 MN. During the construction process of the subway tunnel and the station the groundwater lowering was suspended for 2 years and continued in 1994 until the end of 1996. Figure 2.37a and c shows the variation of the groundwater level and the average measured pile resistances for the inner, middle and outer pile ring with time. Groundwater drawdown is accompanied by an increase of the pile resistances and a groundwater rise by a decrease of the pile resistances, as was found also for other case histories such as the WestendDuo (Section 3.2.2) or the Westhafen Tower.

The CPRF of the Westhafen Tower involved the use of a circular raft with a diameter of 68 m and a total number of 32 bored piles (d_p = 1.5 m, L_p = 12.5 m and L_p = 15 m), where the piles were placed in a circular area with a diameter of 37 m at the centre beneath the superstructure of the tower. The foundation level of the 1.8-m- to 4.2-m-thick raft lies up to 21.4 m deep below ground level in the Frankfurt Formation. Under undisturbed conditions, the groundwater level is situated 3 m below ground level in the quaternary layers, equivalent to the water level at the nearby river Main. The Frankfurt Formation is underlain by the Frankfurt limestone to a depth of approximately 38 m below ground level. Figure 2.38 shows the variation of the measured pile resistances with time. During the construction of the building, the groundwater level had to be drawn down below the final excavation level. When the weight of the building was sufficient to withstand the uplift, the groundwater drawdown was stopped and the groundwater level rose again, yielding an uplift of 570 MN acting on the raft. As a consequence, the pile resistances dropped between 4 MN to 9 MN immediately after the end of the groundwater drawdown.

The numerical study for the Messeturm by Ganal (2024) was carried out by means of 3D FEA with the code Tochnog (Tochnog Professional Company 2022). The time-dependent displacement caused by consolidation processes as well as by the material behaviour of the soil was modelled using coupled pore pressure-displacement analyses. The finite element mesh comprised hexahedral elements with both displacements and pore pressures varying linearly across the elements.

Figure 2.34b shows the finite element mesh of the system where 1/4 of the complete 3D problem was modelled considering the two symmetry planes of the structure. The boundary between the Frankfurt Formation and the Frankfurt Limestone, which represents the bottom of the finite element mesh, was assumed to be 74.8 m below the bottom of the raft. Only the soil below the top surface of the raft is modelled with finite elements. The soil above that level is considered only in respect of the surcharge it provides.

Figure 2.34c shows the details of the piled raft. The thickness of the raft decreases in three steps from the core area (t_r = 6 m) to the edge of the raft (t_r = 3.0 m). The circular piles were replaced by square piles with the same shaft circumference in the analysis. For the modelling of the contact zone between soil and the large diameter bored piles, thin solid continuum elements were introduced instead of special interface elements (Figure 2.34d), with the contact between structure and soil represented as perfectly rough. This means that no relative motion takes place between the nodes of the finite elements that represent the structure and those of the finite elements that represent the soil. The material behaviour in the contact area was simulated by the material behaviour of the soil.

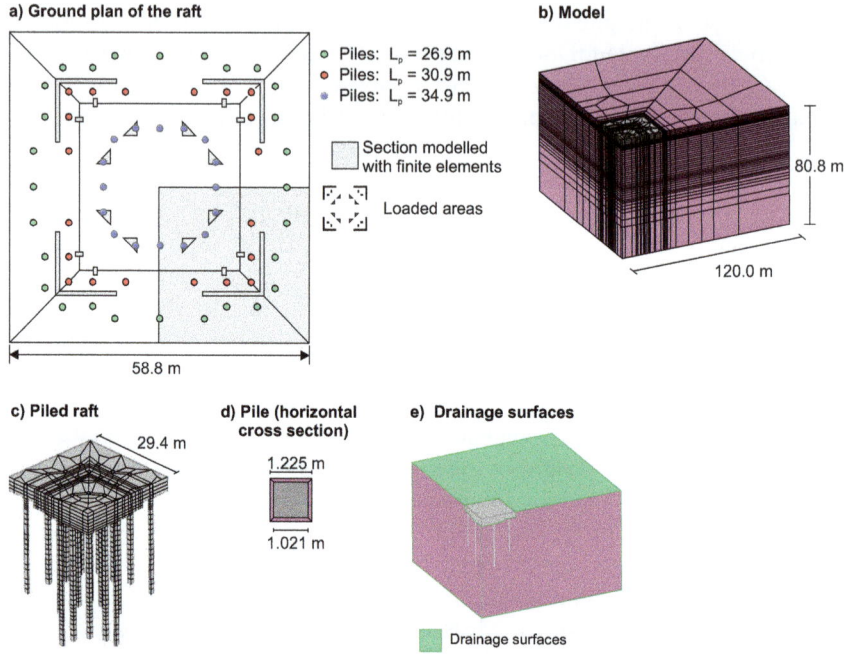

Figure 2.34 Messeturm: Finite element mesh after Ganal (2024).

For the coupled displacement-pore pressure FEA carried out, it is assumed that the pore fluid flow is governed by Darcy's law, that the soil pores are fully saturated ($S_r = 1$) and that the pore fluid (water) is incompressible. Drainage of the soil takes place beneath the raft as well as at every model surface except for the symmetry axes (Figure 2.34e). The piles are assumed to be impermeable. The groundwater level is controlled by a defined hydraulic pressure head at the drainage surfaces. Groundwater level changes are then considered by changing the hydraulic pressure head.

Table 2.8 summarises the step-by-step analysis of the construction process. The piles were modelled "wished-in-place", i.e. changes in the soil surrounding the pile caused by the installation process were not modelled. The total dead weight of the raft of $G_{raft} = 410$ MN was modelled as a gravity load. The total settlement inducing load of the superstructure of $P = G + Q/3 = 1445$ MN was applied on top of the raft in the areas of walls and columns (Figure 2.34a).

The raft and the piles are considered to behave linear-elastically, with the material parameters summarised in Table 2.9.

The FEA on the load-bearing behaviour of the foundation of the Messeturm are carried out with the AVISA model (Tafili and Triantafyllidis 2020) which allows simulation of both creep and alternating loads and is therefore well suited for the current boundary value problem. The model

Table 2.8 Messeturm: Step-by-step analysis of the construction process in the FEA after Ganal (2024)

Step		P [MN]	GW [m.a.s.l.]	Δt [d]	Date
1	In situ stress state	-	92.5	-	-
2	Start of the GW drawdown	-	92.5/90.9	17	26.06.88
3	Excavation to 7.5 m below ground level	-	90.9/89.0	22	13.07.88
4	Pile installation (wished in place)	-	89.9/86.1	32	04.08.88
5	Excavation to 14.0 m below ground level	-	86.1/83.4	52	05.09.88
6	Application of the raft dead weight	410	83.4/83.0	12	29.10.88
7	Application of the raft stiffness	410	83.0	1	10.11.88
8	Loading + GW management	1860	83.0/92.4/83.0	998	11.11.88
9	GW management	1860	83.0/93.2/83.5/92.5	2681	06.08.91
10	Consolidation + creep	1860	92.5	14433	08.12.98

Notes: P: total settlement inducing load at the respective time $P = G_{raft} + G + Q/3$; GW: groundwater level in [m] above sea level; Δt: duration of the process; Date: starting date of the associated analysis step

Table 2.9 Messeturm: Material parameters for piles and raft used in the FEA after Ganal (2024)

Parameter			Raft	Piles
Young's modulus	E	MPa	34,000	25,000
Poisson's ratio	ν	-	0.2	0.2
Total unit weight of moist soil	γ	kN/m³	25	25
Buoyant unit weight	γ'	kN/m³	15	15

formulation under medium large to large strains $1\% < |\varepsilon_1| < 10\%$ follows the framework of hypoplasticity using the total strain rate $\dot{\varepsilon}$, the hypoplastic strain rate $\dot{\varepsilon}^{hp}$ and the viscous (time-dependent) strain rate $\dot{\varepsilon}^{vis}$. The stress rate tensor $\dot{\sigma}$ then obeys the following equation:

$$\dot{\sigma} = m\mathbf{E} : \left(\underbrace{\dot{\varepsilon}}_{linear} - \underbrace{y_h \dot{\varepsilon}^{hp}}_{hypoplastic} - \underbrace{\dot{\varepsilon}^{vis}}_{viscous} \right) \qquad (2.28)$$

with a transversal hypoelastic stiffness tensor as a function of stress $\mathbf{E} = f(\sigma)$. m and y_h are scalar functions that consider the influence of intergranular strain after load reversal by increasing the elastic stiffness tensor and decreasing the hypoplastic strain rate $\dot{\varepsilon}^{hp}$. A detailed discussion of the AVISA constitutive model was given by Tafili and Triantafyllidis (2020) and

Table 2.10 Messeturm: Material parameters for the soil used in the FEA after Ganal (2024)

Parameter			Frankfurt Formation
Unit weight:			
Total unit weight of the moist soil	γ	[kN/m³]	18.5
Buoyant unit weight	γ'	[kN/m³]	8.5
Transversal (hypo)elasticity:			
compression index	λ	[-]	0.108/0.092*
swelling index	κ	[-]	0.029/0.025*
Poisson's ratio	ν_h	[-]	0.23
anisotropic coefficient	α	[-]	1.5
Critical and loading surface:			
critical state slope	M_c	[-]	$0.8334 + 0.0042 \cdot z$
max. void ratio at $p_{ref} = 1$ kPa	e_{i0}	[-]	1.44
loading surface factor	f_{b0}	[-]	1.40
Viscosity:			
viscosity index	I_v	[-]	0.030
viscosity exponent	n_{OCR}	[-]	0.4
Intergranular strain anisotropy (ISA):			
stiffness factor	m_R	[-]	2.5
ISA yield surface radius	R	[-]	$1.8 \cdot 10^{-4}$
ISA hardening parameter	β_0	[-]	0.100
min. ISA exponent	χ_0	[-]	5
max. ISA exponent	χ_{max}	[-]	60
accumulation rate factor	C_a	[-]	0.001

* site specific values established with Eq. 2.32
z depth below ground surface [m]

Tafili (2020), while a more concise description of the model was provided by Tafili et al. (2023).

The material parameters applied for the Frankfurt Formation are summarised in Table 2.10. The base values given for the Frankfurt Formation reflect the parameters of the Frankfurt clay derived from a large number of laboratory tests with samples taken from different sites across the city and the back-analysis of element tests (Ganal 2024, Tafili et al. 2023) without considering embedded sand and limestone bands.

While the material parameters of the Frankfurt clay inside the Frankfurt Formation show only limited scatter locally, the proportion of the much stiffer embedded sand and limestone bands varies from site to site. In order to capture the effect of the greater stiffness of the limestone and sand bands in the FEA, Ganal (2024) proposed to reduce the parameters λ and κ derived from the oedometric tests by exactly the settlement-insensitive portion to establish a system stiffness for the Frankfurt Formation as a whole for the respective project site, using the relationships:

$$\lambda_{site} = (1 - n_{site})\lambda_{base}; \quad \kappa_{site} = (1 - n_{site})\kappa_{base} \qquad (2.29)$$

where λ_{base}, κ_{base} = base value of compression and swelling index; λ_{site}, κ_{site} = site specific value of compression and swelling index; n_{site} = site specific share of embedded sand and limestone bands. For the Messeturm, the total share of limestone and sand of 14.6% (Ganal 2024) results in a reduction factor of $(1 - n_{site}) = 0.854$. This approach was also applied successfully for the back-analysis of the SGZ-Bank/Park Tower (Ganal and Reul 2023, Ganal 2024, Section 4.3.2).

Through the modified formulation of the AVISA model by Ganal (2024) for the viscous strain rate $\dot{\varepsilon}^{vis}$, the proportion of creep settlements is automatically adjusted in the model itself in the same manner as λ:

$$\dot{\varepsilon}^{vis} = I_v \lambda \left(\frac{1}{OCR}\right)^{\frac{1}{I_v}} \dot{\varepsilon}_r \, m \qquad (2.30)$$

where $m = f(\sigma)$ stress-dependent flow rule; $\dot{\varepsilon}_r$ = reference creep rate.

The critical state slope M_c varies linearly with the depth z [m] below ground surface reflecting a certain spatial variability of the clay which may be attributed, for example, to the sedimentation process of the material. The data on the overconsolidation ratio OCR provided by Franke et al. (1985) and Ganal (2024) are plotted in Figure 2.36. The OCR profile used in the FEA is based on Ganal (2024) and was then slightly adjusted on the basis of preliminary FEA. For the initialisation of the horizontal stress by means of the coefficient of earth pressure at rest K_0 (Figure 2.36), the approach of Mayne and Kulhawy (1982) was used, with

$$K_0 = (1 - \sin(\varphi')) \cdot OCR^{\sin(\varphi')} \qquad (2.31)$$

For the numerical simulations it is convenient to replace the heterogeneous Frankfurt Formation (Figure 2.35a) with its embedded limestone and sand bands of varying thickness with a homogenous soil mass and an appropriate equivalent system permeability (Figure 2.35b). Following the approach by Franzen and Reul (2022), taking a permeability of $k_{clay} = 1.4 \cdot 10^{-11}$ m/s for the clay and a significantly higher permeability ($\geq 1 \cdot 10^{-4}$ m/s) for the limestone and sand bands into account, the system permeability k_{sys} can be estimated applying 1-dimensional consolidation theory (Terzaghi and Fröhlich 1936) as:

$$k_{sys} = k_{clay}\left(\frac{h_{sys}}{h_{clay}}\right)^2 \qquad (2.32)$$

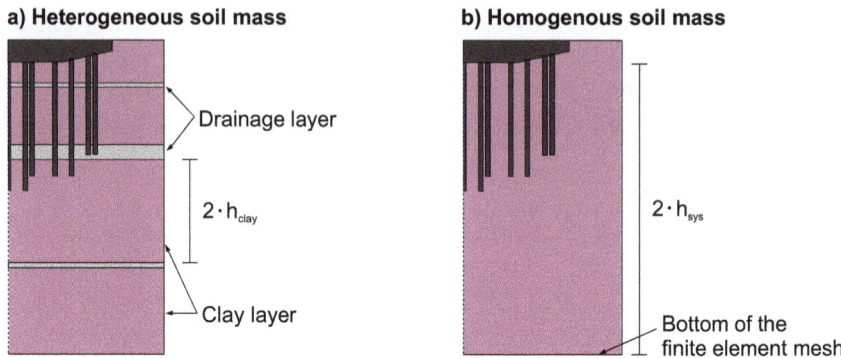

Figure 2.35 Messeturm: System permeability for the Frankfurt after Ganal (2024).

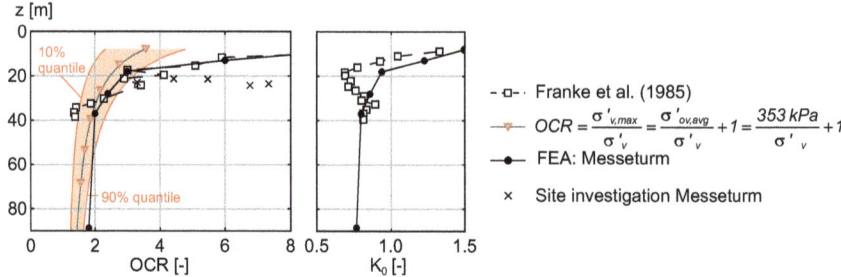

Figure 2.36 Messeturm: Depth profiles of the overconsolidation ratio OCR and the coefficient of earth pressure at rest K_0 after Ganal (2024).

where k_{clay} = permeability of clay; h_{clay} = characteristic length of drainage path within the Frankfurt clay (Figure 2.35a; half distance between neighbouring permeable layers, i.e. limestone and sand bands); h_{sys} = characteristic length of dewatering in the finite element model, where the lower edge of the foundation is considered to be decisive. For the estimation of h_{clay} only limestone and sand bands with a thickness $d \geq 0.5$ m were taken into account.

Evaluation of the nine available borehole logs from the Messeturm site, with lengths between 23.0 m and 63.1 m in the Frankfurt Formation, yielded an average half distance between permeable layers $h_{clay,avg}$ = 4.18 m and maximum half distance $h_{clay,max}$ = 12.85 m. With h_{sys} = 37.4 m the system permeability was estimated by Ganal (2024) to be in the range of k_{sys} = $1.1 \cdot 10^{-9}$ m/s to k_{sys} = $1.2 \cdot 10^{-10}$ m/s which differs slightly from the values provided by Franzen and Reul (2022) where only two borehole logs were evaluated.

Figure 2.37 compares the measured variation of settlements, pile resistances and piled raft coefficient with time with the results of the FEA. Additionally, the development of the groundwater level with time is shown

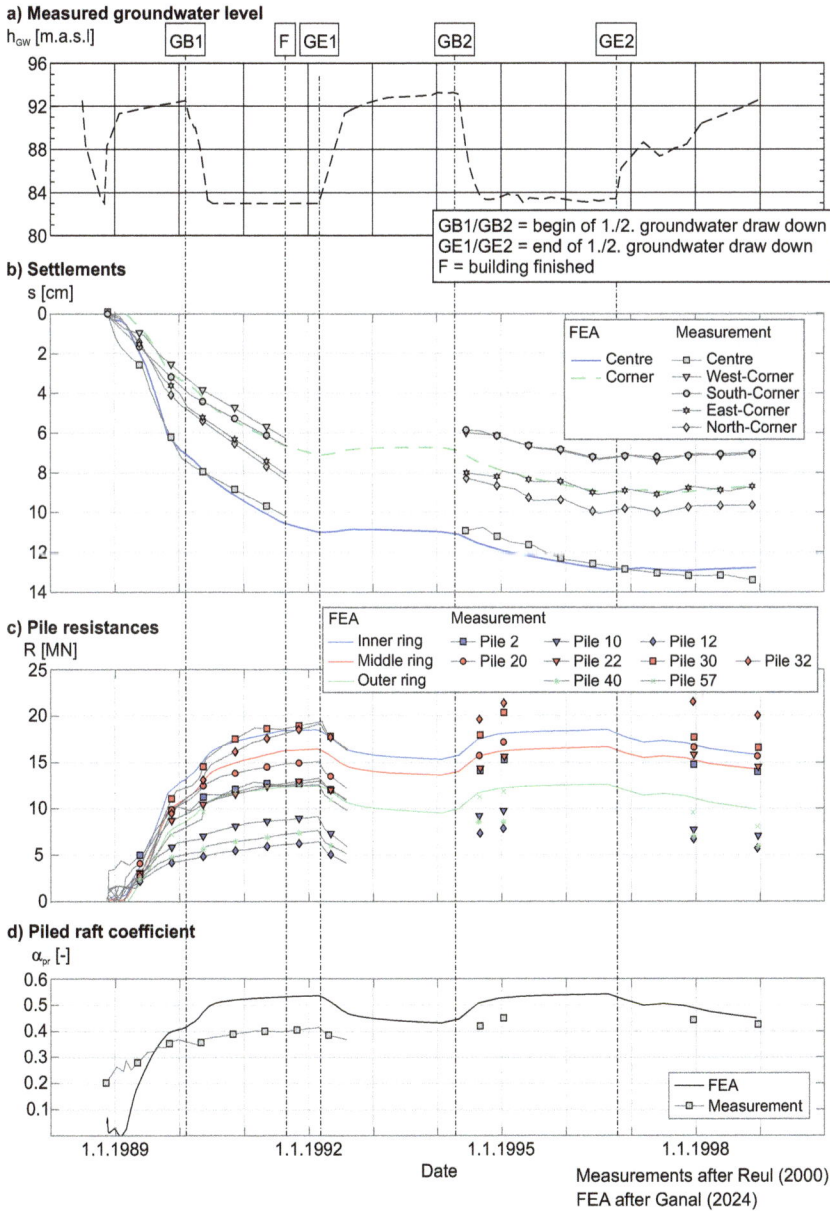

Figure 2.37 Messeturm: Variation of groundwater level, settlements, pile resistances and piled raft coefficient with time.

in Figure 2.37a. Due to the location of the groundwater drawdown for the subway tunnel north-east of the Messeturm mentioned earlier, the settlements at the northern and eastern corner are slightly higher than at the southern and western corner (Figure 2.37b). This influence is also visible in

78 Combined Pile-Raft Foundations

Figure 2.39 where the settlements in a cross section running in an east-west direction are plotted. However, since the finite element model is restricted to 1/4 of the foundation, this effect is not captured in the FEA.

For the piled raft coefficient (Figure 2.37d) the FEA yields $\alpha_{pr} = 0.52$ (groundwater drawdown) and $\alpha_{pr} = 0.43$ (natural groundwater level), respectively. Based on the assumption that the average pile resistance can be derived from the 12 instrumented piles, the piled raft coefficient at the time of the last documented measurement, where the groundwater is situated almost at its natural level, is $\alpha_{pr} = 0.43$.

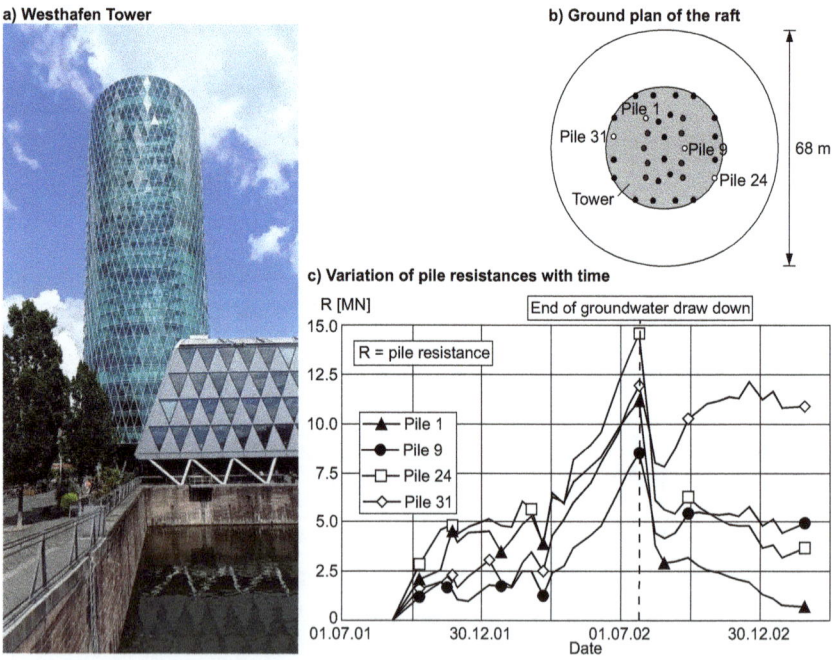

Figure 2.38 Westhafen Tower, Frankfurt.

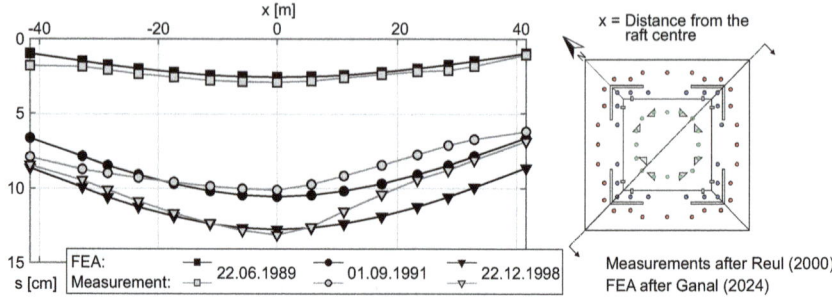

Figure 2.39 Messeturm: Distribution of settlements in the cross section of the raft.

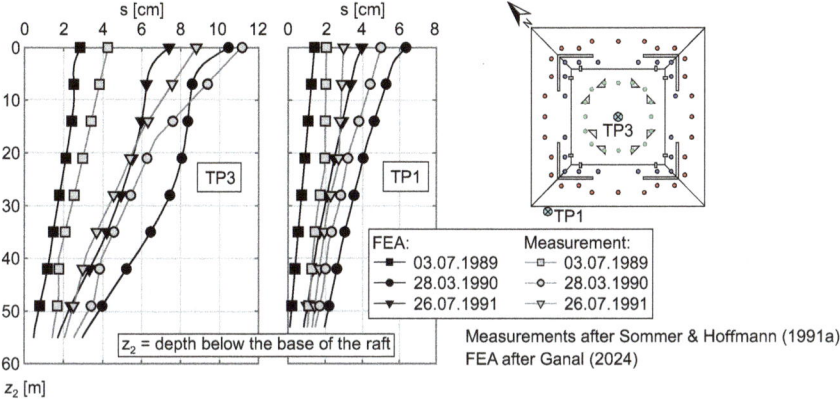

Figure 2.40 Messeturm: Settlement profiles at extensometers TP1 and TP3.

Comparison of the measured and calculated settlement profiles at the extensometers TP1 and TP3 is plotted in Figure 2.40 at three different points in time. For both extensometers reasonable agreement is achieved between measured and calculated settlement profiles.

Figure 2.41 shows the average pile load distribution along the pile shaft for the outer pile ring ($L_p = 26.9$ m), the middle pile ring ($L_p = 30.9$ m) and the inner pile ring ($L_p = 34.9$ m) at three different points in time. While measurements and FEA are in reasonable agreement for the outer and the middle pile ring, the FEA appears to overestimate the pile loads for the inner pile ring. However, uncertainties concerning the estimation of the Young's modulus for the piles need to be kept in mind.

Figure 2.42 shows the results of a sensitivity analysis of the influence of the system permeability and the viscosity index on the long-term settlements

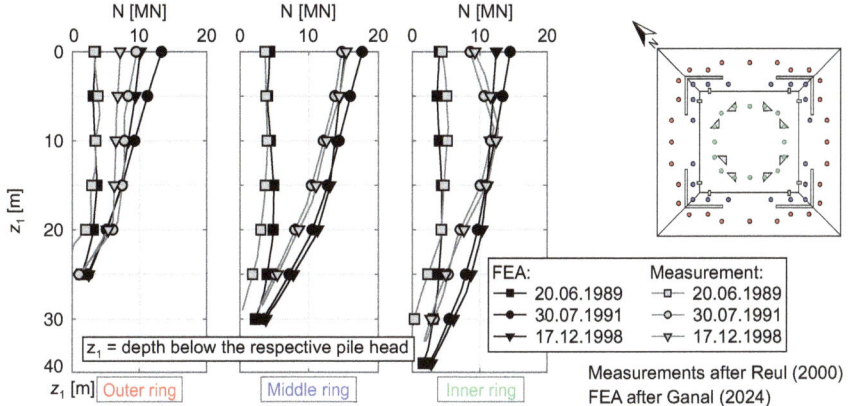

Figure 2.41 Messeturm: Pile load distribution along the pile shaft.

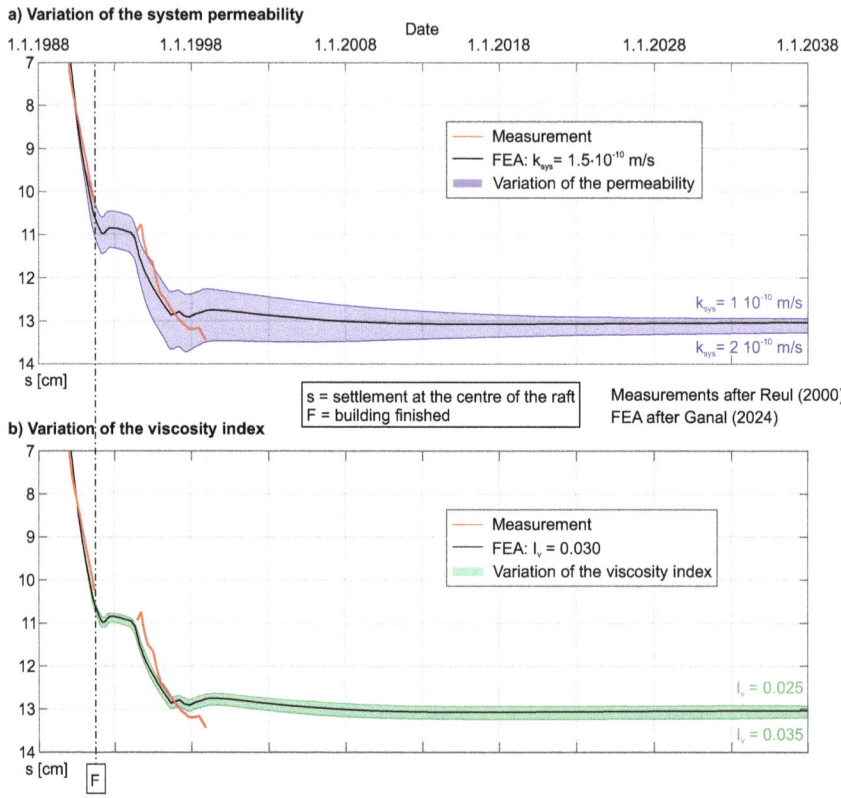

Figure 2.42 Messeturm: Investigation of long-term settlements.

of the foundation. With the system permeability varying in the range between $k_{sys} = 1.0 \cdot 10^{-10}$ m/s and $k_{sys} = 2.0 \cdot 10^{-10}$ m/s (Figure 2.42a), it takes approximately 20 years until the final settlements are reached. Smaller permeabilities lead to higher excess pore pressures during load application, resulting in less initial settlement. Due to the rate-dependent formulation of the AVISA model and the different settlements rates caused by the varying permeabilities, this leads to the final settlements differing by approximately $s = 0.3$ cm (Ganal 2024).

In Figure 2.42b, the viscosity index is varied in the range $I_v = 0.025$ to $I_v = 0.035$. At the completion of the Messeturm, the influence of the viscosity index on the settlement is small. In the first years after completion, different viscosity indices lead to different settlement rates. However, this effect decreases with time. Ganal (2024) found that the influence of the viscosity index on the simulated bearing behaviour of the raft foundation of the SGZ-Bank (Section 4.3.2) was more pronounced than for the two investigated piled rafts of the Messeturm and the Park Tower (Section 4.3.2).

In the case of a raft foundation relatively large loads are transferred to the subsoil close to the foundation level. Despite the high OCR at that depth, it appears that the soil is exposed to stress levels corresponding to primary loading and therefore creep becomes an issue (Ganal 2024). On the contrary, the piles of piled rafts distribute parts of the building loads to larger depth, resulting in lower stress levels and therefore limit creep deformation. Franzen and Reul (2022) also investigated the influence of the viscosity index on the settlement of the Messeturm. Using Mašín's visco-hypoplastic model for clay (Mašín 2014, Jerman and Mašín 2020), a significant influence of the viscosity index on the settlements was observed, although this cannot be confirmed on the basis of the results presented by Ganal (2024).

2.4 OTHER LOADING CONDITIONS

2.4.1 Piled rafts subjected to monotonic vertical and lateral loading

In contrast to piled rafts under predominantly vertical monotonic loading, comparatively few in situ measurements on completed structures and large-scale model tests (field investigations) have been documented for piled rafts under combined vertical and horizontal loading, regardless of whether monotonic or cyclic (Section 2.4.2) actions are involved. Section 4.3.4 provides the detailed case history of the Neue Messehalle 3 in Frankfurt, while other studies of this topic are summarised in Table 2.11.

A number of publications investigated the influence of the passive earth pressure acting on embedded rafts on the overall lateral resistance of the foundation (Beatty 1970, Zafir and Vanderpool 1998, Mokwa and Duncan 2001, Rollins and Sparks 2002) by means of large-scale load tests. Although the plan area of the rafts was relatively small with a maximum reported edge length of 4.1 m (Beatty 1970), the rafts provided between 40% and 50% of the overall lateral resistance of the foundations. Also based on large-scale model tests with relatively small rafts (plan area 2.14 m × 3.05 m), Kim et al. (1979) concluded that the lateral resistance of FPG and piled rafts is similar if more than half of the piles (steel profiles 10BP42; n_p = 6 per raft; L_p = 12.2 m) are battered.

Extensive investigations based on model tests in sand were published by a group of Japanese researchers including Matsumoto (Kanazawa University) and Horikoshi (Taisei Corporation). Watanabe et al. (2001) and Horikoshi et al. (2003a) reported on centrifuge tests on single piles, raft foundations and piled rafts in dry sand. Some of the tests of Watanabe et al. (2001) were also evaluated by Horikoshi et al. (2002a, b). After initially being loaded exclusively vertically, the raft foundation and the piled raft were subsequently loaded monotonically horizontally, with the piled raft mobilising significantly greater resistance than the raft foundation. For the connection

Table 2.11 Investigations on foundations subjected to monotonic vertical and lateral loading

Reference	Methodology	Foundation	Pile type	Soil types	Loading conditions
Beatty (1970) after Mokwa and Duncan (2001)	LST	PR	DP	Soft clay	MLL
Kim and Brungraber (1976), Kim et al. (1979)	LST	FPG, PR, SP	DP	Clay	MLL, MVL
Zafir and Vanderpool (1998) after Mokwa and Duncan (2001)	LST	PR	BP	Silty sand, clayey sand, sandy clay	MLL
Mokwa and Duncan (2001)	LST	PR, SP	DP	Sandy clay, sandy silt, silty sand	MLL
Watanabe et al. (2001), Horikoshi et al. (2002a), Horikoshi et al. (2002b)	CMT	PR, R, SP	MP	Sand	MLL, MVL
Kitiyodom Pastsakorn et al. (2002)	SMT	FPG, PR, R	MP	Sand	MLL, MVL
Rollins and Sparks (2002)	LST	PR	DP	Clay, silt, sand	MLL
Horikoshi et al. (2003a)	CMT	PR, R, SP	MP	Sand	MLL, MVL
Matsumoto et al. (2004a)	SMT	PR	MP	Sand	MLL
Turek (2006)	SMT, FEA	FPG, PR, R	MP	Sand	MLL, MVL
Turek (2006), Section 4.3.4	IMS	PR	BP	Clay	MLL, MVL
Comodromos et al. (2016)	ONA	PR, SP	BP	Silty sand, clay	MLL, MVL
Al-abboodi and Sabbagh (2017)	SMT	PR	MP	Sand	MVL, PL
Deb and Pal (2020)	SMT	PR	MP	Clay, sand	MLL, MVL
Shrestha et al. (2018), Ravichandran et al. (2018), Shrestha and Ravichandran (2019)	FEA, ONA	PR	n.s.	Clay, sand	MLL
Deb and Pal (2020)	SMT	PR	MP	Clay, sand	MLL, MVL

Notes: LST: large-scale load test; SMT: small-scale model test; CMT: centrifuge model test; IMS: in situ measurements on real structures; FEA: finite element analysis; ONA: other numerical or analytical methods; FPG: freestanding pile group; PR: piled raft; SP: single pile; R: raft foundation; BP: bored pile; DP: driven pile; MP: model pile (installation procedure may vary from piles in prototype scale); MLL: monotonic lateral loading (can additionally cause moment loading); MVL: monotonic vertical loading; PL: passive loading due to lateral soil movement; n.s.: not specified

between pile heads and raft hinges and rigid connections were tested by Horikoshi et al. (2003a) with the piled raft with the rigid connection showing a larger stiffness. Matsumoto et al. (2004a) used small-scale model tests on a piled raft under monotonic horizontal and moment loading to investigate the effects of the constraints at the pile head (hinge vs. rigid connection to the raft) on the load-bearing behaviour. In contrast to the investigation by Horikoshi et al. (2003a) here the piled raft with the hinge connection showed at least initially stiffer behaviour than the piled raft with the rigid connection. Matsumoto et al. (2004a) attribute this discrepancy to the different ratios of the pile bending stiffness and the soil modulus in the centrifuge test and the small-scale test. Kitiyodom Pastsakorn et al. (2002) reported results from small-scale model tests on monotonically vertically and horizontally loaded FPG and piled rafts. Even for large horizontal displacements, the contribution of the raft to the total horizontal resistance of the piled raft still amounted to only about 30%.

It is interesting to note that in the last decade with the so-called hybrid mudmats or Hybrid Subsea Foundations (HSF), piled raft-type foundations found their application in offshore engineering. HSF comprising a skirted mudmat and pin-piles at the corners are subsea foundations, e.g. for pipeline end terminations (Figure 2.43), which transfer large horizontal and moment

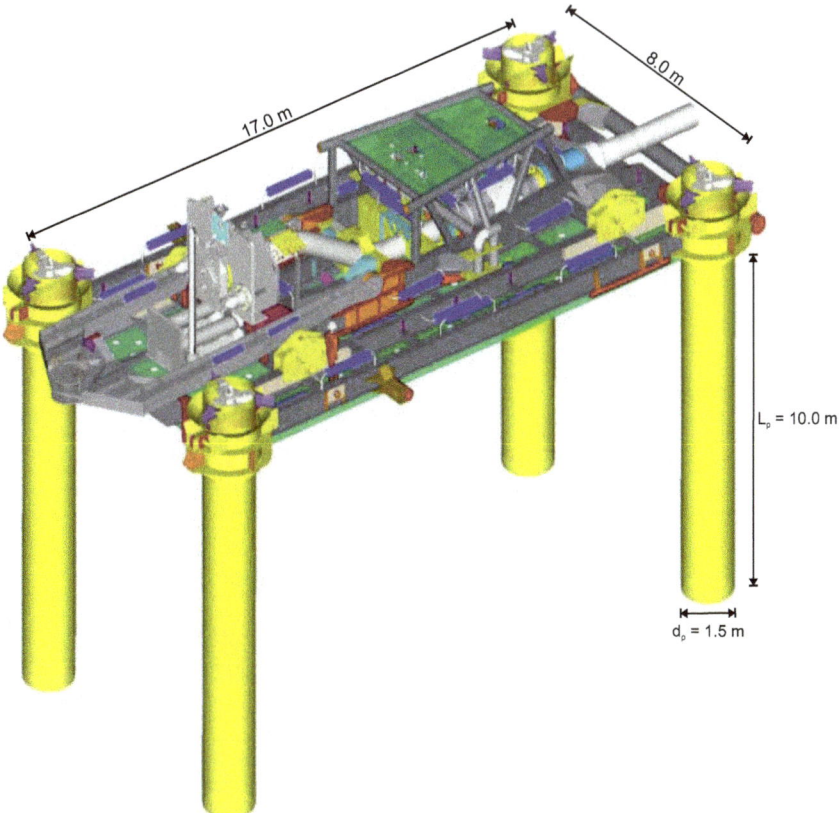

Figure 2.43 Hybrid Subsea Foundation (HSF) for a pipeline end termination structure after Wallerand et al. (2017)

loads in the soft marine sediments. An analytical design approach to assess the capacity of HSF was presented by Dimmock et al. (2013).

2.4.2 Piled rafts subjected to periodic and dynamic loading

2.4.2.1 General remarks

DGGT (2014) distinguishes between cyclic, dynamic and impact loading. Cyclic loading is a recurrent loading where inertial forces of the pile-soil-system do not need to be considered. Examples are slowly changing live loads, changing hydrostatic pressure, water waves, wind and rail and road traffic (depending on the frequency relative to the natural frequency of the structure). The influence of yearly cyclic changes of the temperature on an integral bridge, constructed without joints and bearings with superstructure and substructure being monolithically connected, is shown in Figure 2.44. According to Wenner et al. (2019), the 170-m-long integral railway bridge was instrumented in order to measure the long-term structural behaviour resulting from yearly temperature changes resulting in annual expansion and contraction cycles of the superstructure of $\Delta L = 36$ mm (Figure 2.44).

In contrast, dynamic loading is a recurrent loading where inertial forces of the pile-soil-system need to be considered. Examples are water waves, wind and rail and road traffic (depending on frequency relative to the natural frequency of the structure), machine-induced excitation, earthquakes, construction works and explosions.

Impact loading involves sudden loading that is only effective for a short period of time where, as for dynamic loading, inertial forces of the pile-soil-system need to be considered. Examples are construction works, explosions and vehicle impact due to collision.

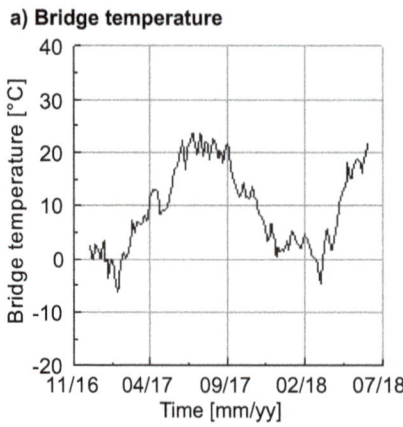
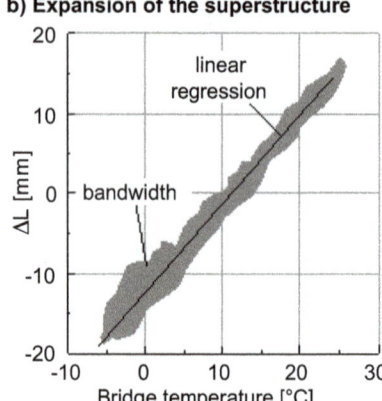

Figure 2.44 Influence of yearly cyclic temperature changes on an integral bridge (Wenner et al. 2019 after Niemann 2020).

Based on DGGT (2014) it is suggested that cyclic loading, rather than dynamic loading, may be considered where the frequency of the applied cyclic loading is less than about 30% of the natural frequency of the foundation system.

EC7-1 distinguishes between common cyclic, dynamic and impact loads, as, for example, caused by standard loading on traffic areas or from site operations, and substantial cyclic, dynamic and impact loads caused, for example, by collisions, pressure waves in the air or in (pore)water or by machine vibrations. While common cyclic, dynamic and impact loads are considered as variable static actions (EC7-1, A.2.4.2.1 A(8a)), in the case of substantial cyclic, dynamic and impact loads it is necessary to assess whether these may be taken into account by static equivalent loads or whether additional investigations are necessary to capture inertia and softening effects or the accumulation of deformation and pore water pressure (EC7-1, A.2.4.2.1 A(8b)).

In the context of pile foundations, loads are considered to be substantial if pile behaviour is influenced by softening effects (shear strength), accumulation of deformation or accumulation of pore pressure. For substantial cyclic, dynamic and impact loading on piles, depending on the boundary conditions the bearing behaviour may deviate considerably from piles subjected to monotonic, static loading. According to EC7-1, A.2.4.2.1 A(8b), this deviation has to be considered in the analysis of pile foundations. According to DGGT (2014), a cyclic axial load can be considered substantial if

$$F_{cyc} > 0.2R_{ult} \text{ or } F'_{cyc} > 0.1R_{ult} \qquad (2.33)$$

where F'_{cyc} = cyclic load amplitude; and F_{cyc} = cyclic load range. For other load types, quantitative criteria whether a load is substantial are not available.

The actions shown in Figure 2.45 for the example of cyclic lateral loading of a single pile represent the ideal case in terms of a harmonic cyclic load. In practice, various periodic loading situations can act on piles including interruptions in the cyclic actions. If different load ranges need to be considered, it is generally assumed that the accumulated pile displacement is independent of the load sequence. An equivalent cyclic load consisting of equivalent effects and a corresponding equivalent number of cycles may be derived (DGGT 2014).

2.4.2.2 Piled rafts subjected to cyclic loading

Table 2.12 summarises the investigations on piled rafts subjected to cyclic loading documented in the literature. Raft foundations, FPG and piled rafts under both monotonic vertical and cyclic lateral loading were the subject of small-scale model tests in dry sand reported by Matsumoto et al. (2010). The focus of those tests, which also varied the constraints at the pile head

86 Combined Pile-Raft Foundations

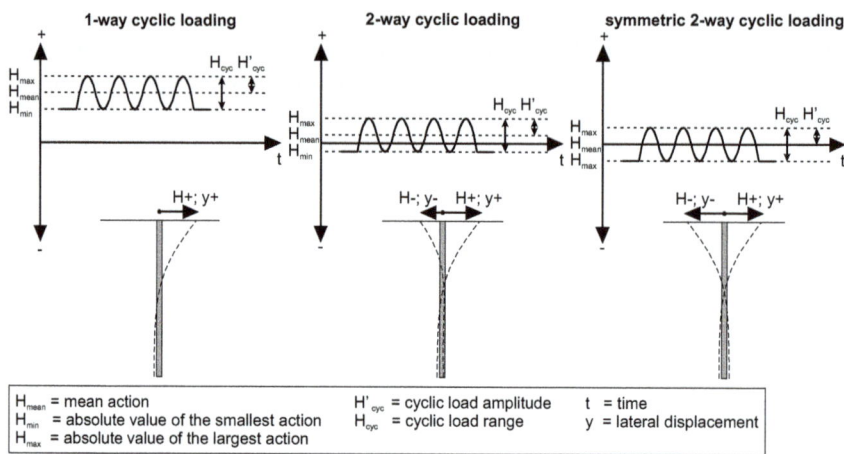

Figure 2.45 Actions from cyclic lateral loading.

Table 2.12 Investigations on foundations subjected to cyclic loading

Reference	Methodology	Foundation	Pile type	Soil types	Loading conditions
Matsumoto et al. (2010)	SMT, ONA	FPG, PR, R	MP	Dry sand	MVL, CLL
Unsever et al. (2013, 2014, 2015b)	SMT, FEA	FPG, PR, R, SP	MP	Dry sand	MVL, CLL
Vu et al. (2017)	SMT	FPG, PR	MP	Dry sand	MVL, CLL
Sawada and Takemura (2014)	CMT	FPG, PR, R	MP	Dry sand	MVL, CLL
Hamada et al. (2015b)	SMT, ONA	FPG, PR, R, SP	MP	Dry sand	MVI, CLL
Louw (2024)	IMS, FEA	PR	BP	Rock	MVL, CLL

Notes: SMT: small-scale model test; CMT: centrifuge model test; IMS: in situ measurements on real structures; FEA: finite element analysis; ONA: other numerical or analytical methods; FPG: freestanding pile group; PR: piled raft; SP: single pile; R: raft foundation; BP: bored pile; MP: model pile (installation procedure may vary from piles in prototype scale); CLL: cyclic lateral loading (can additionally cause moment loading); MVL: monotonic vertical loading.

(varying between hinge, semi-hinged, semi-rigid and rigid connection to the raft), included the contribution of the piles to the load transfer and the internal forces in the piles. Matsumoto et al. (2010) found that the horizontal load proportion carried by the raft as well as the lateral stiffness of the foundation becomes lower as the pile head connection becomes less rigid. The small-scale tests were back-analysed by Matsumoto et al. (2010) by applying a plate-beam spring model for the piled raft.

Unsever et al. (2013, 2014) reported on small-scale model tests comparing the bearing behaviour of single piles, raft foundations, FPG and piled rafts under monotonic vertical and cyclic lateral loading in dry sand where pile configuration for the FPG as well as for the piled raft comprised three piles placed in a row parallel to the lateral loading direction. As expected, due to the contribution of the raft to the load transfer, the piled raft mobilised a larger horizontal resistance than the FPG. Neither the vertical or the horizontal piled raft coefficient was affected significantly during cyclic lateral loading and amounted to $\alpha_{pr,v} \approx 0.7$ and $\alpha_{pr,h} \approx 0.8$, respectively. On the other hand, the cyclic lateral loading yielded an accumulation of vertical displacement which was up to two times larger for the FPG than for the piled raft. Unsever et al. (2015b) presented a back-analysis of the model tests on piled rafts by means of 3D FEA, whereby the material behaviour of the sand was modelled with the elasto-plastic Hardening Soil (HS) model. However, only results for the first load cycle, i.e. de facto for a monotonic loading, were presented; that was possibly due to the choice of the HS model, which is not well suited for the simulation of cyclic material behaviour. Using a similar test set-up, Vu et al. (2017) compared the bearing behaviour of FPG and piled rafts with inclined piles to corresponding foundations with vertical piles. As documented by Unsever et al. (2013, 2014), for the same number of lateral loading cycles the FPG showed a more significant accumulation of settlements than the piled rafts. However, these accumulated settlements caused by cyclic lateral loads could be significantly reduced by inclined piles for the FPG as well as for the piled rafts.

Sawada and Takemura (2014) investigated raft foundations, FPG and piled rafts under monotonic vertical loading and cyclic horizontal and moment loading typical of bridge foundations, using centrifuge model tests in dry sand. As in the small-scale model tests by Unsever et al. (2013, 2014) and Vu et al. (2017), they observed a settlement accumulation due to cyclic lateral loading which was least pronounced for the piled rafts and most pronounced for the FPG.

Hamada et al. (2015b) reported results from small-scale model tests in dry sand on single piles, raft foundations, FPG and piled rafts under cyclic horizontal loading with the main aim to derive simplified equations for the seismic design of piled rafts. A large proportion of the horizontal loads on the piled rafts was thus transferred via friction at the base of the raft, which was constructed realistically as a cast-in-place concrete foundation in the model tests.

Louw (2024) presented measurements for an onshore wind turbine (hub height h_{hub} = 117 m; rotor diameter d_{rotor} = 126 m) in South Africa founded on a piled-raft with the piles socketed in rock. The in situ measurements (but with no details of cumulative settlement) were carried out over a time period of two years, including the construction phase and the first year of operation (Figure 2.46). It is interesting to note that although the piles were socketed in rock they carried only approximately 35% of the total vertical turbine self-weight after installation. For the last measurements available, i.e. after one year of cyclic loading, Louw (2024) reports an increase to 60% for the

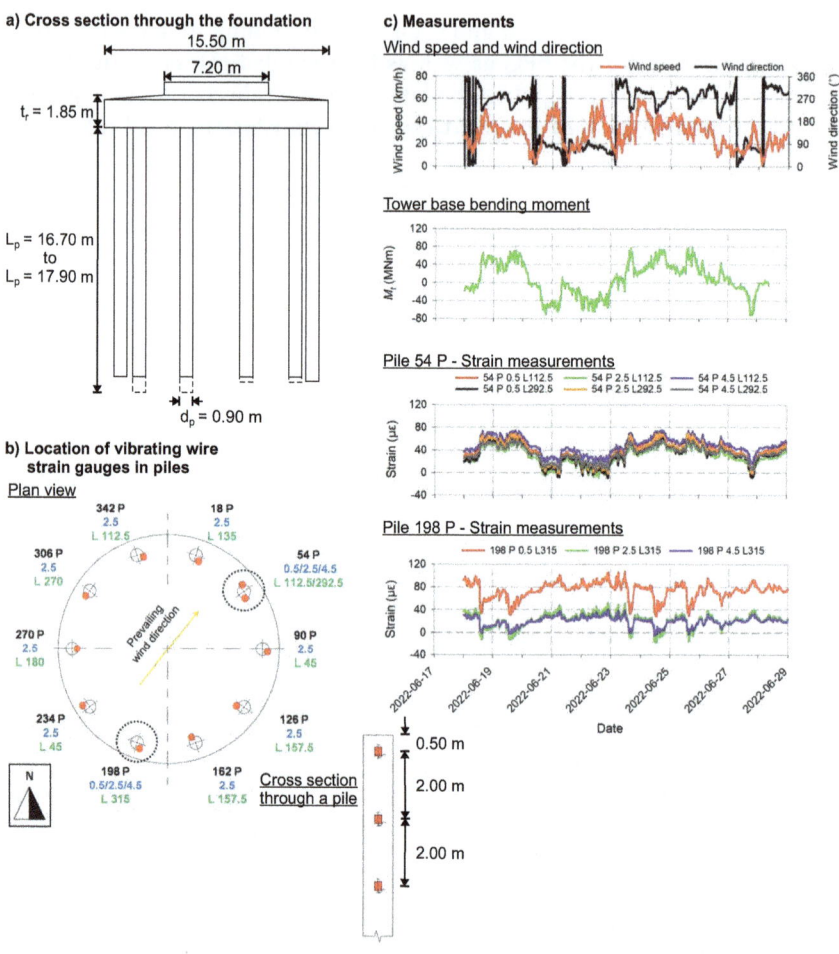

Figure 2.46 Measurements for an onshore wind turbine after Louw (2024).

share of the piles in the transfer of the total vertical turbine self-weight. Additionally Louw (2024) reported a numerical parametric study on piled rafts subjected to monotonic horizontal loads and overturning moments, varying the load magnitude, diameter and thickness of the raft, the pile diameter and stiffness and height of the soil layer. The numerical model based on the finite element method (FEM) was calibrated by means of the back-analysis of the abovementioned in situ measurements.

2.4.2.3 *Piled rafts subjected to dynamic loading*

Table 2.13 gives an overview of the literature on piled rafts subjected to dynamic loading. Mendoza et al. (2000) presented measurements on a bridge foundation in Mexico City, founded in soft clay on a piled raft referred to

Bearing behaviour of piled rafts 89

Table 2.13 Investigations on foundations subjected to dynamic loading

Reference	Methodology	Foundation	Pile type	Soil types	Loading conditions
Mendoza et al. (2000)	IMS*	PR	DP	Soft clay	MVL, SL
Yamashita et al. (2012a, 2018)	IMS*, FEA	PR & DMW	PHC	Sand, clay, silt	MVL, SL
Yamashita et al. (2012b)	IMS*	PR	PHC	Sand, silt	MVL, SL
Yamashita et al. (2013)	IMS*	PR & DMW	PHC	Fill, clay, silt, sand	MVL, SL
Hamada et al. (2015a)	IMS	PR	BP	Sand, clay	MVL, SL
Yamashita et al. (2016)	IMS*	PR & DMW	PHC	Sand, silt	MVL, SL
Yamashita et al. (2019)	IMS*	PR & DMW	PHC	Fill, sand, clay, silt	MVL, SL
Pitteloud and Meier (2019)	IMS*, FEA	PR	BP	Marl, molasse	MVL, DL (wind)
Meymand (1998)	SMT	FPG, PR, SP	MP	Clay	MVL, MLL, DL
Horikoshi et al. (2002a, 2002b)	CMT	PR	MP	Dry sand	DL
Horikoshi et al. (2003b), Eslami et al. 2011), Kumar et al. (2016), Degu et al. (2024)	CMT, FEA	FPG, PR	MP	Dry sand	DL
Matsumoto et al. (2004b)	SMT	PR	MP	Dry sand	DL
Banerjee et al. (2014)	CMT, FEA	PR	MP	Kaolin clay	DL
Takemura et al. (2014)	CMT	PR	MP	Dry & saturated sand	DL
Unsever et al. (2015a), Unsever et al. (2017)	SMT	FPG, PR	MP	Dry & saturated sand	DL
Azizkandi et al. (2017)	SMT	PR	MP	Dry sand	DL
Goh and Zhang (2017)	CMT, FEA	PR	MP	Kaolin clay	DL
Jafarian et al. (2021)	CMT	PR	MP	Saturated sand & gravel	DL
Qin and Ma (2021)	CMT, FEA	PR	MP	Kaolin clay	DL

* see also Appendix B

SMT: small-scale model test; CMT: centrifuge model test; IMS: in situ measurements on real structures; FEA: finite element analysis; FPG: freestanding pile group; PR: piled raft; SP: single pile; R: raft foundation; DMW: deep mixing wall; BP: bored pile; DP: driven pile; PHC: pretensioned spun high-strength concrete pile placed in a borehole; MP: model pile (installation procedure may vary from piles in prototype scale); MLL: monotonic lateral loading (can additionally cause moment loading); MVL: monotonic vertical loading; DL: dynamic loading by means of a shaking table; SL: seismic loading because of an earthquake.

as "pile-box-foundation" (Appendix B.2, Table B.9). During the measurements carried out over a period of 3.5 years, two earthquake events in 1997 were also recorded, so that insights could be gained into the load distribution mechanism between piles and raft under both monotonic static and dynamic actions. According to Mendoza et al. (2000), pile loads decreased during the earthquake events but recovered few days afterwards to their pre-earthquake level. The seismic loads on piles located near the foundation edge were reported to be higher than in the central parts of the foundation.

Extensive measurements of structures founded on piled rafts in Japan, some of which were carried out over several years, were documented by Yamashita et al. (2012a, 2012b, 2013, 2016, 2019) and Hamada et al. (2015a). Structures for which settlement measurements are available are also listed in Appendix B.2, Table B.7. During these long-term measurements, various earthquake events such as the Tohoku earthquake (September 11 2011) were recorded with the resulting dynamic actions and their effects on the load-bearing behaviour of the piled rafts. For some structures, mixed-in-place (MIP) walls were placed under the rafts in addition to the piles to prevent soil liquefaction due to earthquakes. In the publications MIP walls are frequently termed deep mixing walls (DMW). Yamashita et al. (2018) discussed the influence of soil improvement consisting of DMW on the bearing behaviour of structures founded on piled rafts (Yamashita et al. 2012a) under earthquake loading on the basis of 3D FEA. The material behaviour of the silty sands prone to liquefaction and silts was simulated using an elasto-plastic material model (Tsujino et al. 1994), while for the DMW a nonlinear elastic material model was applied. With the 3D FEA, in which the earthquake loading was considered from various acceleration profiles of real earthquakes, Yamashita et al. (2018) showed that the DMW can reduce the settlements as well as the bending moments in the piles.

Pitteloud and Meier (2019) presented results from high-frequency monitoring of the piled raft foundation of a high-rise building in Basel, Switzerland (Appendix B.2, Table B.6) under dynamic wind loading. Measurements taken during a windstorm event in November 2015 indicated that the forces from the wind loads were primarily transferred by the building cores yielding a much more pronounced variation of the pile resistances of the central piles than of piles located outside the core area of the building. From these measurements they estimated a natural frequency of the building for excitation in north-south direction of $f_n \approx 0.35$ Hz. Moreover, the measurements are subjected to back-analysis using 3D FEA, applying the Hardening Soil small strain model for the soil, which comprised marl and molasse.

The piled raft model of Watanabe et al. (2001) (Section 2.4.1) was subjected to dynamic horizontal loading in the centrifuge by Horikoshi et al. (2002a, 2002b), which caused permanent settlement and horizontal displacement. Horikoshi et al. (2003b) discussed the influence of the constraints at the pile head (hinge vs. rigid connection to the raft) on the horizontal resistance of the foundation under and dynamic horizontal loads As observed by Horikoshi et al. (2003a), for monotonic horizontal loading the piled raft with the rigid connection showed stiffer behaviour than the piled raft with the hinge connection. From comparison with a FPG (rigid connection), Horikoshi et al. (2003b) concluded that the contact of the raft with the soil reduced the inclination and bending moments of the rafts of the piled raft. The centrifuge tests by Horikoshi et al. (2003b) were the subject of numerical back-analyses by means of 3D FEA by Eslami et al. (2011) and Kumar et al. (2016) applying simple linear elastic-ideal plastic soil models and by Degu et al. (2024) applying the HS small strain model (Benz 2007) which is better suited to capture cyclic soil behaviour.

For otherwise identical test boundary conditions, Unsever et al. (2015a) subject the FPG and the piled raft in sand from Unsever et al. (2013, 2014) (Section 2.4.2) to dynamic horizontal loading in shaking table tests. While the inclination behaviour of FPG and piled raft were considered similar by Unsever et al. (2015a), they found the settlements of the piled raft to be approximately half of the settlements of the FPG. Additionally, Unsever et al. (2017) investigated the FPG and the piled raft in saturated sand, applying dynamic horizontal loading by means of a shaking table. Although, even for the piled raft, the horizontal resistance decreased due to liquefaction, the piled raft proved to be more effective in reducing settlements in liquefied soil than the FPG.

Takemura et al. (2014) compared the load-bearing behaviour of raft foundations and piled rafts under a liquid-filled tank under dynamic loading in centrifuge model tests. The tests were carried out both in dry sand and in water-saturated sand in order to investigate liquefaction effects by means of a shaking table. Under horizontal acceleration typical for an earthquake loading, the piled raft was able to reduce uneven settlements compared to the raft foundation. However, the effectiveness of the piled raft was less pronounced in the saturated sand which Takemura et al. (2014) attribute to a decrease of effective stress caused by liquefaction which in turn corresponds to a significant decrease of the piled raft coefficient

Jafarian et al. (2021) reported results from two centrifuge model tests on piled rafts in water-saturated sand and gravel with and without gravel drains. In test no. 1 the whole soil body in the strong box consisted of sand prone to liquefaction, while in test no. 2 a layer of gravel was placed on top of the sand just beneath the raft. Under the dynamic loads applied by means of a shaking table, the drains were able to reduce both the excess pore water pressures and the settlements.

Meymand (1998) presented results from small-scale model tests on vertically and horizontally loaded single piles, FPG and piled rafts in artificial soft clay made from bentonite, kaolin and fly ash. Horizontal dynamic loading was applied by means of a shaking table. It was found, for example, that the shear stresses mobilised at the base of the raft contributed more significantly to the horizontal resistance of the foundation than the passive earth pressure acting at the raft sides. Moreover, a piled raft that had been first subjected to monotonic lateral loading showed lesser seismic response than an identical piled raft that had not been pre-loaded, suggesting, according to Meymand (1998), that pre-loading remoulded the soil and base-isolated the foundation.

Banerjee et al. (2014) and Goh and Zhang (2017) investigated piled rafts in kaolin under dynamic loading in centrifuge tests, where the seismic profiles applied with the shaking table were derived from earthquake events in Singapore. The tests by Banerjee et al. (2014) focused especially on the influence of pile geometry and pile stiffness. By means of a back-analysis of the centrifuge tests, Banerjee et al. (2014) validated a 3D FE model. The material behaviour of the kaolin under cyclic loading was modelled using a non-linear constitutive law that accounted for both hysteresis effects and cycle-dependent degradation of the shear modulus (Banerjee 2010). Based on the model tests and numerical simulations, Banerjee et al. (2014) defined an active pile length that distinguished between stiff and flexible piles, resulting in different mechanisms for the horizontal load transfer into the soil. Qin and Ma (2021) also conducted centrifuge tests in kaolin, comparing the bearing behaviour of FPG (termed high-raft case in the publication) and piled rafts under dynamic loads. In the centrifuge model tests as well as in additional accompanying 3D FEA the piles of the FPG showed larger bending moments than the piles of the piled raft.

The influence of the superstructure was the focus of a number of experimental investigations. Using the piled raft model of Matsumoto et al. (2004a) (Section 2.4.1), Matsumoto et al. (2004b) reported results of dynamic shaking table tests where the resonance frequency of the superstructure decreased with increasing height of the centre of gravity. For frequencies lower than the resonance frequency, they found the bearing behaviour of the piled raft similar to monotonic horizontal loading. Azizkandi et al. (2017) reported shaking table tests on small-scale models of piled rafts in dry sand showing larger bending moments in the piles if the mass of a superstructure is considered in the model. The influence of the superstructure was investigated by Goh and Zhang (2017) in the abovementioned centrifuge tests by varying the vertical load on the raft. Goh and Zhang (2017) also use a 3D FE model to back-analyse their centrifuge tests, also applying the constitutive law proposed by Banerjee (2010), which was calibrated with laboratory tests. Based on this, Goh and Zhang (2017) conducted a numerical parametric study and derived empirical correlations for estimating the maximum bending moments in the piles and the maximum acceleration of the raft.

Helpful tools to estimate the dynamic behaviour of piled rafts are the solutions provided by Gazetas (1991) to establish stiffness and damping of raft foundations and pile groups subjected to dynamic loading. A summary of the current knowledge on the bearing behaviour and the design of foundations under dynamic loading and seismic events was provided by, among others, Poulos (2017) and Katzenbach et al. (2017).

REFERENCES

ABAQUS Theory Manual - Version 5.8 (1998). Hibbitt, Karlsson & Sorensen, Pawtucket R.I.

Al-abboodi, I., Sabbagh, T. (2017). Model tests on piled raft subjected to lateral soil movement. *International Journal of Geotechnical Engineering*, 12, 4, 357–367.

Alnuaim, A. M., El Naggar, M. H., El Naggar, H. (2015). Performance of micropiled raft in clay subjected to vertical concentrated load - centrifuge modelling. *Canadian Geotechnical Journal*, 52, 12, 2017–2029, https://doi.org/10.1139/cgj-2014-0448

Azizkandi, A. S., Baziar, M. H., Razmi, B. (2017). Experimental study on seismic response of structure with piled raft foundation. *Proceedings of the 19th International Conference on Soil Mechanics and Geotechnical Engineering*, 835–838.

Banerjee, S. (2010). Centrifuge and numerical modelling of soft clay–pile–raft foundations subjected to seismic shaking. Dissertation, National University of Singapore.

Banerjee, S., Goh, S. H., Lee, F. H. (2014). Earthquake-induced bending moment in fixed-head piles in soft clay. *Géotechnique*, 64, 6, 431–446.

Beatty, C. I. (1970). Lateral test on pile groups. *Foundation Facts*, 6, 1, 18–21.

Benz, T. (2007). Small-strain stiffness of soils and its numerical consequences. PhD thesis, Universität Stuttgart, Institut für Geotechnik, Stuttgart.

Bhartiya, P., Basu, D., Chakraborty, T. (2022). Time-dependent response of rectangular piled rafts in clayey soils. *ASCE Journal of Geotechnical and Geoenvironmental Engineering*, 148. https://doi.org/10.1061/(ASCE)GT.1943-5606.0002758

Borel, S. (2001). Comportement et dimensionnement des fondations mixtes. Ph.D. thesis de ENPC, Spécialité Géotechnique, Paris.

Brand, E. W., Muktabhant, C., Taechathummarak, A. (1972). Load tests on small foundation in soft clay. *Proceedings of the ASCE Conference on Performance of Earth and Earth Supported Structures*, Purdue University, Vol. 1, Part 2, 903–928.

Comodromos, E. M., Papadopoulou, M. C., Laloui, L. (2016). Contribution to the design methodologies of piled raft foundations under combined loadings. *Canadian Geotechnical Journal*, 53, 4, 559–577.

Conte, G., Mandolini, A., Randolph, M. F. (2003). Centrifuge modelling to investigate the performance of piled rafts. *Proceedings of the Conference Deep Foundations on Bored and Auger Piles*, Ghent, 379–386.

Cooke, R. W. (1986). Piled raft foundations on stiff clays- a contribution to design philosophy. *Géotechnique*, 36, 2, 169–203.

Cooke, R. W., Bryden-Smith, D. W., Gooch, M. N., Sillett, D. F. (1981) Some observations of the foundation loading and settlement of a multi-storey building on a piled raft foundation in London Clay. *Proceedings of the ICE, Part 1*, 70, 433–460.

Cui, C., Luan, M., Zhao, Y. (2009). Time-dependent behavior of piled-raft on soil foundation with reference to creep and consolidation. *Electronic Journal of Geotechnical Engineering*, 14, 1–14.

Davis, E. H., Taylor, H. (1962). The movement of bridge approaches and abutments on soft foundation soils. *Proceedings of the 1st Biennial Conference* Aust. Road Res. Board, 740.

Deb, P., Pal, S. K. (2020). Nonlinear analysis of lateral load sharing response of piled raft subjected to combined V-L loading. *Marine Georesources & Geotechnology*, 39, 8, 994–1014.

Degu, Y., Reul, O., Worku, A., Tschuchnigg, F. (2024). Calibration of a numerical model for the investigation of piled rafts under dynamic loading. *Proceedings of the 18th European Conference on Soil Mechanics and Geotechnical Engineering*, Lisbon (accepted for publication)

de Sanctis, L., Mandolini, A. (2006). Bearing capacity of piled rafts on soft clay soils. *Journal of Geotechnical and Geoenvironmental Engineering*, 132, 12, 1600–1610.

Deutsche Gesellschaft für Geotechnik, (DGGT). Recommendations on Piling (EA-Pfähle). Ernst & Sohn, 2014.

Dimmock, P., Clukey, E. C., Randolph, M. F., Gaudin, C., Murff, J. D. (2013). Hybrid subsea foundations for subsea equipment. *Journal of Geotechnical and Geoenvironmental Engineering, ASCE*, 139, 12, 2182–2192.

DIN EN 1997-1:2014-03 (2014). Eurocode 7: Geotechnical design – Part 1: General rules. German version EN 1997-1:2004 + AC:2009 + A1:2013

Eslami, M. M., Aminikhah, Ahmadi, M. M. (2011). A comparative study on pile group and piled raft foundations (PRF) behavior under seismic loading. *Computational Methods in Civil Engineering*, 2, 2, 185–199.

Fattah, M. Y., Al-Mosawi, M. J., Al-Zayadi, A. A. (2014). Contribution to long term performance of piled raft foundation in clayey soil. *International Journal of Civil Engineering and Technology*, 5, 7, 130–148.

Fioravante, V., Giretti, D., Jamiolkowski, M. B. (2008). Physical modelling of piled rafts. *Proceedings of the Deep foundations on bored and auger piles*, Ghent, Belgium, 241–248.

Franke, E., Lutz, B. (1994) Pfahl-Platten-Gründungs-Messungen. Forschungsabschlussbericht zum Forschungsauftrag Fr 600 - 11/1.

Franke, E., Mader, H., Schetelig, K., Schneewolf, T. (1985). Anisotropie des Eigenspannungszustandes der wechsel-lagernden Locker- und Festgesteinsschichten des Frankfurter Raumes, Ingenieurgeologische Probleme im Grenz-bereich zwischen Locker- und Festgesteinen, 399–416.

Franzen, A., Reul, O. (2022). Numerical investigation of the long-term settlement behaviour of piled rafts in overconsolidated clay, *Proceedings of the 20th International Conference on Soil Mechanics and Geotechnical Engineering*. Sydney, 3445–3450.

Fraser, R. A., Wardle, L. J. (1976). Numerical analysis of rectangular rafts on layered foundations. *Géotechnique*, 26, 4, 613–630.

Ganal, A. (2024). Time dependent bearing behaviour of foundations subjected to alternate loading in overconsolidated clay. Ph.D. thesis, Schriftenreihe Geotechnik, Universität Kassel, Heft 30.

Ganal, A., Reul, O. (2023). Back analysis of long-term measurements of a high-rise building founded on a raft foundation in over-consolidated clay. *Proceedings of the 10th European Conference on Numerical Methods in Geotechnical Engineering*, London, https://doi.org/10.53243/NUMGE2023-116

Ganal, A., Jacobsz, S. W., Reul, O. (2022). Centrifuge tests on foundations under alternating loads in overconsolidated clay. *Proceedings of the 10th International Conference on Physical Modelling in Geotechnics*, Daejeon, 816–819.

Garcia, F., Lizcano, A., Reul, O. (2006). Numerical modelling of the case history of a piled raft with a viscohypoplastic model. In: *Numerical Modelling of Construction Processes in Geotechnical Engineering for Urban Environment* – Triantafyllidis, T. (ed.), 265–271, Taylor & Francis Group, London, ISBN: 0415397480.

Gazetas, G. (1991). Foundation vibrations. In: *Foundation Engineering Handbook* – Fang, H. S. (ed.), 2nd edition, 563–593, Chapman & Hall.

Goh, S. H., Zhang, L. (2017). Estimation of peak acceleration and bending moment for pile-raft systems embedded in soft clay subjected to far-field seismic excitation. *ASCE Journal of Geotechnical and Geoenvironmental Engineering*, 143, 11, https://doi.org/10.1061/(ASCE)GT.1943-5606.0001779

Hamada, J., Aso, N., Hanai, A., Yamashita, K. (2015a). Seismic performance of piled raft subjected to unsymmetrical earth pressure based on seismic observation records. *Proceedings of the 6th International, Conference on Earthquake Geotechnical Engineering*.

Hamada, J., Tsuchiya, T., Tanikawa, T., Yamashita, K. (2015b). Lateral loading tests on piled rafts and simplified method to evaluate sectional forces of piles. *Geotechnical Engineering Journal SEAGS & AGSSEA*, 46, 2, 29–42.

Hansbo, S. (1993). Interaction problems related to the installations of pile groups. *Proceedings of the Conference Deep Foundations on Bored and Auger Piles*, Ghent, 59–66, Rotterdam: Balkema.

Hemsley, J. A. (1998). *Elastic analysis of raft foundations*. Thomas Telford, London.

Hoang, L. T., Matsumoto, T. (2020). Long-term behavior of piled raft foundation models supported by jacked-in piles on saturated clay. *Soils and Foundations*, 60, 198–217. https://doi.org/10.1016/j.sandf.2020.02.005

Horikoshi, K., Randolph, M. F. (1996). Centrifuge modelling of piled raft foundations on clay. *Géotechnique*, 46, 4, 741–752.

Horikoshi, K., Randolph, M. F. (1997). On the definition of raft-soil stiffness ratio. *Géotechnique*, 47, 5, 1055–1061.

Horikoshi, K., Randolph, M. F. (1998). A contribution to the optimum design of piled rafts. *Géotechnique*, 48, 2, 301–317.

Horikoshi, K., Randolph, M. F. (1999). Estimation of overall settlement of piled rafts. *Soils and Foundations*, 39, 2, 59–68.

Horikoshi, K., Matsumoto, T., Watanabe, T., Fukuyama, H. (2002a). Performance of piled raft foundations subjected to seismic loads. *Proceedings of the International Workshop on Foundation Design Codes and Soil Investigation in View of International Harmonization and Performance Based Design*, IWS Kamakura 2002, 1–8.

Horikoshi, K., Watanabe, T., Fukuyama, H., Matsumoto, T. (2002b). Behavior of piled raft foundations subjected to horizontal loads. *Proceedings of the 1st International Conference on Physical Modelling in Geotechnics*, St. John's, 715–721.

Horikoshi, K., Matsumoto, T., Hashizume, Y., Watanabe, T., Fukuyama, H. (2003a). Performance of piled raft foundations subjected to static horizontal loads. *International Journal of Physical Modelling in Geotechnics*, 2, 37–50.

Horikoshi, K., Matsumoto, T., Hashizume, Y., Watanabe, T. (2003b). Performance of piled raft foundations subjected to dynamic loading. *International Journal of Physical Modelling in Geotechnics*, 2, 51–62.

ISSMGE Technical Committee TC 212 (2013). ISSMGE Combined Pile-Raft Foundation Guideline. Report of the ISSMGE Technical Committee TC 212 – Deep Foundations, ed. Katzenbach, R., Choudhury D, Technische Universität Darmstadt.

Jafarian, Y., Fallahzedeh, M., Lee, C., Haddad, A. Hedayati, J. (2021). Centrifuge modeling for seismic performance of floating piled raft with and without drainage wells in liquefiable site. *ASCE, International Journal of Geomechanics*, 21, 5. https://doi.org/10.1061/(ASCE)GM.1943–5622.0001994

Jerman, J., Mašín, D. (2020). Hypoplastic and viscohypoplastic models for soft clays with strength anisotropy. *International Journal of Numerical Analysis Methods in Geomechanics*, 44, 1396–1416. https://doi.org/10.1002/nag.3068

Katzenbach, R., Arslan, U., Moormann, C., Reul, O. (1998). Piled raft foundation - Interaction between piles and raft. *Proceedings of the International Conference on Soil-Structure Interaction in Urban Civil Engineering, Darmstadt Geotechnics*, 2, 4, 279–296.

Katzenbach, R., Leppla, S., Choudhury, D. (2017). *Foundation systems for high-rise structures*. CRC Press, London.

Kim, J. B., Brungraber, R. J. (1976). Full-scale lateral load tests of pile groups. *Journal of the Geotechnical Engineering Division, ASCE*, 102, GT1, 87–105.

Kim, J. B., Singh, L. P., Brungraber, R. J. (1979). Pile cap soil interaction from full-scale lateral load tests. *Journal of the Geotechnical Engineering Division, ASCE*, 105, GT5, 643–653.

Kitiyodom Pastsakorn, K., Hashizume, Y., Matsumoto, T. (2002). Lateral load tests on model pile groups and piled raft foundations in sand. *Proceedings of the 1st International Conference on Physical Modelling in Geotechnics*, St. John's, 709–714.

Kumar, A., Choudhury, D., Katzenbach, R. (2016). Effect of earthquake on combined pile–raft foundation, *International Journal of Geomechanics*, 16, 5, 1–16.

Liu, J. L., Yuan, Z. L., Zhang, K. P. (1985). Cap-pile-soil interaction of bored pile groups. *Proceedings of the XIth ICSMFE*, San Francisco, Vol. 3, 1433–1436.

Liu, J., Huang, Q., Li, H., Hu, W. L. (1994). Experimental research on bearing behaviour of pile groups in soft soil. *Proceedings of the 13th ICSMFE*, 2, 535–538.

Louw, H. (2024). Soil-structure interaction of horizontally loaded piled-raft foundations. PhD thesis, University of Pretoria.

Mašín, D. (2014). Clay hypoplasticity model including stiffness anisotropy. *Géotechnique*, 64, 232–238. https://doi.org/10.1680/geot.13.P.065

Matsumoto, T., Fukumura, K., Pastsakorn, K., Horikoshi, K., Oki, A. (2004a). Experimental and analytical study on behaviour of model piled rafts in sand subjected to horizontal and moment loading. *International Journal of Physical Modelling in Geotechnics*, 4, 3, 1–19.

Matsumoto, T., Fukumura, K., Horikoshi, K., Oki, A. (2004b). Shaking table tests on model piled rafts in sand considering influence of superstructures. *International Journal of Physical Modelling in Geotechnics*, 4, 3, 21–38.

Matsumoto, T., Nemoto, H., Mikami, H, Yaegashi, K., Arai, T., Kitiyodom, P (2010). Load tests of piled raft models with different pile head connection conditions and their analyses. *Soils and Foundations*, 50, 1, 63–81.

Mayne, P. W., Kulhawy, F. H. (1982). K0-OCR relationships in soil, Japan Society of Civil Engineers, *Journal of Geotechnical Engineering*, 108, 851–872.

Mendoza, M. J., Romo, M. P., Orozco, M., Dominguez, L. (2000). Static and seismic behavior of a friction pile-box foundation in Mexico City clay. *Soils and Foundations*, 40, 4, 143–154.

Meymand, P. J. (1998). Shaking table scale model tests of nonlinear soil-pile-superstructure interaction in soft clay. PhD thesis, University of California, Berkeley.

Modak, R., Singh, B (2023). Numerical study on settlement-dependent variation of raft-soil-pile interactions for large piled raft in clay soil. *Ocean Engineering*, 281. https://doi.org/10.1016/j.oceaneng.2023.115011

Mokwa, R. L., Duncan, J. M. (2001). Experimental evaluation of lateral-load resistance of pile caps. *ASCE Journal of Geotechnical and Geoenvironmental Engineering*, 127, 2, 185–192.

Niemunis A. (2003). Extended hypoplastik models for soils. Schriftenreihe des Institutes für Grundbau und Bodenmechanik der Ruhr-Universität Bochum, Band 34.
Niemann, C. (2020). Pile groups under cyclic, lateral loading. Schriftenreihe Geotechnik Universität Kassel, Heft 27.
Optum CE (2022). Optum G3 2021 2.1.6.
Phung, D. L. (1993). *Footings with settlement-reducing piles in non-cohesive soil.* Department of Geotechical Engineering, Chalmers University of Technology, Gothenburg. Dissertation.
Pitteloud, L., Meier, J. (2019). High-frequency monitoring results of a piled raft foundation under wind loading. *International Journal of Geotechnical and Geological Engineering*, 13, 3, 90–102.
Prakoso, W. A., Kulhawy, F. H. (2001). Contribution to piled raft optimum design. *Journal of Geotechnical and Geoenvironmental Engineering, ASCE*, 127, 1, 17–24.
Poulos, H. G. (2000). Practical design procedures for piled raft foundations. In J. A. Hemsley (Ed.), *Design applications of raft foundations* (pp. 425–467). Thomas Telford.
Poulos, H. G. (2001). Piled-raft foundation: design and applications. *Géotechnique*, 51, 2, 95–113.
Poulos, H. G. (2017). *Tall building foundation design.* CRC Press, London.
Qin, H., Ma, K. (2021). Dynamic behaviour difference between high- and low-raft forms of piles in earthquakes. *Geotechnical Research*, 8, 3, 85–92.
Randolph, M. F. (1994). Design Methods for pile groups and piled rafts. *Proceedings of the 13th International Conference on Soil Mechanics and Foundation Engineering*, New Delhi, 5, 61–82.
Ranganatham, B. V., Kaniraj, S. R. (1978). Settlement of model pile foundations in sand. *Indian Geotechnical Journal*, 8, 1, 1–26.
Ravichandran, N., Shrestha, S., Piratla, K. (2018). Robust design and optimization procedure for piled-raft foundation to support tall wind turbine in clay and sand. *Soils and Foundations*, 58, 744–755.
Reul, O. (2000). In-situ-Messungen und numerische Studien zum Tragverhalten der Kombinierten Pfahl-Plattengründung. Mitteilungen des Institutes und der Versuchsanstalt für Geotechnik der Technischen Universität Darmstadt, Heft 53.
Reul, O. (2001). Numerical study on the bearing behaviour of piled rafts subjected to nonuniform vertical loading. Research Report Geo: 03294, The University of Western Australia, Department of Civil and Resource Engineering.
Reul, O. (2002). Study of the influence of the consolidation process on the calculated bearing behaviour of a piled raft. *Proceedings of the 5th NUMGE*, Paris, 383–388.
Reul, O. (2004). Numerical study of the bearing behaviour of piled rafts. *International Journal of Geomechanics*, 4, 2, 59–68.
Reul, O., Randolph, M. F. (2003). Piled rafts in overconsolidated clay – Comparison of in-situ measurements and numerical analyses. *Géotechnique*, 53, 3, 301–315.
Reul, O., Randolph, M. F. (2004). Design strategies for piled rafts subjected to non-uniform vertical loading. *ASCE Journal of Geotechnical and Geoenvironmental Engineering*, 130, 1, 1–13.
Rincón, R. E., Cunha, R. P., Caicedo, H. B. (2020). Analysis of settlements in piled raft systems founded in soft soil under consolidation process. *Canadian Geotechnical Journal*, 57, 537–548. https://doi.org/10.1139/cgj-2018-0702

Rollins, K. M., Sparks, A. (2002). Lateral resistance of full-scale pile cap with gravel backfill. *ASCE Journal of Geotechnical and Geoenvironmental Engineering*, 128, 9, 711–723.

Russo, G. (1998). Developments in the analysis and design of piled rafts. *Workshop prediction and performance in geotechnical engineering*, Napoli, 279–327.

Sales, M. M. (2000). Anàlise do comportamento de sapatas estaqueadas. Ph.D. thesis em Geotecnia, Univ. de Brasilia.

Sawada, K., Takemura, J. (2014). Centrifuge model tests on mechanical behavior of piled raft foundation in sand subjected to horizontal and moment loads. In *Proceedings of the 8th International Conference on Physical Modelling in Geotechnics*, Perth (Gaudin, C. and White, D. (eds)) Taylor & Francis, London, UK, 1, 637–644.

Sheil, B. (2017). Numerical simulations of the reuse of piled raft foundations in clay. *Acta Geotechnica*, 12, 1047–1059. https://doi.org/10.1007/s11440-017-0522-8

Shrestha, S. Ravichandran, N. (2019). 3D Nonlinear finite element analysis of piled-raft foundation for tall wind turbines and its comparison with analytical model. *Journal of GeoEngineering*, 14, 4, 259–276, http://doi.org/10.6310/jog.201912_14(4).5

Shrestha, S., Ravichandran, N., Rahbari, P. (2018). Geotechnical design and design optimization of a pile-raft foundation for tall onshore wind turbines in multilayered clay. *International Journal of Geomechanics*, 18, 2, 04017143, https://doi.org/10.1061/(ASCE)GM.1943-5622.0001061

Sloan, S. W. (2013). Geotechnical stability analysis. *Géotechnique*, 63, 7, 531–572.

Sommer, H. (1993). Development of locked stresses and negative shaft resistance at the piled raft foundation - Messeturm Frankfurt/Main. *Proceedings of the Deep Foundations on Bored and Auger Piles*, 347–349, Rotterdam: Balkema.

Sommer, H., Hoffmann, H. (1991a). Load-settlement behaviour of the fairtower (Messeturm) in Frankfurt/Main. *Proceedings of the 4th International Conference on Ground Movements and Structures*, 612–627, London: Pentech Press.

Sommer, H., Hoffmann, H. (1991b). Last-Verformungsverhalten der Gründung des Messeturmes Frankfurt/Main. Festkolloquium 20 Jahre Grundbauinstitut Prof. Dr.-Ing. H. Sommer und Partner, 63–71.

Sommer, H., Katzenbach, R., DeBeneditiis, C. (1990). Last-Verformungsverhalten des Messeturms Frankfurt/Main. 21.Baugrundtagung der Deutschen Gesellschaft für Geotechnik in Karlsruhe, 371–380.

Sommer, H., Tamaro, G., DeBeneditis, C. (1991). Messe Turm, foundations for the tallest building in Europe. *Proceedings of the 4th International Conference on Piling and Deep Foundations*, 139–145, Rotterdam: Balkema

Tafili, M. 2020. On the behaviour of cohesive soils: Constitutive description and experimental observations. Schriftenreihe. des Institutes für Bodenmechanik und Fels-mechanik am KIT 186.

Tafili, M., Triantafyllidis, T. (2020). AVISA: Anisotropic visco-ISA model and its performance at cyclic loading. *Acta Geotechnica*, 15, 2395–2413.

Tafili, M., Ganal, A., Wichtmann, T., Reul, O. (2023). On the AVISA model for clay – Recommendations for calibration and verification based on the back analysis of a piled raft. *Computers and Geotechnics*, 154, 105–126.

Takemura, J., Seki, S., Yamada, M. (2014). Dynamic response and settlement behavior of piled raft foundation of oil storage tank. In *Proceedings of the 8th International Conference on Physical Modelling in Geotechnics*, Perth (Gaudin, C. and White, D. (eds)) Taylor and Francis, London, UK, 1, 613–620.

Terzaghi, K., Fröhlich, O. K. (1936). *Theorie der Setzung von Tonschichten – Eine Einführung in die analytische Tonmechanik*; Deuticke, Leipzig und Wien.

Thaher, M. (1991). Tragverhalten von Pfahl-Platten-Gründungen im bindigen Baugrund, Berechnungsmodelle und Zentrifugen-Modellversuche. Schriftenreihe des Institutes für Grundbau und Bodenmechanik der Ruhr-Universität Bochum, Vol. 15.

Thaher, M.; Jessberger, H. L. (1991). The behavior of pile-raft foundations, investigated in centrifuge model tests. *Proceedings of the International Conference Centrifuge - Centrifuge 91*, Boulder, 225–234.

Tochnog Professional Company (2022). Tochnog Professional FEA. https://www.tochnogprofessional.nl

Tran, T. V., Teramoto, S., Kimura, M., Boonyatee, T., Vinh, L. B. (2012). Effect of ground subsidence on load sharing and settlement of raft and piled raft foundations. *International Journal of Civil and Environmental Engineering*, 6, 2, 120–127.

Tsujino, S., Yoshida, N., Yasuda, S. (1994). A simplified practical stress-strain model in multi-dimensional analysis. *Proceedings of the International Symposium on Pre-failure Deformation Characteristics of Geomaterials*, Sapporo, 463–468.

Turek, J. (2006). Beitrag zur Klärung des Trag- und Verformungsverhaltens horizontal belasteter Kombinierter Pfahl-Plattengründungen. Mitteilungen des Institutes und der Versuchsanstalt für Geotechnik der TU Darmstadt, Heft 72.

Ueshita, K., Meyerhof, G. G. (1968). Surface displacement of an elastic layer under uniformly distributed loads. Highway Research Record, No. 228, 1–10.

Unsever, Y. S., Kawamori, M., Matsumoto, T., Shimono, S. (2013). Cyclic horizontal load tests of single pile, pile group and piled raft in model dry sand. *Proceedings of the 18th Southeast Asian Geotechnical & Inaugural AGSSEA Conference*, Singapore, https://doi.org/10.3850/978-981-07-4948-4_044

Unsever, Y. S., Matsumoto, T., Shimono, S., Ozkan, M. Y. (2014). Static cyclic load tests on model foundations in dry sand. *Geotechnical Engineering Journal SEAGS & AGSSEA*, 45, 2, 40–51.

Unsever, Y. S., Matsumoto, T., Shimono, S. (2015a). Shaking table tests of piled raft and pile group foundations in dry sand. *Proceedings of the 6th International Conference on Earthquake Geotechnical Engineering*.

Unsever, Y. S., Matsumoto, T., Özkan, M. Y. (2015b). Numerical analyses of load tests on model foundations in dry sand. *Computers and Geotechnics*, 63, 255–266.

Unsever, Y. S., Matsumoto, T., Esashi, K., Kobayashi, S. (2017). Behaviour of model pile foundations under dynamic loads in saturated sand. *Bulletin of Earthquake Engineering*, 15, 1355–1373.

Vu, A.-T., Matsumoto, T., Yoshitani, R., Nguyen, T.-L. (2017). Behaviour of pile group and piled raft foundation models having batter piles. *Journal of Earth Engineering*, 2, 1, 27–40.

Wallerand, R., Kay, D., Cafi, M., Dimmock, P., Randolph, M. F. (2017). Hybrid subsea foundations – from research to project application. *Proceedings of the 8th International Conference Offshore Site Investigation and Geotechnics*, Society for Underwater Technology, London, 2, 802–809.

Watanabe, T., Fukuyama, H., Horikoshi, K., Matsumoto, T. (2001). Centrifuge modeling of piled raft foundations subjected to horizontal loads. *Proceedings of the 5th International Conference on Deep Foundation Practice incorporating Piletalk*, Singapore, 371–378.

Wenner, M., Seidl, G., Garn, R., Marx, S. (2019). Langzeitverhalten einer 170 m langen integralen Eisenbahnbrücke. *Bautechnik*, 96, 2, 120–132.

Yamashita, K., Hamada, J., Onimaru, S., Higashino, M. (2012a). Seismic behavior of piled raft with ground improvement supporting a base-isolated building on soft ground in Tokyo. *Soils and Foundations*, 52, 5, 1000–1015.

Yamashita, K., Hashiba, T., Ito, H. (2012b). Settlement and load sharing behaviour of a piled raft subjected to strong seismic motion. *Proceedings of the 11th Australia New Zealand Conf. on Geomechanics*, 1514–1519.

Yamashita, K., Wakai, S., Hamada, J. (2013). Large-scale piled raft with grid-form deep mixing walls on soft ground. *Proceedings of the 18th International Conference on Soil Mechanics and Geotechnical Engineering*, 2637–2640.

Yamashita, K., Hamada, J., Tanikawa, T. (2016). Static and seismic performance of a friction piled raft combined with grid-form deep mixing walls in soft ground. *Soils and Foundations*, 56, 3, 559–573.

Yamashita, K., Shigeno, Y., Hamada, J., Chang, D.-W. (2018). Seismic response analysis of piled raft with grid-form deep mixing walls under strong earthquakes with performance-based design concerns. *Soils and Foundations*, 58, 65–84.

Yamashita, K., Tanikawa, T., Uchida, A. (2019). Long-term behaviour of piled raft with DMW grid on reclaimed land. *Geotechnical Engineering Journal of the SEAGS & AGSSEA*, 50, 3.

Zafir, Z., Vanderpool, W. E. (1998). Lateral response of large diameter drilled shafts: I-15/US 95 load test program. *Proceedings of the 33rd Engineering Geology and Geotechnical Engineering Symposium*, University of Nevada, 161–176.

Chapter 3

Design, construction and monitoring of CPRF

3.1 LIMIT STATE APPROACH AND TECHNICAL REGULATIONS FOR THE DESIGN PROCESS

3.1.1 Overall factor of safety

Traditionally, in civil engineering safety against failure of a structure was evaluated deterministically and by means of calculating a global ratio of minimum favourable resistance to maximum unfavourable forces, moments or stresses. For example, for a foundation the design criteria to be fulfilled for a certain failure mode, single pile under axial or lateral loading, footing under combined loading, etc. are defined as follows:

$$FS \leq \frac{R_u}{P} \tag{3.1}$$

where FS = global factor of safety; R_u = ultimate capacity of the foundation for the considered mode of failure and P = maximum load for the considered mode of failure.

As pointed out by Poulos et al. (2001), factors of safety were usually based on experience and precedent, although some attempts were made in the latter part of the century to relate safety factors to statistical parameters of the ground and the foundation type. Typical overall factors of safety in geotechnical engineering as documented, for example, by Meyerhof (1995) are summarised in Table 3.1.

3.1.2 Limit state design

3.1.2.1 General remarks

The overall factor of safety approach does not take into account the very different variance of the variables involved. To overcome this restriction, limit state design approaches, such as the Eurocode in particular, apply different partial factors of safety for actions E, e.g. forces, moments,

Table 3.1 Typical overall factors of safety after Meyerhof (1995)

Failure type	Item	Factor of safety
Shearing	Earthworks	1.3–1.5
	Earth retaining structures, excavations, offshore foundations	1.5–2.0
	Foundations on land	2.0–3.0
Seepage	Uplift, heave	1.5–2.0
	Exit gradient, piping	2.0–3.0
Ultimate pile loads	Load tests	1.5–2.0
	Dynamic formulae	3.0

temperature-induced strains and resistances R. Then the design criterion which has to be met is defined as follows:

$$E_d = \gamma_E \cdot E_k \leq \frac{R_k}{\gamma_R} = R_d \qquad (3.2)$$

where γ_E = partial factor of safety for the actions; γ_R = partial factor of safety for the resistances; E_k = characteristic action; E_d = design action; R_k = characteristic resistance and R_d = design resistance.

A detailed discussion of the limit state design approach defined in the Eurocode 7 (EC7) for geotechnical engineering is given, for example, by Frank et al. (2004). Orr (2007) summarises the historical development towards EC7 and its implementation.

Partial factors of safety will vary depending on the limit state/failure mode, e.g. single pile under axial or lateral loading, footing under combined loading, and the design scenario, i.e. usual utilisation scenarios, temporary utilisation scenarios such as an excavation pit, extraordinary scenarios such as fire, explosions or impact and earthquake scenarios. In theory the values of the respective factors of safety are established based on a probabilistic concept assuming that actions and resistances occur as random, variable quantities which can be described by a normal distribution and allowing a certain likelihood of the failure of the structure. However, it was realised, for example, during the implementation of EC7, that a purely probabilistic approach is difficult to establish in geotechnical engineering for the following reasons (e.g. Schuppener 2012):

- In practice, there are severe limits to the statistical assessment of the subsoil, because the extent of site explorations and soil mechanical investigations is usually insufficient to apply a probabilistic analysis.
- Unlike other branches of civil engineering and the construction materials applied there, the subsoil itself cannot be manufactured according to a recipe with well-defined properties. As a result, the coefficient of

variation of the properties of the subsoil is much larger than for other construction materials.
- Human errors in planning and execution, which are the main cause of damage, cannot easily be taken into account in probabilistics.

For these reasons, for example during the transition to EC7, for the conversion of traditional geotechnical standards to the limit state concept with its application of partial factors of safety, it was decided to maintain the safety level of the previous overall factor of safety concept as far as possible and to make changes only in justified exceptional cases (e.g. Schuppener 2012). Therefore, the partial factors of safety were selected so that a design based on the limit state concept results in approximately the same dimensions of structures as a design based on the standards of the overall factor of safety concept (e.g. Schuppener 2012).

Following the safety concept of the Eurocode, the characteristic value (indicated with an index k) corresponds to the 5% fractile of a quantity. For example, the value of a resistance is considered to be characteristic if there is only a 5% probability that even smaller values can occur during the lifetime of a structure. However, for geotechnical engineering, the EC7 code suggests another approach to establish the characteristic value acknowledging that due to the non-homogenous nature of soil the behaviour is determined by the mean behaviour of the relevant zone of soil (Orr 2007). According to Eurocode EC7 the characteristic value of a geotechnical parameter should be established as the mean value (50% fractile) at the 95% confidence level. Bond (2011) describes a procedure for determining the characteristic value of a geotechnical parameter under these constraints.

To establish the design resistance there are two approaches:

1. The resistance, e.g. for a pile under axial loading, is computed applying characteristic values for the shear strength (e.g. φ'_k, c'_k, $s_{u,k}$) and is then divided by an appropriate partial factor of safety.
2. The resistance is computed applying design values for the shear strength; i.e. the shear strength parameters are divided by appropriate factors of safety (sometimes referred to as "material factor").

3.1.2.2 Ultimate limit state (ULS)

The ultimate limit state (ULS) of a structure is defined by its collapse or failure. For the check for ULS Eq. (3.2) has to be evaluated for the relevant failure mode or ultimate limit state. For example EC7 distinguishes between five different ultimate limit states, namely:

EQU: Loss of overturning equilibrium of the structure, considered as a rigid body, on the ground in which the strength of structural materials and the ground do not contribute in providing resistance.

STR: Internal failure or excessive deformation of the structure or structural elements, including footings, piles, basement walls, etc. in which the strength of structural materials is significant in providing resistance.

GEO: Failure or excessive deformation of the ground, in which the strength of soil or rock is significant in providing resistance.

UPL: Loss of equilibrium of the structure or the ground due to uplift by water pressure (buoyancy) or other vertical actions.

HYD: Hydraulic heave, internal erosion and piping in the ground caused by hydraulic gradients.

In the design process, for example for a foundation, the check for the ULS includes the proof that the design load is not larger than the design value of the bearing capacity of the foundation which in terms of EC7 corresponds to the limit state GEO.

3.1.2.3 Serviceability limit state (SLS)

The serviceability limit state (SLS) of a structure is defined by a performance which no longer meets the initial requirements. In the case of foundations, the SLS usually is defined by excessive settlements and differential settlements. Based on the work by Burland and Wroth (1975), Poulos et al. (2001) list the following parameters connected with settlement and differential settlement as depicted in Figure 3.1:

- Overall settlement
- Tilt, both local and overall
- Angular distortion (or relative rotation) between two points
- Relative deflection (for walls and panels).

Based on a literature review, Poulos et al. (2001) distilled allowable values for the above quantities (Table 3.2).

In the framework of the EC7 the check for SLS is carried out by demonstrating:

$$E_d \leq C_d \tag{3.3}$$

where E_d = design value of the effects of action, e.g. deformations and differential settlements (please note the definition differing from Eq. (3.2)) and C_d = limiting value of the specific effect.

It is interesting to note that the CPRF guideline (ISSMGE TC 212 2013) suggests applying characteristic loads for the check for SLS (or, in other words, to set the partial factors of safety for actions equal to 1).

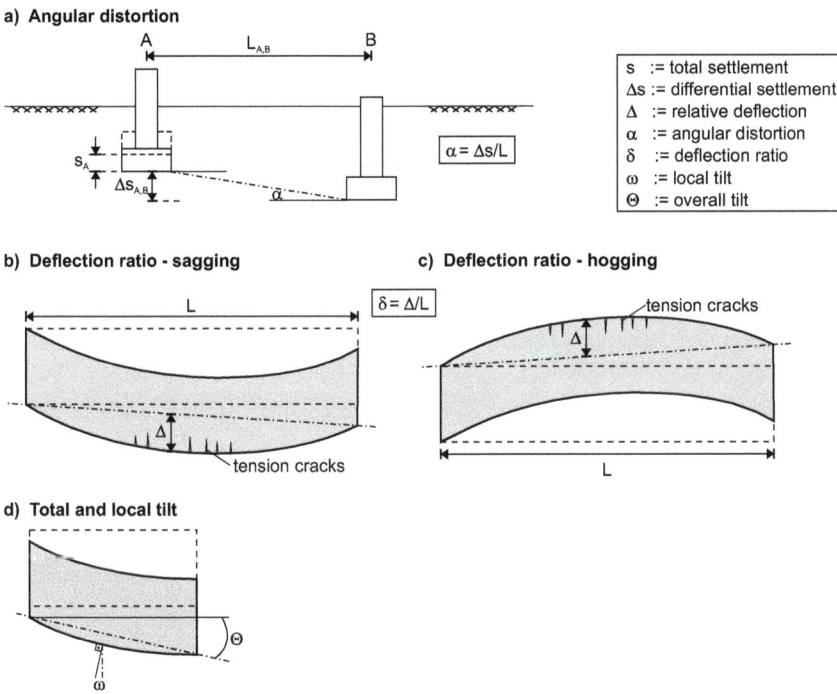

Figure 3.1 Definitions of differential settlement and distortion for framed and load-bearing wall structures adapted from Poulos et al. (2001)

3.1.3 Combined pile-raft foundation (CPRF) according to ISSMGE Technical Committee TC 212 (2013)

3.1.3.1 General remarks

The experience gained with piled rafts in Frankfurt in the 1980s and 1990s (Section 1.3 & Chapter 4) led to the development of the German "KPP-Richtlinie" (DGGT 2001), a guideline for the design, dimensioning and construction of piled rafts. Piled rafts designed according to this guideline are termed Combined Pile-Raft Foundations (CPRF; in German: Kombinierte Pfahl-Plattengründung/KPP). The design philosophy of the German "KPP-Richtlinie" was adopted more or less one-to-one in the *ISSMGE Combined Pile-Raft Foundation Guideline* (ISSMGE TC 212 2013).

The design approach is valid for piled rafts under predominantly vertical loading. According to ISSMGE TC 212 (2013), the guideline shall not be used in cases where layers of relatively small stiffness (e.g. soft cohesive and organic soils) are situated closely beneath the raft. This restriction is motivated by the fact that in such a situation the piles will carry the main proportion of the load (indicated by high piled raft coefficients) so that design

Table 3.2 Summary of criteria for settlement and differential settlement of structures after Poulos et al. (2001)

Type of structure	Type of damage/concern	Criterion	Limiting value(s)
Framed buildings and reinforced load bearing walls	Structural damage Cracking in walls and partitions Visual appearance Connection to services	Angular distortion Angular distortion Tilt Total settlement	1/150 to 1/250 1/500 (1/1000 to 1/1400 for end bays) 1/300 50 mm to 75 mm (sand) 75 mm to 135 mm (clay)
Tall buildings	Operation of lifts & elevators	Tilt after lift installation	1/1200 to 1/2000
Structures with unreinforced load bearing walls	Cracking by sagging Cracking by hogging	Deflection ratio Deflection ratio	1/2500 (length L/height H = 1) 1/1250 (L/H = 5) 1/5000 (L/H = 1) 1/2500 (L/H = 5)
Bridges – general	Ride quality Structural distress Function	Total settlement Total settlement Horizontal movement	100 mm 63 mm 38 mm
Bridges – multiple span	Structural damage	Angular distortion	1/250
Bridges – single span	Structural damage	Angular distortion	1/200

as a pile foundation is more appropriate. In the KPP-Richtlinie (DGGT 2001), the scope of application is explicitly quantified as follows:

$$\alpha_{pr} \stackrel{!}{<} 0.9 \tag{3.4}$$

$$\frac{E_{upper}}{E_{lower}} \stackrel{!}{\geq} 0.1 \tag{3.5}$$

where α_{pr} = piled raft coefficient and E_{upper}, E_{lower} = Young's modulus of the upper and lower soil layers, respectively, as depicted in Figure 3.2.

A more rational approach might also take the thickness of the different layers in to account, e.g.:

$$\frac{E_{upper}}{h_{upper}} \stackrel{!}{\geq} 0.1 \frac{E_{lower}}{h_{lower}} \tag{3.6}$$

According to ISSMGE TC 212 (2013), the various interactions highlighted in Figure 1.9 must be appropriately considered in the computational model

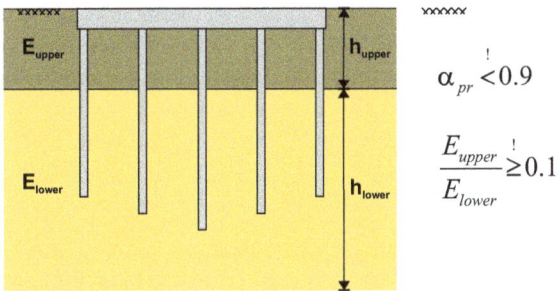

Figure 3.2 Limitations of the application of the design guideline according to the KPP-Richtlinie (DGGT 2001).

applied in the design of the foundation. This requires a realistic geometric model of the foundation elements and the soil continuum and appropriate material models for structure and soil as well as for the contact behaviour between soil and foundation. An indication of the quality of the computational model is its ability to simulate the bearing behaviour of a single pile, i.e. shearing of the pile shaft-soil interface and inelastic compression at the pile base. Therefore, ISSMGE TC 212 (2013) recommends carrying out pile load tests at the site or under comparable soil conditions to validate the computational model. If necessary, time-dependent behaviour of soil and structure has to be considered. The stiffness of the superstructure and its influence on the bearing behaviour of the CPRF have to be considered within the analysis.

3.1.3.2 Ultimate limit state (after ISSMGE TC 212 2013)

According to ISSMGE TC 212 (2013), a proof of the ultimate limit state (ULS), i.e. the external and internal bearing capacity, has to be carried out for a CPRF. The <u>external bearing capacity</u> of the foundation elements interacting with the surrounding soil shall be verified in accordance with ISSMGE TC 212 (2013) as follows:

$$E_d = E_{G,k} \cdot \gamma_G + E_{Q,k} \cdot \gamma_Q \leq \frac{R_{1,tot,k}}{\gamma_R} = R_{1,tot,d} \qquad (3.7)$$

where E_d = design value of the effect of actions; $E_{G,k}$, $E_{Q,k}$ = characteristic values of the effect of permanent and variable actions, respectively; $R_{1,tot,k}$, $R_{1,tot,d}$ = characteristic value and design value, respectively, of the total resistance of the foundation in the ULS and γ_G, γ_Q, γ_R = partial factors of safety according to relevant design codes (e.g. Eurocode EC7).

ISSMGE TC 212 (2013) suggests evaluating the total resistance of the foundation $R_{1,tot,k}$ from the resistance-settlement curve established with an adequate computational model (see above) as the value for which the settlements start to increase visibly. The analysis is carried out taking characteristic material parameters for soil and structural elements into account. It is important to note that no proof (or limiting ratio of mobilised resistance and bearing capacity) is required for individual piles. For certain so-called "Simple cases" which are characterised by a simple geometry (e.g. uniform pile length, rectangular or circular raft), homogeneous subsoil conditions and symmetric, static loading of the raft, ISSMGE TC 212 (2013) allows the total resistance of the piled raft to be calculated by means of the resistance of the raft foundation alone.

The <u>internal bearing capacity</u> describes the bearing capacity of the foundation components piles and raft as well as the superstructure for design cases such as piles under compression, bending and shear or bending of the raft. A sufficient safety against material failure has to be proven for all structural elements according to the relevant design codes (e.g. Eurocode EC 2).

3.1.3.3 Serviceability limit state (after ISSMGE TC 212 2013)

Just as for the ULS, according to ISSMGE TC 212 (2013), a proof of the serviceability limit state (SLS) has to be carried out for the external serviceability and the internal serviceability. The <u>external serviceability</u> of the foundation elements interacting with the surrounding soil shall be verified in accordance with ISSMGE TC 212 (2013) as follows:

$$E_{2d} = E_{2k} \cdot \leq C_d \tag{3.8}$$

where E_{2k}, E_{2d} = characteristic value and design value, respectively, of the effect of actions relevant for the SLS and C_d = limiting design value of the relevant SLS criterion.

ISSMGE TC 212 (2013) requires the effects of actions relevant for the SLS to be established by means of an adequate computational model (see above). The analysis is carried out adopting characteristic material parameters for soil and structural elements. During the life span of the building the effects E_2, such as settlements s_2, differential settlements Δs_2 and angular distortions α, have to be smaller than the limiting design value of the relevant serviceability criterion. The limiting design values of the relevant serviceability criteria C_d are defined taking into account the sensitivity of the structure as well as adjacent buildings and infrastructure to deformations and especially to differential settlements (ISSMGE TC 212 2013).

The <u>internal serviceability</u> describes the serviceability of the foundation components (piles and raft(s)) as well as the superstructure for design cases such as the crack width of concrete members or the maximum deflection of the raft. The serviceability has to be proven for all structural elements according to the relevant design codes (e.g. Eurocode EC2).

From experience gained from a large number of case histories (e.g. Chapter 4) it can be assumed that for piled rafts of large plan area, as typical for high-rise buildings where loading is dominated by vertical monotonic conditions, the SLS, rather than the ULS, will be decisive in the design process.

The CPRF guideline employs the limit state design approach (Section 3.1.2). However, it is also possible to use overall factors of safety.

3.1.4 Loads

Usual sources of loading relevant for foundation design which are provided by the structural engineer include the following (Poulos 2017):

- Dead loads
- Live loads
- Wind loads
- Earthquake loads
- Loads arising from earth pressure (relevant for the design of basement walls)
- Load arising from ground movements
- Loads from other sources, such as snow and ice

A detailed summary of building loads relevant for foundation design is given by Poulos (2017).

3.1.5 Design process – interaction between structural and geotechnical engineering

It is still common practice for structural engineers to carry out the foundation design with an analytical model with the raft placed on Winkler springs representing the soil and the piles simplified as elastic springs (Figure 3.3). The design of CPRF therefore requires an interactive design process between

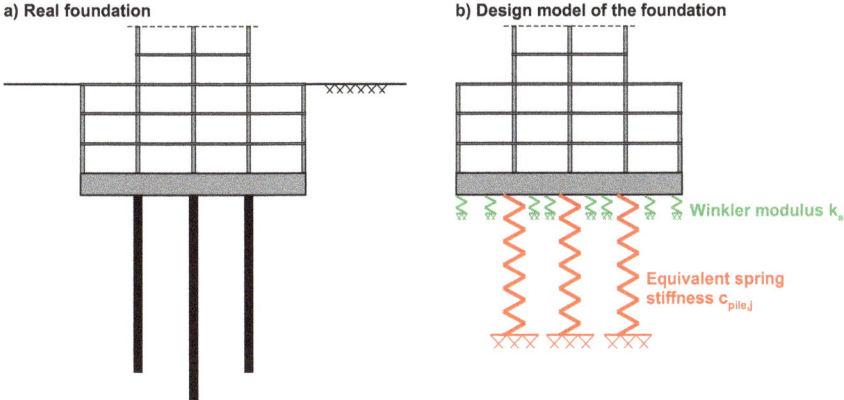

Figure 3.3 Reality vs. design model.

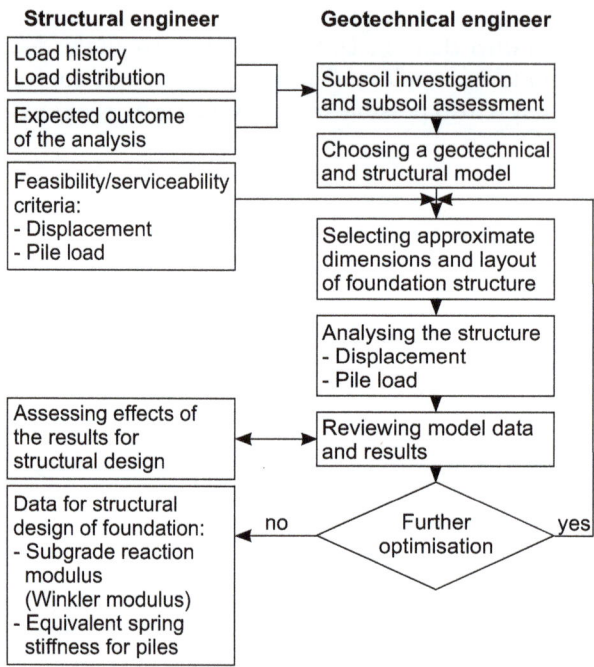

Figure 3.4 Interaction between structural engineer and geotechnical engineer in the design process of CPRF. (After Katzenbach et al. 2003)

structural engineering and geotechnical engineering, as illustrated by the flow diagram in Figure 3.4. The calculation of deformations and pile loads for complex foundation systems under consideration of the nonlinear behaviour of soil requires the application of numerical analysis. From the results of this analysis, spring stiffness to model the behaviour of the piles and subgrade reaction modulus to model the response of the raft are derived for the structural design analysis of the foundation. A detailed discourse of the geotechnical aspects involved in the design and construction of high-rise buildings is given, for example, by Arslan and Ripper (2003) and Poulos (2017).

3.2 ANALYSIS METHODS

3.2.1 General remarks

This section presents analysis methods to investigate the bearing behaviour of the piled raft in the serviceability limit state (SLS) and the (ULS) (Section 3.1.3). Available research on the bearing behaviour of piled rafts (Chapter 2) indicates that in most cases the SLS is the relevant state for design. This chapter therefore focuses mainly on analysis methods to establish deformations of the foundation and parameters relevant for the dimensioning of the foundation and the superstructure by the structural engineer, namely the load share between piles and raft, Winkler modulus and equivalent pile spring stiffness.

Analysis methods for investigating internal bearing capacity and internal serviceability of the raft and piles and for the dimensioning of these structural elements are beyond the scope of this book.

3.2.2 Analysis methods for investigating the SLS

3.2.2.1 Classification of analysis methods

Analysis methods for investigating the bearing behaviour of piled rafts might be classified into the following groups:

1. Empirical correlations
2. Methods replacing the piled raft by an equivalent raft or an equivalent pier
3. Methods based on elasticity theory
4. Numerical analysis based on the Boundary Element Method (BEM)
5. Numerical analysis based on the Finite Element Method (FEM) and the Finite Difference Method (FDM)

For a summary and discussion of analysis methods of groups 2, 3 and 4, one may refer, for example, to Clancy and Randolph (1993, 1996), Randolph (1994), Poulos et al. (1997), Poulos (2017) and Viggiani et al. (2012).

3.2.2.2 Empirical correlations

3.2.2.2.1 General remarks

Empirical correlations to estimate settlements of pile groups and piled rafts were proposed, for example, by Skempton (1953), Vesic (1969), Cooke (1986) and Mandolini et al. (2005). Although these approaches generally are limited to certain subsoil conditions or foundation configurations, they can be useful to get an approximate estimate of the settlements in an early stage of the design process.

Empirical correlations frequently apply group settlement ratios R_s to derive the settlements of pile groups (FPG) and piled rafts from the settlement of a single pile (e.g. Skempton 1953, Vesic 1969, Mandolini et al. 2005) with dimensions equal to the dimensions of the piles in the group:

$$R_s = \frac{s}{s_s} \qquad (3.9)$$

where s = average settlement of a piled foundation and s_s = settlement of a single pile with dimensions equal to the dimensions of the piles in the group under the average working load of the group.

3.2.2.2.2 Vesic (1969)

Based on small-scale model tests on pile groups and piled rafts in medium dense sand, Vesic (1969) proposes the following estimation:

$$s = R_s \cdot s_s = \sqrt{\frac{B}{d_p}} \cdot s_s \qquad (3.10)$$

where B = breadth of the pile group measured from pile axis to pile axis$_x$ and d_p = pile diameter. It has to be noted that in the model tests the piles were forced into the soil in entire groups by jacking.

3.2.2.2.3 Mandolini et al. (2005)

Mandolini et al. (2005) suggest the following expressions to establish the upper limit and the best estimate of the average settlement of a piled foundation:

$$s_{max} = R_{s,max} \cdot s_s = \frac{0.5}{R} \cdot \left(1 + \frac{1}{3R}\right) \cdot n_p \cdot s_s \qquad (3.11)$$

$$s = R_s \cdot s_s = 0.29 \cdot n_p \cdot R^{-1.35} \cdot s_s \qquad (3.12)$$

with

$$R = \sqrt{\frac{n_p \cdot e}{L_p}} \qquad (3.13)$$

where s_{max}, s = upper limit and best estimate, respectively, of the average settlement of a piled foundation; $R_{s,max}$, R_s = upper limit and best estimate, respectively, of the group settlement ratio; s_s = settlement of a single pile under the average working load of the group; R = aspect ratio after Randolph and Clancy (1993); n_p = number of piles; e = pile spacing and L_p = pile length.

3.2.2.2.4 Cooke (1986)

Cooke (1986) gives the following empirical relationship between settlement and width of a deep foundation, taking into account the structures summarised in Appendix B, Table B.3 (exception: Queen Elizabeth Conference

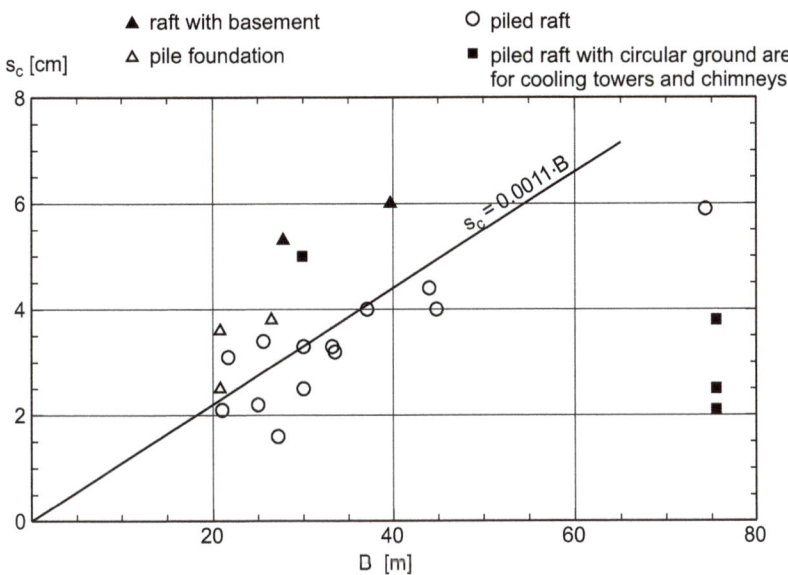

Figure 3.5 Observed settlements s_c of structures on deep foundations plotted against the equivalent breadth B (after Cooke 1986).

Centre; Victoria Street Redevelopment, Block C) and a number of other deep foundations (Figure 3.5):

$$s_c = 0{,}0011 \cdot B = 0.0011 \cdot \frac{I_{m,R}}{I_{m,Q}} \cdot b \tag{3.14}$$

where s_c = settlement at the end of consolidation; B = breadth of the equivalent square foundation; $I_{m,R}$ = influence factor for the mean settlement under a rectangular footing (after Poulos and Davis 1974, p.58); $I_{m,Q}$ = influence factor for the mean settlement under a square footing (after Poulos and Davis 1974, p.58) and b = breadth of the deep foundation.

3.2.2.3 Numerical analysis based on the Finite Element Method (FEM)

3.2.2.3.1 General remarks

The Finite Element Method is well suited to model and investigate the bearing behaviour of piled rafts as can be concluded from Chapter 2. In this chapter, analysis by means of FEM implies that subsoil, piles and raft are modelled with finite elements. Methods where only the raft and the piles, but not the subsoil, are modelled with finite elements (e.g. Clancy and Randolph 1993, Ta and Small 1996) are not discussed here. The following recommendations concerning the modelling of piled rafts with FEM are mainly based on a guideline by the German Geotechnical Society (DGGT 2014). Compared to FEM there are only few examples available where the bearing behaviour of piled rafts was investigated with the finite difference method (FDM) (e.g. Comodromos et al. 2016). However, it is believed that the following remarks in general also apply to analysis with the FDM.

3.2.2.3.2 Model section and discretisation (DGGT 2014)

Boundary value problems such as piled rafts usually consist of systems whose structural properties, i.e. loading and geometry, change in both horizontal and vertical directions. Accordingly, three-dimensional numerical models are usually used to address such problems. To carry out the FEA, the geometric situation of the real foundation configuration must be suitably modelled. In general, only a confined model section is taken into account with the size of the model chosen in such a way that the influence of the foundation at the edges of the model section is negligible. Figure 3.6 shows the model section for a piled raft. The boundary conditions are preferably defined in such a way that the horizontal displacement is set to zero at the lateral edges and the vertical displacement is set to zero at the bottom edge. For boundary value problems with large horizontal actions, considerably larger model sections may be necessary. In case of doubt, the adequate size of the model must be verified by means of preliminary analyses.

The investigation of a continuum with the FEM always represents an approximation. The finer the discretisation and/or the higher the order of the (polynominal) shape functions for the displacements in the elements, the better the results will agree with exact solutions for the same constitutive model and material parameters. How large the individual elements may become in order to still obtain sufficiently accurate results depends essentially on the type of elements used and the shape functions selected. In general, the size of the elements may increase if higher-order shape

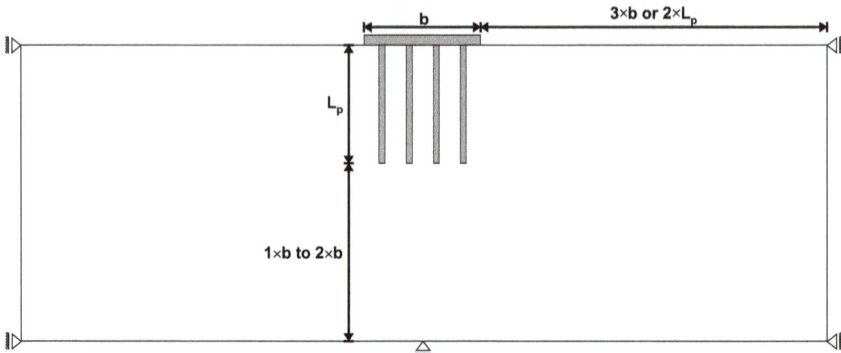

Figure 3.6 Model size and boundary conditions after DGGT (2014)

functions are used and/or if the element is located in areas where only small stress and deformation gradients are to be expected. For a detailed discussion on mesh discretisation, reference can be made to the literature on FEM with focus on application in geotechnical engineering (e.g. Smith and Griffiths 1998, Potts and Zdravkovic 1999).

In the case of foundations, the composition of the subsoil, the structural components to be taken into account and the contact between the subsoil and the structure are to be simulated in detail. The numerical analyses are usually used to investigate the deformations that occur in the SLS. The choice of a suitable constitutive model for the soil and the selection of appropriate material parameters are of great importance for these questions (see below). If numerical analyses are used in specific cases to establish the bearing capacity of a foundation, i.e. to investigate the ULS, the areas in which high stress gradients occur or where the shear strength of the soil is reached must be modelled with a particularly fine mesh.

The FE mesh should be made relatively fine at the edges of rafts and at the contact zone between piles and soil. The contact zone between soil and pile should be modelled with interface elements or contact surfaces. When investigating SLS, the application of thin continuum elements at the pile shaft usually also leads to reasonable results.

A full-scale simulation of the super-structure in the model is only carried out in exceptional cases due to the large effort required for this. For the majority of geotechnical problems, it is only necessary to capture the foundation stiffness or structure stiffnesses, respectively, and the distribution of the loads acting on the foundation. For this purpose, rafts are modelled with continuum elements or, if available, with special plate elements. The use of plate elements can be beneficial because, for example, a variation of the foundation stiffness does not require any modification of the FE mesh. Instead, the plate elements can be assigned a corresponding thickness or a flexural stiffness, which realistically captures the overall stiffness of the structure. A further advantage of the plate elements is that usually the internal forces and moments are calculated automatically in the program requiring no additional evaluation. When selecting suitable plate elements, the application limits of the underlying plate theory must be taken into account.

Piles are usually modelled with continuum elements, whereby it is generally acceptable to transform the circular cross section of the piles into a square equivalent cross section. If the shaft area of the square equivalent cross section is identical with the actual pile dimensions, the pile base area is underestimated compared to reality. In several program systems special pile elements termed "embedded piles" are available, which are superimposed as linear elements on the subsoil (see below).

A discussion of the influence of the mesh refinement on the results of the FEA of piled rafts carried out with the program ABAQUS can be found in Reul and Randolph (2002). Figure 3.7a and 3.7b show the system configuration studied in the scope of this work. The piled raft consists of five piles

116 Combined Pile-Raft Foundations

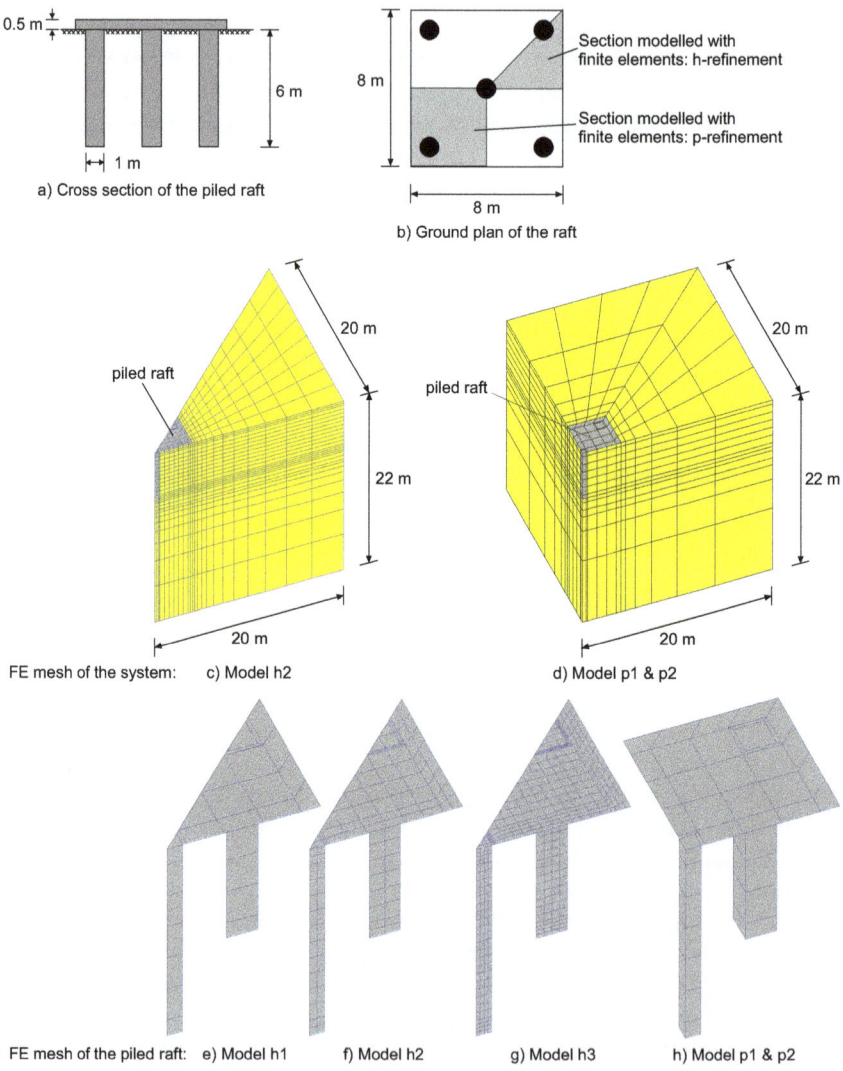

Figure 3.7 Study on the influence of the mesh refinement: System configuration and FE mesh. (Adapted from Reul and Randolph 2002)

(L_p = 6 m, d_p = 1 m) and a square raft (B = 8 m, t_r = 0.5 m). The various finite element meshes are plotted in Figure 3.7c to 3.7h. In the finite element models, the circular piles were replaced by square piles with the same shaft circumference. In the analyses a uniformly distributed load was applied stepwise on the raft. For the h-refinement (reduction of the subdivision size), three finite element meshes were investigated where 1/8 (three symmetry planes) of the system was modelled. The number of first-order elements increases by a factor of about eight between mesh h1 and h2 and again between mesh h2 and h3. For the p-refinement (increase of the order of the

Table 3.3 Study on the influence of the mesh refinement: Material parameters used in the FEA

Parameter		Unit	Soil	Raft	Piles
Young's modulus	E	MPa	14.5+1.75·z*	30000	30000
Poisson's ratio	ν	—	0.25	0.2	0.2
Buoyant unit weight	γ'	kN/m³	9	—	15
Coefficient of earth pressure at rest	K_0	—	0.658	—	—
Slope of the conus yield surface in the p-t plane	β	°	37.67	—	—
Intersection of the conus yield surface with the t-axis	d	kPa	42.42	—	—
Shape parameter of the conus	K	—	0.795	—	—
Shape parameter of the cap	R	—	0.1	—	—

* z = depth below ground surface in m

polynomial trial function approximation), 1/4 (two symmetry planes) of the system is modelled and first-order elements (mesh p1) and second-order elements (mesh p2) were applied. For the modelling of the raft first-order (h-refinement, p1) and second-order (p2) shell elements with reduced integration were used. The material behaviour of the soil was modelled with an elasto-plastic cap model (Chapter 2), while the raft and piles are considered to behave linear-elastically. The material parameters used in the FEA are summarised in Table 3.3.

Figure 3.8a shows the average settlement versus the applied load for the different models investigated. Although all curves are noticeably nonlinear, failure of the piled raft is not observed for any of the models up to the maximum applied load of 34 MN, which corresponds to the ultimate capacity of the equivalent raft foundation. For the investigated mesh refinements, the average settlement increases with increasing number of DOFs (Figure 3.8b), while the differential settlement (Figure 3.8c) and the maximum positive bending moment per unit length (Figure 3.8e) decrease for finer meshes. For the latter two parameters the influence of the mesh refinement is less pronounced for higher load levels. The range of results for the average settlement and the piled raft coefficients is well within the bandwidth of results that can be expected if the deviation of in situ soil parameters is considered. For the system configuration studied, the coarse mesh gives a conservative estimation of the bending moments and differential settlements. Due to the nonlinear pile resistance-settlement behaviour, the piled raft coefficient decreases with increasing load level (Section 2.2). While the piled raft coefficient decreases with increasing number of DOFs for the h-refinement, it increases slightly for the p-refinement. It has to be noted that the piled raft coefficient is not obtained directly from the analyses, but has to be calculated from the pile resistances which themselves are calculated from the vertical stresses in the finite elements that represent the piles. For the h-refinement less favourable triangular prisms are applied in

118 Combined Pile-Raft Foundations

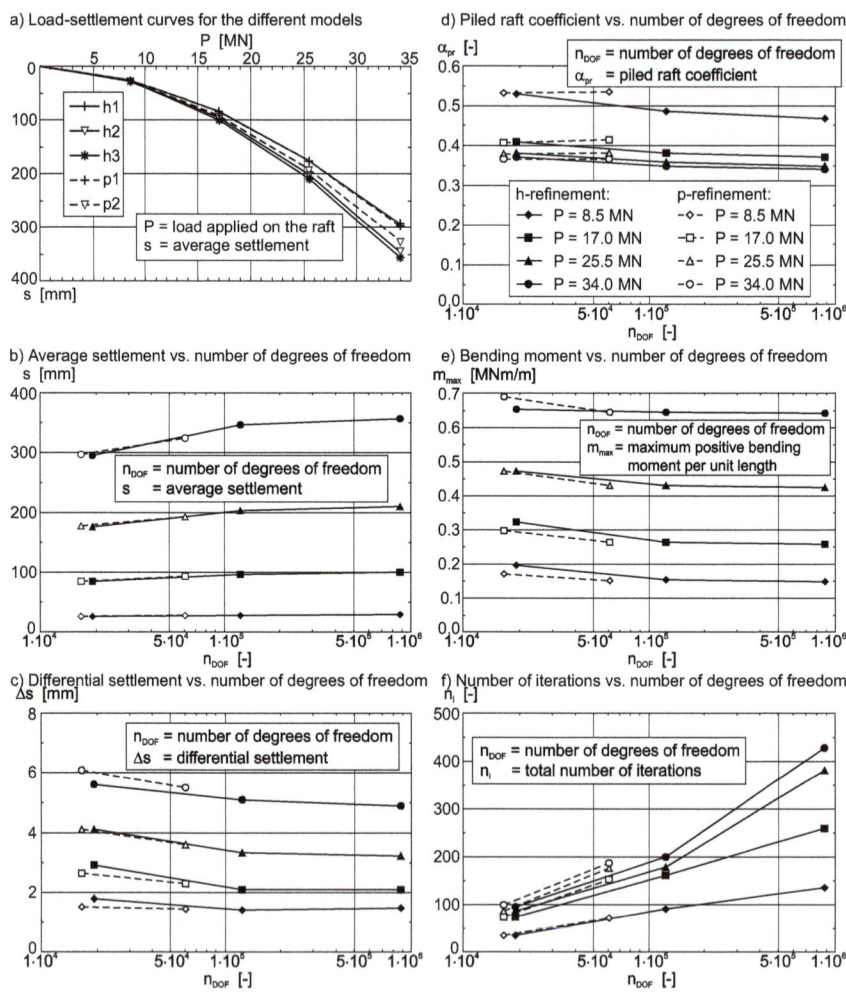

Figure 3.8 Study on the influence of the mesh refinement: Results. (Adapted from Reul and Randolph 2002)

particular to model the piles, which might have an influence on the results. Nevertheless, the decrease of the piled raft coefficient with increasing load level is far more significant than the variation of this parameter with the number of DOFs for h- as well as p-refinement. Generally, for all parameters investigated, the difference in the results between model h1 and model h2 is far more significant than between model h2 and model h3, which indicates convergence of the solution with increasing number of DOFs. For the same number of DOFs h-refinement and p-refinement yield similar results. The number of iterations and, therefore, the cost of the analyses increase considerably between model h2 and model h3, especially for higher load levels (Figure 3.8f). From the results of this study it is concluded that even with the coarser finite element meshes, reasonable results can be achieved.

3.2.2.3.3 Modelling of piles as embedded piles

In several program systems special pile elements termed "embedded piles" are available, which are superimposed as 1D elements on the subsoil (e.g. Sadek and Shahrour, 2004, Engin et al. 2007). The advantage of this approach is that piles are not discretised by means of volume elements but are replaced by an approximation applying beam elements that can take the behaviour of a pile penetrating a finite element in any orientation into account and thus do not affect significantly the number of DOFs of the FE mesh. This concept usually includes the possibility to define a maximum shear strength at the pile shaft and a maximum base resistance per unit area. Figure 3.9 shows a single vertical embedded pile (EP) subjected to a point load at the pile head with embedded interface elements connecting the virtual nodes inside the solid soil elements and the nodes of the embedded beam. A detailed discussion of the capabilities and limitations of the embedded pile approach is given, for example, by Tschuchnigg (2013) and Tschuchnigg and Schweiger (2015).

3.2.2.3.4 Constitutive models (DGGT 2014)

Essential requirements for constitutive models are the simulation of the non-linear stress-strain behaviour of the soil during initial loading and the variation (generally increase) of stiffness with depth. If significant unloading and reloading occur during the construction process, the selected constitutive model must be able to reproduce the different deformation behaviour of the soil during initial loading, reloading and unloading.

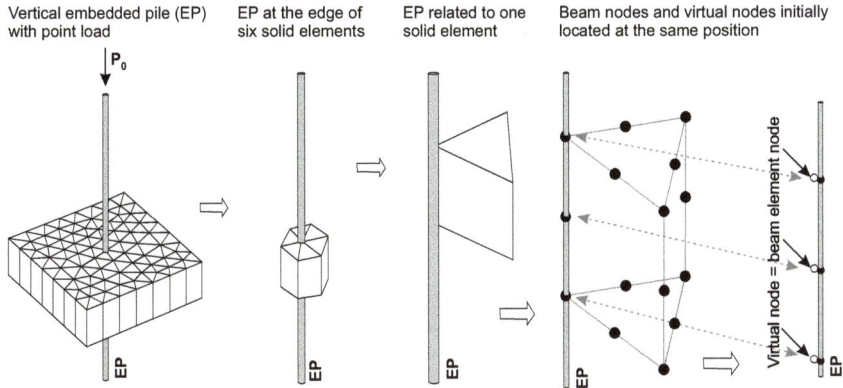

Figure 3.9 Discretisation of a single vertically loaded embedded pile (EP) after Tschuchnigg (2013)

The analyses should take account of foundations being loaded but gradually over time, rather than suddenly, in accordance with the progress of construction. Accordingly, an incremental analysis in which the structural weight and live loads are applied step by step is carried out. For the investigation of the long-term bearing behaviour of foundations in cohesive soils, creep due to the viscosity of the soil may have to be taken into account.

The material behaviour of soils should be simulated with elasto-plastic models or with more advanced approaches such as for example hypoplasticity. However, in the case of predominantly monotonic load paths and appropriately calibrated material parameters, certain analysis results, such as settlements and piled raft coefficient, can also be established with reasonable accuracy using linear-elastic-ideal-plastic models.

Within any given soil stratum, an increase of soil stiffness with depth can be simulated by selecting a suitable depth-dependent approach in the constitutive model or by taking into account the increased soil stiffness at small strains. As an approximation, the change in material properties with depth can also be simulated by introducing several soil layers, each of which is assigned properties corresponding to the mean depth.

Generally, in numerical analyses, linear-elastic material behaviour can be assigned to the structural components including the piles. A strength criterion for the construction materials (usually reinforced concrete) is generally not required, since the analyses of the internal bearing capacity and serviceability of the structural components are performed separately.

Interface elements or contact surfaces with elasto-plastic constitutive models can be used in the transition of structures to the subsoil, especially in modelling the shaft friction of piles.

Experience shows that even with careful identification of the material parameters, the actual settlement behaviour of foundations can only be predicted to a limited extent. For this reason, the constitutive models applied should be calibrated by means of the back-analysis of measurements if suitable data is available.

3.2.2.3.5 Simulation of the construction progress (DGGT 2014)

In order to correctly predict the deformations of a foundation, it is necessary to determine the stress conditions in the subsoil in a realistic way. For this purpose, it is necessary to model the stress state prevailing before the start of the construction. In general, the first analysis step serves to simulate the initial stress state.

In further analysis steps, the construction of the excavation pit, including any lowering of the groundwater level, is simulated. If the influence of the excavation pit support on the bearing behaviour of the foundation is of minor importance, the soil above the foundation level can be taken into account approximately by its self-weight (Figure 3.10).

Design, construction and monitoring of CPRF 121

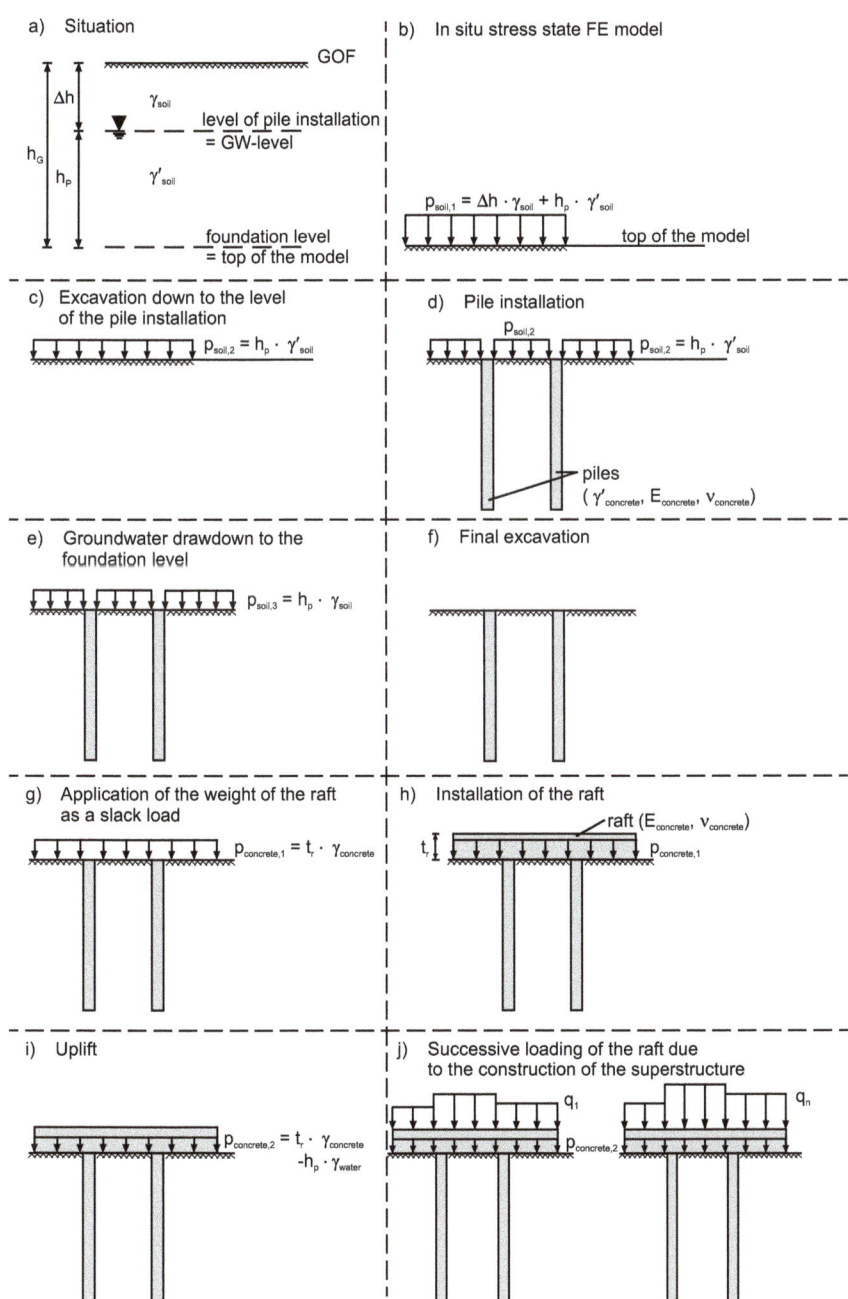

Figure 3.10 Step-by-step analysis of the construction process.

Subsequently, the foundation is included in the model, whereby it may be necessary to proceed in stages, i.e. piles and raft successively. The piles are usually installed "wished-in-place", i.e. ignoring any installation-related changes in the surrounding soil. This simplification is a reasonable approximation for bored piles. The numerical simulation of installation processes such as pile driving is not yet state of the art in engineering practice. The effect of shaft grouting can be simulated approximately by expanding the pile elements, although careful calibration based on data from pile load tests is required.

Finally, the structural weight and the live loads are applied in stages. The number of load steps depends on the magnitude of the load and the amount of nonlinearity in deformation behaviour of the specific soil.

For specific applications, it may be necessary to simulate the removal of the excavation pit support and the rise of the groundwater after the end of the drawdown. For these construction stages, appropriate analysis steps must be included. Furthermore, it must be taken into account that in these cases modelling of the raft alone may not be sufficient. In addition to the excavation pit support, the basement of the structure must also be modelled in a simplified way up to the level of the ground surface.

If interim stages are of interest in cohesive soil, the time-dependent deformation behaviour due to consolidation must be taken into account by coupled pore water pressure-deformation analyses.

If necessary, the construction of existing neighbouring structures must be simulated in preceding analysis steps, whereby the same procedure as described above can be followed.

3.2.2.4 Comparison of analysis methods

For the Westend 1 in Frankfurt, Reul and Randolph (2003) present the results of a study comparing settlements, maximum and minimum pile loads and piled raft coefficients derived from various analysis methods with measurements incorporating the investigations by Poulos et al. (1997). The 90 m × 100 m large office building Westend 1 was constructed between 1990 and 1993. The 208 m high tower and the 60 m high low-rise section of the building complex are founded on two separated rafts. The piled raft of the tower consists of a 47 m × 62 m large raft with a thickness of 3 m to 4.65 m and 40 bored piles with a length of L_p = 30 m and a diameter of d_p = 1.3 m. The bottom of the raft lies 14.5 m below ground level (Figure 3.11a). The groundwater level is situated 7 m below ground level in the quaternary layers. The top surface of the tertiary Frankfurt Formation (Section 4.2.2), which has a thickness of at least 63 m in the vicinity of Westend 1, lies 8.5 m below ground level. The layout of the measurement devices, which consisted of 6 instrumented piles, 13 contact pressure cells, 5 pore pressure cells, 1 multi-point borehole extensometer and 2 combined inclinometers/multi-point borehole extensometers, is shown in Figure 3.11b.

Just as in the creep pile concept (Section 1.4), in the design process of the foundation it was assumed that the piles would fail, i.e. would reach their

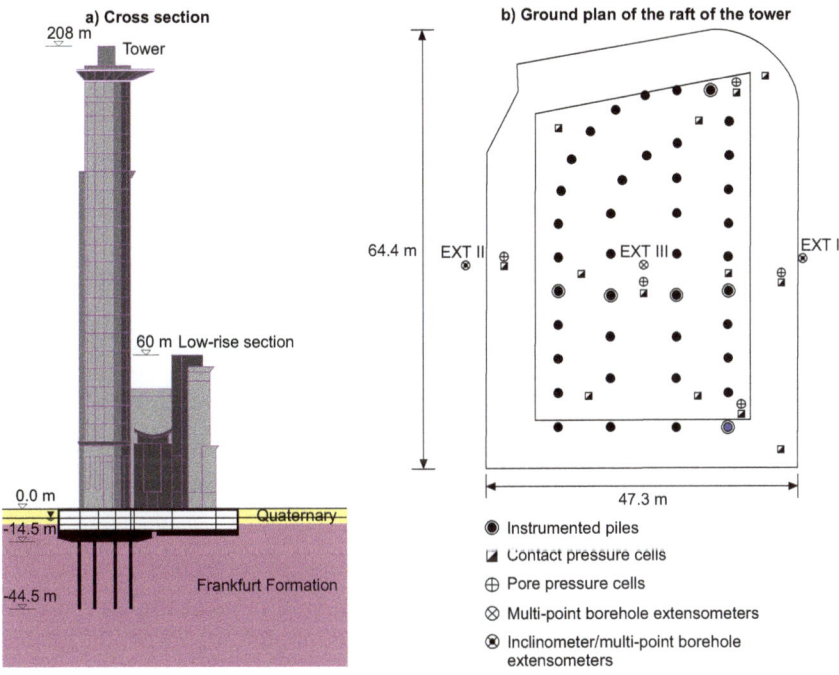

Figure 3.11 Westend I, Frankfurt (Germany)

ultimate capacity, under working load (Franke et al. 1994) with the piles carrying 30% of the settlement inducing load (Wittmann and Ripper 1990).

Reul and Randolph (2003) present the results of a detailed numerical back-analysis of the measurements on the Westend 1 with a three-dimensional finite element model where the nonlinear material behaviour of the Frankfurt Formation was simulated with an elasto-plastic cap-model (Section 2.2.1), while the raft and piles were considered to behave linear-elastically. The material parameters used in the FEA are summarised in Section 2.2.1. A detailed description of the step-by-step analysis of the construction process of the Westend 1 simulated in the FEA is given by Reul and Randolph (2003).

Figure 3.12 shows the comparison of the measured centre settlement, the maximum pile resistance, the minimum pile resistance and piled raft coefficient with the results of the finite element analyses (FEA) and the following analysis methods:

1. Simplified hand calculation method (Poulos and Davis 1980)
2. Strip on springs (Poulos 1991)
3. Plate on springs (Poulos 1994)
4. Combined finite element and boundary element method (Ta and Small 1996)

124 Combined Pile-Raft Foundations

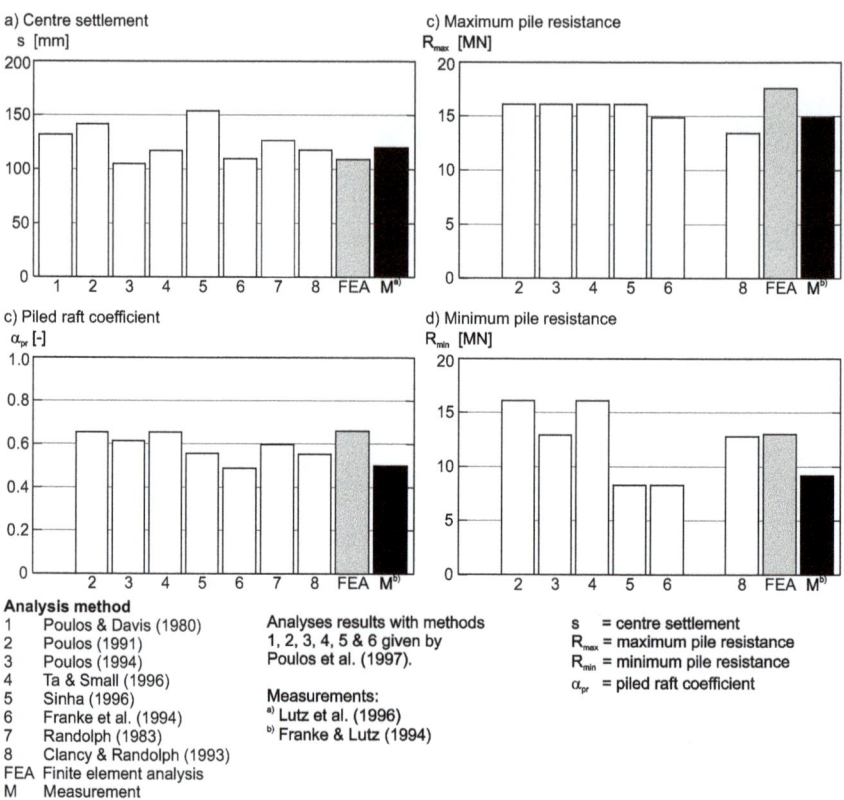

Figure 3.12 Westend 1: Comparison of different analysis methods and measurements. (After Reul and Randolph 2003)

5. Combined finite element and boundary element method (Sinha 1996)
6. Combined finite element and boundary element method (Franke et al. 1994)
7. Flexibility matrix method (Randolph 1983)
8. Load transfer approach for individual piles combined with elastic interaction between piles and raft (Clancy and Randolph 1993)
9. 3D finite element analysis

The results calculated with the first six analysis methods are given by Poulos et al. (1997). For analysis method No. 8 described by Clancy and Randolph (1993), the program HyPR was applied with the parameters of the soil, the raft and the piles summarised in Table 3.5. A detailed description of how the input parameters of approaches 7 and 8 were derived is given by Reul and Randolph (2003). It has to be noted that the subsoil conditions and soil parameters described by Reul and Randolph (2003) and applied in the approaches 7 to 9 may not necessarily comply with the assumptions made for the approaches 1 to 6.

Table 3.5 Comparison of analysis methods: Parameters of the soil, the raft and the piles used in the HyPR analysis

Parameter			Soil	Raft	Piles
Young's modulus	E	MPa	90.1	34000	22000
Poisson's ratio	ν	—	0.15	0.2	—
Ultimate shaft resistance of piles	R_{s1}	MN	—	—	11.0
Ultimate base resistance of piles	R_{b1}	MN	—	—	4.4

The measured centre settlement amounts to 120 mm, 2.5 years after completion of the shell of the building (Lutz et al. 1996) while the settlement attained from the finite element analysis is $s = 109$ mm. The centre settlement calculated with the method by Sinha (1996) is significantly larger than the results achieved with all the other analysis methods. The measured minimum and maximum pile resistances of $R_{min} = 9.2$ MN and $R_{max} = 14.9$ MN, respectively, are taken from Franke and Lutz (1994). Under the assumption that the average resistance of the 6 instrumented piles is equal to the average resistance of the whole pile group, the piled raft coefficient can be derived from the measured pile resistances to give $\alpha_{pr} = 0.50$, while the FEA yields a piled raft coefficient of $\alpha_{pr} = 0.66$. Most of the analysis methods give pile resistances larger than the measured values and therefore overestimate the piled raft coefficient. The calculated maximum pile resistances are generally close to the measured value. The FEA shows the largest deviation and overestimates the maximum pile resistance by 18%. The minimum pile resistance calculated with the methods of Poulos (1991) and Ta and Small (1996) is 75% larger than the measured value. Overall, methods that yield the closest match to the measured settlements tend to overestimate the proportion of load carried by the piles, while (with the exception of Franke et al. 1994) close agreement with the measured pile loads leads to overestimation of settlement.

3.2.3 Analysis methods for investigating the ULS

For investigating the ULS, i.e. the bearing capacity of the foundation, generally FEM and FDM are well suited as long as the influence of the mesh refinement on the results, which is much more pronounced than in the SLS, is considered appropriately. Another option is the application of finite element limit analysis (FELA) to calculate upper and lower bounds of the failure load (Section 2.2.7). Sloan (2013) provides a detailed overview of the FELA including its application on geotechnical boundary value problems and software codes such as Optum CE (Optum CE 2022) incorporate such methodology.

However, it is in most cases the SLS rather than the ULS that will be the limit state decisive for the design of CPRF. Therefore it is believed that in order to comply with the requirements of design codes requesting a proof for the ULS it is sufficient to simply consider the raft alone with classical approaches for shallow foundations.

3.3 REQUIREMENTS FOR THE SITE INVESTIGATION

An appropriate site investigation is necessarily required for the design and dimensioning of CPRF (ISSMGE TC 212 2013). In order to define scope and objectives of a site investigation program, it is helpful to first identify which areas of the subsoil are relevant for the bearing behaviour of pile groups and piled rafts. Pile capacity (ULS) depends on effective stress and fabric conditions at the pile-soil-interface while the deformation response (SLS) is influenced primarily by soil conditions away from the pile (Figure 3.13). Since SLS is the main design criterion for CPRF in most cases, the focus of the site investigation must be on establishing a realistic stiffness profile of the soil. Furthermore, the other parameters applied in the analysis methods for the constitutive models, especially the shear parameters of the soil, need to be determined. A reasonable site investigation program for a CPRF will therefore comprise the following:

- Drilled boreholes to identify the main geological stratification and the groundwater conditions, to take samples for laboratory testing and to provide a location for in situ testing techniques inside the borehole.
- Classical in situ testing techniques such as cone/piezocone penetration tests (CPT/CPTU), standard penetration tests (SPT), dynamic probing (DP) and pressuremeter tests (PMT) to estimate ground stiffness and strength profiles. It is advisory to calibrate and verify results achieved with these techniques by means of ground profiles derived from drilled boreholes in reasonable proximity.

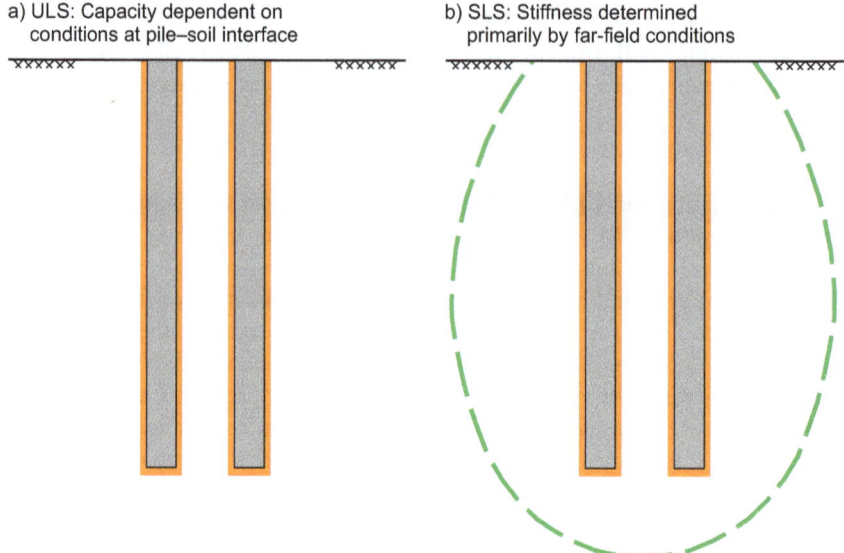

Figure 3.13 Pile group capacity and stiffness after Randolph (2003).

- Geophysical techniques such as seismic borehole tests or seismic CPTs. Again, it is advisory to calibrate and verify results achieved with these techniques by means of ground profiles derived from drilled boreholes.
- Laboratory tests on borehole samples to establish basic soil properties such as water content, particle size distribution or plasticity index as well as soil strength and soil stiffness, for example, by means of triaxial tests and oedometric tests, respectively.

The required depth of the site investigation depends on the geological stratification, the ground area of the foundation and the magnitude of the load which has to be transferred to the ground. As an example, Figure 3.14 shows the investigation depth required by the Eurocode EC7-2. The spacing between boreholes suggested by Eurocode EC7-2 for rising structures and industrial buildings amounts to 15 m–40 m

A comprehensive overview on site investigation techniques appropriate for deep foundations is provided, for example, by Fleming et al. (2009) and Poulos (2017).

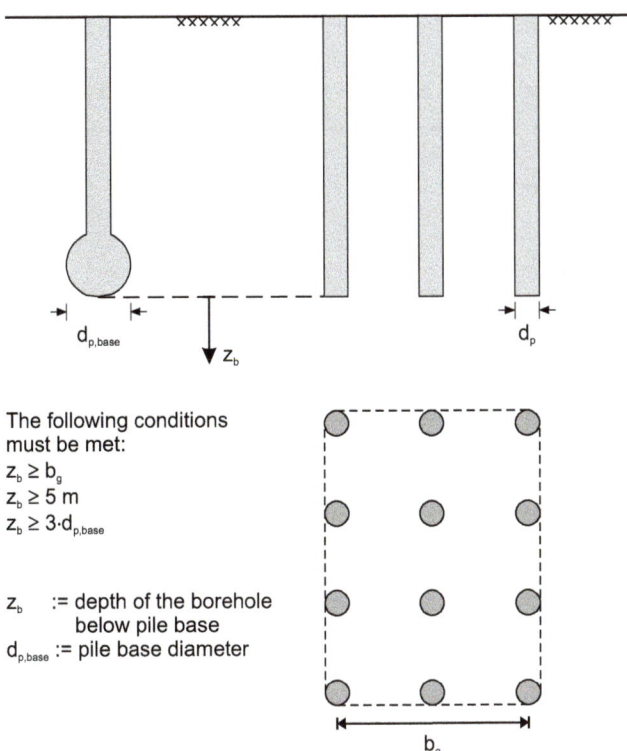

Figure 3.14 Investigation depth for piled foundations after Eurocode EC7-2.

3.4 CONSTRUCTION

3.4.1 Construction of CPRF according to the CPRF guideline (ISSMGE TC 212 2013)

According to the CPRF guideline (ISSMGE TC 212 2013), the construction of a CPRF has to be supervised by a geotechnical expert particularly qualified on this subject with respect to the ground engineering aspects. This applies to the construction both of the piles and of the foundation level.

For the piles, the supervision would include the actual installation process as well as an adequate quality control of the completed pile (Section 3.4.3). To allow for the load transfer of the raft to the soil it is also important to replace spots of soft soil at the foundation level, for example, with lean concrete or similar material to the blinding layer.

3.4.2 Pile types

There are no obvious restrictions for certain pile types for their application for CPRF. However, especially for high-rise buildings, mainly large-diameter bore piles were executed so far (Chapter 4). This is certainly due to the fact that in order to minimise the stresses in the raft a large-diameter pile placed beneath a highly loaded column load is most effective. For displacement piles, especially with relatively close spacing, further research with the focus on the interaction between piles and raft may be needed. Comprehensive overviews on pile types and piling methods are provided, for example, by Tomlinson and Woodward (2007), Fleming et al. (2009) and Viggiani et al. (2012).

3.4.3 Pile integrity tests

Especially for CPRF with highly loaded piles, as is typical for high-rise buildings, quality control of the piles is a major factor to ensure a satisfactory performance of the foundation. To check on pile quality, i.e. proper installation of the piles, the following integrity tests are frequently applied:

- Seismic tests (low-strain dynamic tests).
- Sonic single and multiple borehole tests (cross-hole sonic logging – CSL).
- Radiometric tests.
- Thermal profiling.

An overview of methods on pile integrity testing is provided, for example, by Fleming et al. (2009) and Poulos (2017).

3.5 MONITORING

3.5.1 Monitoring of CPRF according to the CPRF guideline (ISSMGE TC 212 2013)

According to the CPRF guideline (ISSMGE TC 212 2013), monitoring is an elementary and indispensable component of the safety concept. The monitoring comprises geotechnical and geodetic measurements at the new building and also at adjacent structures and is used for the following purposes:

- Verification of the numerical model used in the design process.
- In-time detection of possible critical states.
- Quality assurance.
- Conservation of evidence.

The measurements are supposed to give information on the load distribution between raft and piles and the settlements of the structure. In simple cases, settlements measurements are sufficient. The results of the measurements can then be used to verify, calibrate and improve numerical models for the analysis of foundations and also deep excavations which in general are also the subject of in situ measurements, especially in an urban environment. Figure 3.15 outlines the interplay between in situ measurements, numerical modelling and design.

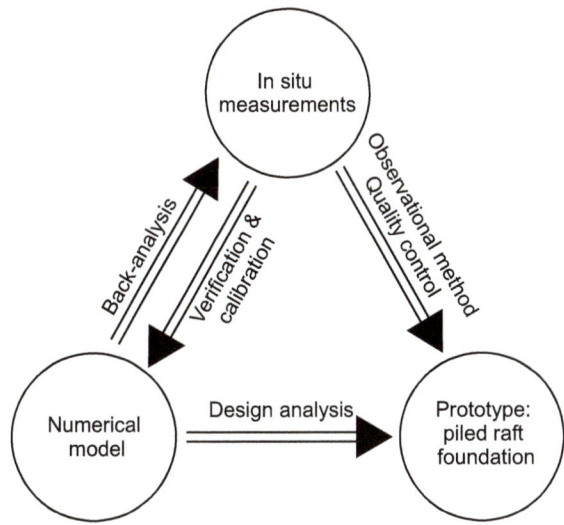

Figure 3.15 Interplay between in situ measurements, numerical model and design.

130 Combined Pile-Raft Foundations

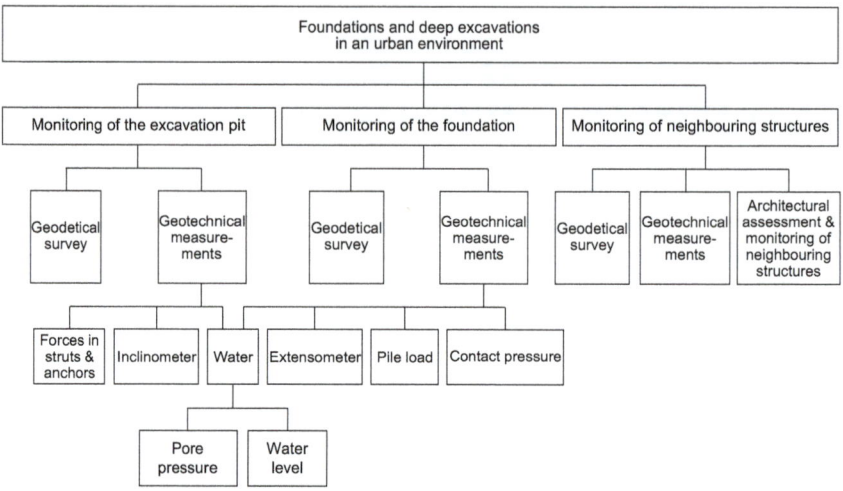

Figure 3.16 Measurements on foundations and deep excavations in an urban environment.

3.5.2 Measurement devices

3.5.2.1 General remarks

In an urban environment, it is recommended to incorporate close monitoring of the excavation pit and its retaining structure, of the foundation itself as well as of buildings and structures in the vicinity of the construction site (Figure 3.16). Typical measurement devices include load cells at the pile head, load cells at the pile base, strain gauges along the pile shaft, contact pressure cells under the raft, pore pressure cells under the raft, multi-point borehole extensometers and inclinometers. An overview of measurement devices is given in the following.

3.5.2.2 Load cells at pile head and pile base

Typical load cells consist of an oil-filled pressure pad made of steel (Figure 3.17a). By measuring the fluid pressure acting in the pressure pad, the external stress acting normal to the pressure pad plane is determined. The fluid pressure of the oil is measured by a piezoelectric sensor and/or a hydraulic compensating valve. For measuring pile head loads, the diameter of the pressure pad, which is laid in a mortar bed on the pile head and surrounded by Styrofoam rings, is smaller than the pile diameter to protect the transducer valve. At the connection of the pile head to the foundation raft, no moments or shear forces are to be transmitted. The transition between pile head and pressure pad is therefore designed as a shear force joint so that the pile head load can be measured directly (Figure 3.17b).

A load cell at the pile base (Figure 3.17c) is usually equipped with a mounting bracket, foam rubber ring and a concreted tip. The load cell will be attached to the base of the reinforcement cage before installation.

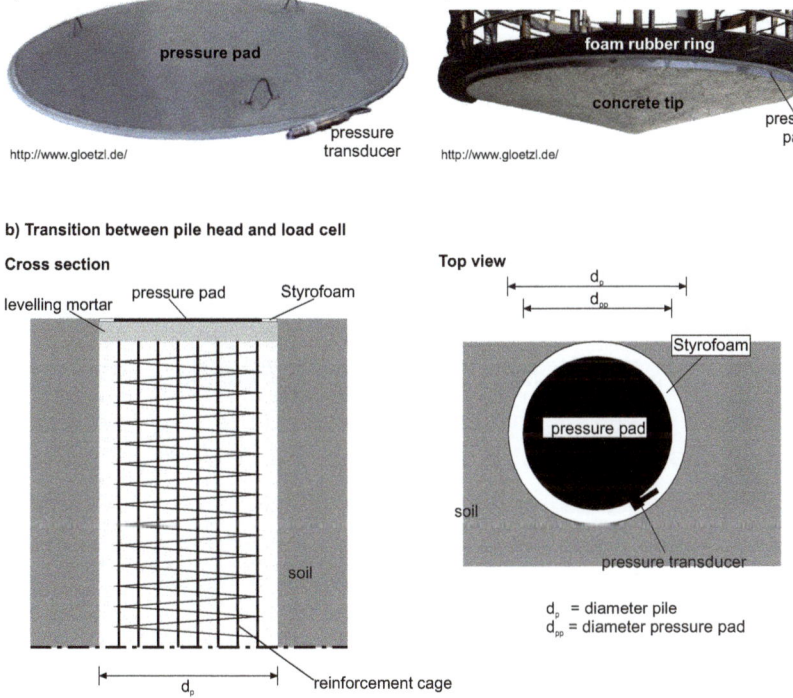

Figure 3.17 Load cells at pile head and pile base.

3.5.2.3 Strain gauges

Strain measurements in piles nowadays are frequently carried out by means of strain gauges based on the vibrating wire technology (Figure 3.18a). If some bending of the piles has to be expected, it is useful to have at least three strain gauges in each cross section (Figure 3.18b). The vibrating wire strain gauge is connected to the reinforcement cage. Based on strain measurement in the pile it is possible to establish the axial load distribution along the pile:

$$N = \varepsilon_a \cdot E_{pile} \cdot A_{pile} \tag{3.15}$$

where N = axial pile load; ε_a = measured axial strain; A_{pile} = cross section area of the pile; E_{pile} = Young's modulus of the pile

The average shaft friction between two measurement planes is then calculated with

$$q_{s(i,i+1)} = \frac{\Delta N_{i,i+1}}{A_{s(i,i+1)}} \tag{3.16}$$

where $q_{s(i,i+1)}$ = average shaft friction between the two measurement planes i and $i+1$; $\Delta N_{i,i+1}$ = difference of the axial pile load between the two

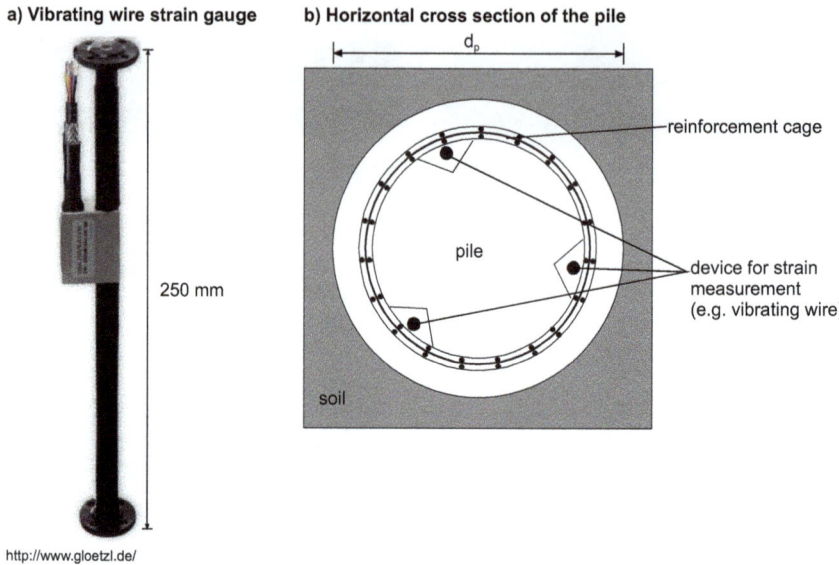

Figure 3.18 Strain measurements by means of vibrating wire strain gauges.

measurement planes i and $i+1$; $A_{s(i,i+1)}$:= pile shaft area between the two measurement planes i and $i+1$.

However, it has to be noted that for bored piles the modulus of elasticity of the pile concrete frequently is smaller than the respective theoretical values as will be discussed in Chapter 4 in the context of the case history of Park Tower. This circumstance results in a degree of uncertainty in the determination of the axial pile loads and subsequently shaft friction profiles based on strain measurements.

3.5.2.4 Fibre optic measurements

It can be expected that in the foreseeable future fibre optic measurements will become increasingly important allowing, for example, continuous strain measurements along a pile shaft. Distributed fibre optics sensing technique takes advantage of the sensitivity of an optical fibre with respect to ambient parameters such as temperature or strain (Soga et al. 2015). The ambient parameters to which the fibre is subjected influence the scattering of a laser light signal travelling throughout the glass material in an optical fibre. The analysis of the scattered light gives information on the magnitude of the ambient parameter such as strain. Soga et al. (2015) provide an overview on the technique of fibre optic measurements and its applications in geotechnical engineering.

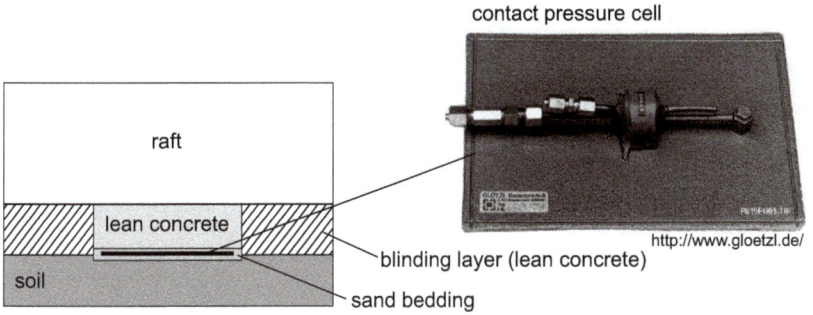

Figure 3.19 **Contact pressure cell.**

3.5.2.5 Contact pressure cells

The total contact stresses under the raft are measured with contact pressure cells. Typical contact pressure cells comprise a pressure pad made of steel filled with hydraulic oil, having a size of up to 40 cm × 40 cm and a thickness of approximately 1.2 cm and are installed in a gap in the blinding layer as shown in Figure 3.19. Similar to the pressure pads used for pile load measurements, the fluid pressure of the oil is measured by a piezoelectric sensor and/or a hydraulic compensating valve. According to Schwab et al. (1991), the measurement inaccuracy for the contact pressure measurements is in the order of 10–20%. At the same time, the ratio of the sum of the contact pressure cell area to the total area of the raft is only about 0.05%, so that a determination of the share of piles and raft in the load transfer, which is carried out exclusively by measuring contact stresses, can lead to significant misinterpretations.

3.5.2.6 Pore pressure transducers

Pore pressure transducers are used to determine the pore water pressure beneath the raft. The pore water pressure is measured either by a piezoelectric sensor or by a hydraulic compensating valve. Figure 3.20 shows the hydraulic valve version. With the injection tip (Figure 3.20a) it is possible to push the cell into the soil without the need to create an oversized hole first which needs to be refilled with an appropriate material later. There are also combined contact pressure (earth pressure) – pore pressure devices available. The effective settlement-inducing contact stresses are determined from the difference between the total contact stresses determined with the contact pressure cells and the neutral stresses (pore pressure) determined with the pore water pressure transducers. Even when placed in a soil layer of low permeability, pore pressure cells can be used to monitor changes in the groundwater level. If the pore pressure cell is in contact with a high-permeability

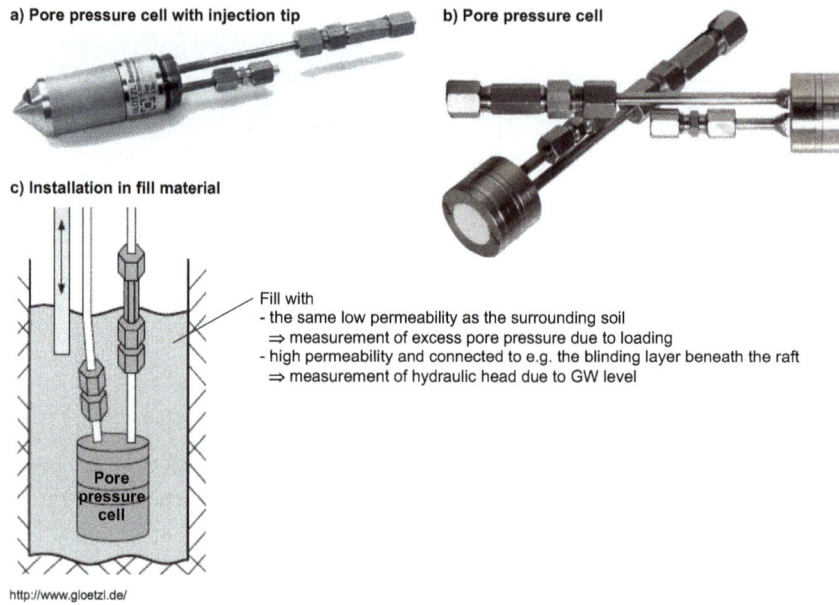

Figure 3.20 Pore pressure transducer.

layer, for example, via the blinding concrete layer beneath the raft, the measured hydraulic head will correspond to the groundwater level.

3.5.2.7 Multi-point borehole extensometer

Extensometers are devices used to measure displacements in the ground along the respective borehole axis. With rod extensometers the displacements of the individual anchor points arranged at different depths are measured in relation to the extensometer head. The individual extensometer rods, e.g. made of glass fibre, are guided from the anchoring point through the backfilled borehole (the grouted annulus) to the extensometer head, protected by a casing tube (Figure 3.21).

3.5.2.8 Inclinometer

Inclinometers are used to measure the lateral deformations of piles, excavation pit walls or of the subsoil. For inclinometer measurements, a mobile measuring probe equipped with inclination sensors is lowered down a guide tube installed in the structure of interest or the subsoil, and the inclination angle between the current probe position and the vertical is measured. From this, the position of the guide tube is determined as a three-dimensional polygon line. Frequently, accelerometers are used as inclination sensors.

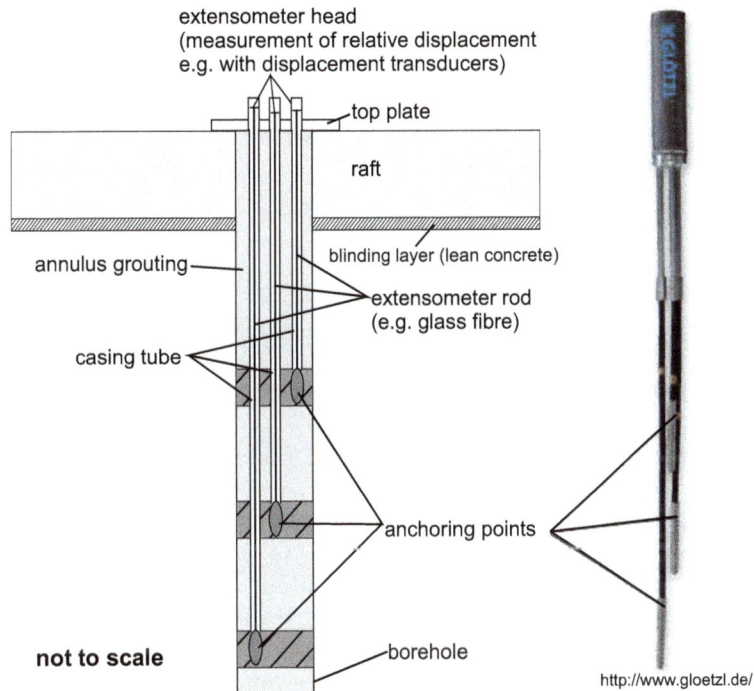

Figure 3.21 Multi-point borehole extensometer.

3.5.2.8.1 Geodetic measurements of horizontal and vertical raft movements

Even if no geotechnical measurements were carried out, at least for high-rise buildings the geodetic measurement of horizontal and vertical deformations of the foundation can be considered common practice. However, one of the crucial aspects in the interpretation of the data is the starting point of the measurements, i.e. whether the immediate settlements of the foundation are included or not. This is of special importance if time-dependent deformations due to consolidation and creep are to be expected. To record all time-dependent deformations the location of the measurement points has to be adjusted in the course of the construction process, as indicated in Figure 3.22. After installation of each new point, the data have to be transferred from the old point.

3.5.3 Pile load tests

According to the CPRF guideline (ISSMGE TC 212 2013), no proof load tests are required for individual piles. Pile load tests are therefore not necessary to establish ultimate shaft friction or base resistance per unit area for

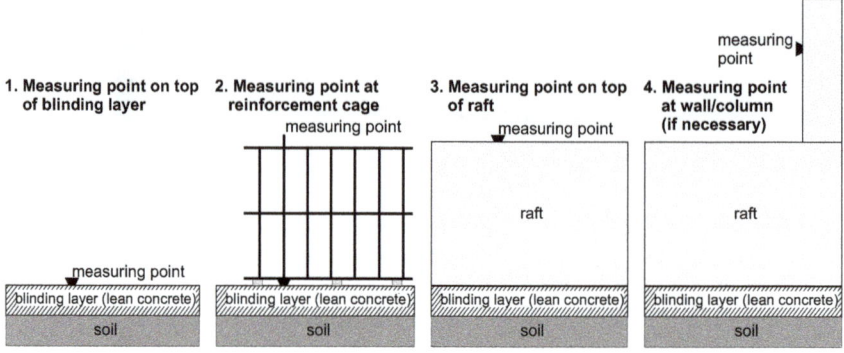

Figure 3.22 Transfer of geodetical measuring points in the construction process.

pile design. However, pile load tests can be valuable to calibrate and verify the computational model, including soil parameters for the constitutive models applied, in particular evaluation of the shear stiffness of the soil. For this purpose it must be carefully evaluated if the installation process of the pile is of importance; i.e. for a displacement pile the modelling of the pile installation as wished-in-place (Section 3.2.2.3) and subsequently increasing the soil stiffness to match the test results might result in serious overestimation of soil stiffness and consequently underestimation of the settlement of the piled raft.

Generally, pile load tests can be categorised between static and dynamic testing methods with dynamic tests mainly applied for establishing pile capacity. Available pile testing methods comprise the following:

- Static axial load tests in compression and tension.
- Static bidirectional load tests (Osterberg cell test).
- Static lateral load tests.
- Statnamic load tests.
- Dynamic load tests (high strain tests).

Since resistance settlement curves are required to verify the load-displacement behaviour, usually static tests are the method of choice in this particular case. For large-diameter piles, bidirectional load tests are of particular merit since they are capable of applying extremely high loads without the need to for reaction piles. For the instrumentation of the test piles the same measurement devices as described above for the actual foundation are used. A discussion of the various methods of pile testing is beyond the scope of this book, and the reader is referred to, for example, Fleming et al. (2009), Viggiani et al. (2012) and Poulos (2017).

3.6 SUGGESTIONS FOR THE DESIGN OF CPRF UNDER MAINLY VERTICALLY LOADING

Based on the experience gained from in situ measurements on completed structures (Chapter 4) and the research on the bearing behaviour of piled rafts (Chapter 2), the following suggestions for the design of CPRF under mainly vertically loading can be derived:

- A more cost-efficient utilisation of the piles of the CPRF, i.e. closer equalisation of the pile forces, can be achieved by maintaining reasonably large pile spacings. The choice of a suitable pile spacing depends, among other things, on the pile length and pile diameter as well as the load level of the foundation, although the positioning of the building columns may also dictate the pile spacing.
- Generally, for the same total pile length, smaller average settlements are achieved with longer piles rather than with a larger number of piles.
- For a raft under uniform loading or core-edge loading, the differential settlements can be most efficiently reduced by installation of piles only under the central area of the raft.
- For a raft subjected to uniform loading the installation of piles seems not to reduce the bending moments compared to the unpiled raft, although suitable located and dimensioned piles may help to reduce differential settlements.
- The arrangement of the piles under column loads leads to a reduction of the bending and shear stresses within the raft.
- The installation of piles with nonuniform length under the loaded areas of the structure allows an efficient reduction of mean settlements, differential settlements and bending moments.
- The overall stiffness of a piled raft decreases with increasing load level.
- The differential settlements are much more sensitive to the raft-soil stiffness ratio and the load configuration than the average settlements.
- The calculation of bending moments is very sensitive to whether or not nonlinear raft-soil interactions are considered.
- The installation of piles under the loaded areas, together with nonuniform pile lengths, appears to yield the minimum total pile length.

However, it has to be pointed out that each project has its own individual constraints, which require the design of an optimised foundation by comparing variants. The raft foundation should always be included in the consideration of variants as a benchmark for the optimisation of the foundation design.

REFERENCES

Arslan, U., Ripper, P. (2003). Geotechnical Aspects of the Planning and Building of High-rises. *High-Rise Manual – Typology and Design, Construction and Technology*, 58–75, Basel: Birkhäuser.

Bond, A.J. (2011). A procedure for determining the characteristic value of a geotechnical parameter. *Proceedings of the 3rd International Symposium on Geotechnical Safety and Risk*, Bundesanstalt für Wasserbau, 419–426.

Burland, J.B., Wroth, C.P. (1975). Settlement of buildings and associated damage. *Settlement of Structures*, Pentech Press, London, 611–654.

Clancy, P., Randolph, M.F. (1993). An approximate analysis procedure for piled raft foundations. *International Journal for Numerical and Analytical Methods in Geomechanics*, 17, 849–869.

Clancy P., Randolph M.F. (1996). Simple design tools for piled raft foundations. *Géotechnique*, 46, 2, 313–328.

Comodromos, E.M., Papadopoulou, M.C., Laloui, L. (2016). Contribution to the design methodologies of piled raft foundations under combined loadings. *Canadian Geotechnical Journal*, 53, 4, 559–577.

Cooke, R.W. (1986). Piled raft foundations on stiff clays- a contribution to design philosophy. *Géotechnique*, 36, 2, 169–2034.

Deutsche Gesellschaft für Geotechnik (DGGT), Arbeitskreis Pfähle (2001). Richtlinie für den Entwurf, die Bemessung und den Bau von Kombinierten Pfahl-Plattengründungen (KPP) – KPP-Richtlinie.

Deutsche Gesellschaft für Geotechnik (DGGT), Arbeitskreis Numerik in der Geotechnik (2014). *Empfehlungen des Arbeitskreises Numerik in der Geotechnik – EANG*. Berlin: Ernst & Sohn.

Engin, H.K., Septanika, E.G., Brinkgreve, R.B.J. (2007). Improved embedded beam elements. In Pande & Pietruszczak (eds), Numerical Models in Geomechanics; *Proceedings of the 10th International NUMOG Symposium*, Rhodes, Greece, 475–480, Rotterdam: Balkema.

Eurocode EC2-1-1 (2011). Design of concrete structures- Part 1-1: General rules – Rules for buildings, bridges and civil engineering structures. EN 1992-1-1:2011.

Eurocode EC7 (2014). Geotechnical design – Part 1: General rules. EN 1997-1:2004 + AC:2009 + A1:2013.

Eurocode EC7 (2010). Geotechnical design – Part 2: Ground investigation and testing. EN 1997-2:2007 + AC:2010.

Fleming, W.G.K., Weltman, A. J., Randolph, M.F., Elson, W.K. (2009). *Piling Engineering*. 3rd ed.; Taylor & Francis Group.

Frank, R., Bauduin, C., Driscoll, R., Kavvdas, M., Krebs Ovesen, N., Orr, T., Schuppener, B. (2004). Designers guide to EN 1997-1. Eurocode 7: Geotechnical design Part 1: General rules. Thomas Telford London.

Franke, E., Lutz, B. (1994). Pfahl-Platten-Gründungs-Messungen. Forschungsabschlußbericht zum Forschungsauftrag Fr 600–11/1.

Franke, E., Lutz, B., El-Mossallamy, Y. (1994). Measurements and Numerical Modelling of High Rise Building Foundations on Frankfurt Clay. *Proceedings of the Conference on Vertical and Horizontal Deformations of Foundations an Embankments*, Texas, ASCE Geotechnical Special Publication No. 40, Vol. 2, 1325–1336.

ISSMGE Technical Committee TC 212 (2013). ISSMGE Combined Pile-Raft Foundation Guideline. Report of the ISSMGE Technical Committee TC 212 – Deep Foundations, ed. Katzenbach, R., Choudhury D, Technische Universität Darmstadt.

Katzenbach, R., Schmitt, A., Turek, J. (2003). Introduction. *Interaction between structural and geotechnical engineers*, 11–14, London: Thomas Telford.

Lutz, B., Wittmann, P., El Mossallamy, Y., Katzenbach, R. (1996). Die Anwendung von Pfahl-Plattengründungen - Entwurfspraxis, Dimensionierung und Erfahrungen mit Gründungen in überkonsolidierten Tonen auf der Grundlage von Messungen. *Vorträge der Baugrundtagung 1996 in Berlin*, 153–164, Essen: DGGT.

Mandolini, A. Russo, G., Viggiani, C. (2005). Pile foundations: experimental investigations, analysis and design: state of the art report. *Proceedings of the 16th International Conference Soil Mechanics and Geotechnical Engineering*, Osaka, 1, 177–213.

Meyerhof, G.G. (1995). Development of geotechnical limit state design. *Canadian Geotechnical Journal*, 32, 128–136.

Optum CE (2022). Optum G3 2021 2.1.6.

Orr, T.L.L. (2007). The development and implementation of Eurocode 7. Transactions of Engineers Ireland.

Poulos, H.G. (1991). Analysis of piled strip foundations. *Proceedings of the Computer Methods and Advances in Geomechanics*, Rotterdam: Balkema.

Poulos, H.G. (1994). An approximate numerical analysis of pile-raft interaction. *International Journal for Numerical and Analytical Methods in Geomechanics*, 18, 73–92.

Poulos, H.G. (2017). *Tall Building Foundation Design*. London: CRC Press.

Poulos, H. G., Davis, E. H. (1974). *Elastic solutions for soil and rock mechanics*. New York: Wiley & Sons.

Poulos, H.G., Davis, E.H. (1980). *Pile foundation analysis and design*. New York: John Wiley & Sons.

Poulos, H. G., Small, J. C., Ta, L. D., Sinha, J., Chen, L. (1997). Comparison of some methods for analysis of piled rafts. *Proceedings of the 14th ICSMFE*, Hamburg, 2, 1119–1124, Rotterdam: Balkema.

Poulos, H.G., Carter, J.C., Small, J.C. (2001). Foundations and retaining structures – Research and practice. *Proceedings of the 15th ICSMFE*, Istanbul, Vol. 4, 2527–2606, Rotterdam: Balkema.

Potts, D. M., Zdravkovic, L. (1999). *Finite element analysis in geotechnical engineering: Theory and application*. Thomas Telford.

Randolph, M.F. (1983). Design of piled raft foundations. *Proceedings of the International Symp. on Recent Developments in Laboratory and Field Tests and Analysis of Geotechnical Problems*, Bangkok, 525–537.

Randolph, M.F. (1994). Design Methods for pile groups and piled rafts. *Proceedings of the 13th ICSMFE*, New Delhi, Vol. 5, 61–82, Rotterdam: Balkema.

Randolph M.F. (2003). Science and empirism in pile foundation design. *Géotechnique*, 53, 10, 847–875.

Randolph, M.F., Clancy, P. (1993). Efficient design of piled rafts. *Proceedings of the II International Geotech. Seminar on Deep Foundations on Bored and Auger Piles*, Ghent, Belgium, 119–130.

Reul, O., Randolph, M.F. (2002). Study of the influence of finite element mesh refinement on the calculated bearing behaviour of a piled raft. *Proceedings of the 8th International Symposium on Numerical Models in Geomechanics*, 259–264.

Reul, O., Randolph, M.F. (2003). Piled rafts in overconsolidated clay – Comparison of in-situ measurements and numerical analyses. *Géotechnique*, 53, 3, 301–315.

Sadek, M., Shahrour, I. (2004). A three dimensional embedded beam element for reinforced geomaterials. *International Journal for Numerical and Analytical Methods in Geomechanics*, 28, 9, 931–946.

Schuppener, B. (2012). Grundlagen der geotechnischen Bemessung. In: Schuppener, B. (ed): *Kommentar zum Handbuch Eurocode 7* – Geotechnische Bemessung, 39–69, Berlin: Ernst & Sohn.

Schwab, H., Günding, N., Lutz, B. (1991). Monitoring pile raft soil interaction. *Proceedings of the 3rd International Symposium on Field Measurements in Geomechanics*, Oslo, 1–11, Rotterdam: Balkema.

Sinha, J. (1996). Piled Raft Foundations Subjected to Swelling and Shrinking Soils. PhD thesis, University of Sydney.

Skempton, A.W. (1953). Discussion contribution: Piles and pile foundations, settlement of pile foundations. *Proceedings of the 3rd ICSMFE*, Zürich, Vol. 3, 172.

Sloan, S. W. (2013). Geotechnical stability analysis. *Géotechnique*, 63, 7, 531–572.

Smith, I.M., Griffiths, D.V. (1998). *Programming the finite element method.* 3rd ed., John Wiley & Sons.

Soga, K., Kwan, V., Pelecanos, L., Rui, Y., Schwamb, T., Seo, H., Wilcock, M. (2015). The Role of Distributed Sensing in Understanding the Engineering Performance of Geotechnical Structures. *Proceedings of the 16th ECSMGE*, ICE Publishing, 13–48.

Ta, L.D., Small, J.C. (1996). Analysis of piled raft systems in layered soils. *International Journal for Numerical and Analytical Methods in Geomechanics*, 20, 57–72.

Tomlinson, M., Woodward, J. (2007). *Pile design and construction practice.* 5th ed., CRC Press, https://doi.org/10.4324/9780203964293

Tschuchnigg, F. (2013). 3D Finite Element Modelling of Deep Foundations Employing an Embedded Pile Formulation. Gruppe Geotechnik Graz, Heft 50.

Tschuchnigg, F., Schweiger, H.F. (2015). The embedded pile concept – Verification of an efficient tool for modelling complex deep foundations. *Computers and Geotechnics*, 63, 244–254.

Vesic, A.S. (1969). Experiments with instrumented pile groups in sand. *Proceedings of the Performance of Deep Foundations*, ASTM, STP 444, 177–222.

Viggiani, C., Mandolini, A., Russo, G. (2012). *Piles and pile foundations*. CRC Press.

Wittmann, P., Ripper, P. (1990). Unterschiedliche Konzepte für die Gründung und Baugrube von zwei Hochhäusern in der Frankfurter Innenstadt. Vorträge der Baugrundtagung 1990 in Karlsruhe, 381–397, Essen: DGEG.

Chapter 4

Case histories

4.1 GENERAL REMARKS

This chapter presents a number of detailed case studies of piled raft design and performance. The main focus is on "modern" (essentially this century) design of CPRFs, based on the German KPP-Richtlinie (DGGT/DIBt 2001) or the *ISSMGE Combined Pile-Raft Foundation Guideline* (ISSMGE TC 212 2013) guideline. However, a brief review is given initially of early developments on the stiff clays of London and Frankfurt where the design approach followed a more conventional treatment of the foundations as piled rafts, i.e. taking account of load being transferred directly to the underlying soil by the pile cap.

Towards the end of the 20th century, the design of piled foundations started to take account of the load transferred directly between the pile cap and the soil. The pile group itself was still designed conventionally, but the additional bearing capacity of the pile cap was also considered, essentially following the generic approach of "piled rafts". These early developments for the stiff, overconsolidated formations in the London and Frankfurt regions are discussed first. The chapter then goes on to document later case histories where the designs followed more formally the CPRF approach.

Appendix B summarises the key design and performance details of all the piled rafts and CPRFs that the authors have been involved in or have come across in the literature, divided according to different regions of the world. A number of those cases, in particular those in which the first author had direct involvement, are discussed in detail here.

4.2 EARLY DEVELOPMENTS IN OVERCONSOLIDATED LONDON CLAY AND FRANKFURT FORMATION

4.2.1 Overconsolidated London Clay

The term London Clay is used to refer to both the London Clay itself and the overconsolidated, stiff-plastic Woolwich and Reading Beds. A compilation of soil properties for the London Clay can be found, for example, in

Skempton and DeLory (1957), Skempton and Henkel (1957), Ward et al. (1958) and Ward et al. (1965).

The 90-m-high Hyde Park Cavalry Barracks (Hooper 1973), built from 1967 to 1970, was already introduced in Section 1.3 as one of the earliest documented case histories of a piled raft.

Green and Hight (1976) and Hight and Green (1976) described the geotechnical instrumentation and measurement results obtained on the foundation of the approximately 70-m-high Dashwood House in London. The structure is founded on a 32.6 m × 33.8 m raft (t_r =1.5 m) and 462 piles (d_p = 0.49 m, L_p = 15 m). Beneath the raft, there is first an approx. 1-m-thick gravel layer, followed by the London Clay. The piles are arranged under the raft at a uniform axial spacing of $e = 3.1d_p$. Based on measurements of the pile resistances of 9 piles and measurements of the contact normal stresses by means of contact pressure cells, Green and Hight (1976) determined a piled raft coefficient of $\alpha_{pr} = 0.66$ after completion of the construction, with the largest pile resistance of $R = 0.69$ MN measured at a corner pile. The measurements indicate a dependence of the pile resistances on the distance to the main load from the rising structure on the raft. Piles in the vicinity of the principal loads exhibit significantly higher pile resistances than piles located at a greater distance from walls and columns. Contact stresses were higher under the edge of the raft than under the centre of the raft. The settlements amounted to $s = 3.3$ cm after completion of the structure.

Hooper (1979) described the foundation of the 185-m-high National Westminster Bank Tower, comprising measurements on 4 instrumented piles, 20 contact pressure cells and an extensometer. The tower, which was founded on an approximately circular raft ($r = 27$ m; $t_r = 2.0$ m to 4.5 m) and 375 piles ($d_p = 1.22$ m, $L_p = 26.5$ m), some of which were widened to 2.14 m at the pile base, settled by a maximum of $s = 4$ cm by the end of construction. At that time, the piles accounted for 75% of the load transfer.

The approximately 42-m high-rise building at Stonebrigde Park in London features a foundation design similar to Dashwood House with a total of 351 piles ($d_p = 45$ cm, $L_p = 13$ m) arranged with a uniform axial spacing of $e = 3.6d_p$ under the 19.2 m × 43.3 m large raft ($t_r = 0.9$ m). Cooke et al. (1981) presented the measurement results of the foundation instrumented with 8 monitoring piles, 11 contact pressure cells, and an extensometer at the centre of the foundation in London Clay. With the monitoring piles equipped with load cells at the pile head and at the pile base, ratios of pile resistances of 2.0 : 1.5 : 1.0 were determined for corner, edge, and inner piles, with the shaft resistance of the corner piles being about three times higher than the shaft resistance of the inner piles. Based on the measurements, Cooke et al. (1981) reported an increase of the piled raft coefficient from $\alpha_{pr} = 0.55$ in the initial phase of the construction process to $\alpha_{pr} = 0.75$ after completion of the construction. 4 years after completion of the structure, the maximum settlements amounted to $s = 1.7$ cm. For the building at Stonebrigde Park there are also measurements available that were carried out during the demolition of the building complex (Butcher et al. 2006).

Burland and Kalra (1986), Kalra and Willows (1986), and Price and Wardle (1986) presented the geotechnical aspects, design and construction, and results of geotechnical measurements at the Queen Elizabeth II Conference Centre in London. The 34.5-m-high structure, which in some areas was built over an existing basement structure, was constructed within the protection of a diaphragm wall, applying the top-down method. During excavation, the slabs were temporarily supported by steel columns founded on 21 pile-like members widened to up to 4 m in diameter and 4.4 m in length. A total of 8 piles (d_p = 1.8 m, L_p = 16 m), one of which was equipped with a load cell at the pile head, were installed under the highly loaded, permanent structural columns to reduce settlement. In addition, 4 contact pressure cells were installed under the 2-m-thick raft. Based on the geotechnical measurements, Price and Wardle (1986) determined a piled raft coefficient of α_{pr} = 0.3 at the end of construction, with the contribution of the piles to the load transfer increasing as construction progressed. The maximum settlement at completion of the conference building amounted to approximately s = 2 cm.

The approximately 49-m-high Victoria Street Redevelopment, Block C, was also built using the top-down method (Hooper 1979). Bored piles with lengths between 13.7 m and 16.8 m were located under the 1.5-m-thick raft. Shortly before the end of construction, the piled raft coefficient reached a value of α_{pr} = 0.25.

The piled rafts in the London Clay presented in this section, which are equipped with geotechnical measuring equipment, are summarised in Appendix B, Table B.3 along with other piled rafts for which settlement measurements have been published.

4.2.2 Overconsolidated Frankfurt Formation

The subsoil conditions in Frankfurt am Main, Germany, are characterised mainly by tertiary soils and rock. They consist of the Frankfurt Formation at the top underlain by the rocky Frankfurt Limestone. According to Kümmerle and Seidenschwann (2009), the geologically correct term for the Frankfurt Formation is Hydrobia layers named after the frequently occurring hydrobia fossils, a genus of small brackish water snails. The Hydrobia layers are subdivided into the Upper Hydrobia (Frankfurt Formation) and Lower Hydrobia (Wiesbaden Formation). Since from a geotechnical point of view there is no relevant difference between the Upper and Lower Hydrobia, the entire layer is referred to simply as the Frankfurt Formation in this book. In the older literature, the Frankfurt Formation was generally termed "Frankfurt Clay", e.g. Sommer and Hoffmann (1991), Sommer et al. (1991) and Reul and Randolph (2003). However, here the term "Frankfurt clay" will be used only for the clayey soils of this layer.

The Frankfurt Formation consists of a heterogeneous sequence of clay and clayey marl, layers of silt and sand of varying thickness (fossil sands), and non-horizontal limestone and dolomite banks, which affect the deformation behaviour of the Frankfurt subsoil. Evaluation of borehole logs from various construction sites in Frankfurt indicates a proportion of sand layers in the overall profile in a range of 5%–24% (Ganal 2024). For the limestone and dolomite banks a proportion between 3% and 24% has been evaluated (Ganal 2024).

Breth (1970) and Amann et al. (1976) described the geotechnical properties of the stiff to semisolid clay and clayey marl predominant in the Frankfurt Formation. Moormann (2002) gave a comprehensive documentation and statistical evaluation of the laboratory tests on samples of clay and clayey marls of the Frankfurt Formation available at the Technical University of Darmstadt, Germany. Ganal (2024) extended this database and carried out an extensive laboratory program to establish the parameters for the visco-hypoplastic AVISA model (Tafili & Triantafyllidis 2020) as summarised by Ganal et al. (2024). Butler (1975) compared the material properties of Frankfurt clay with the London Clay. Table 4.1 summarises typical properties of the clay and clayey marl of the Frankfurt Formation.

The rocky Frankfurt Limestone consists of limestone and dolomite beds, algal limestone reefs, marly calcareous sands and clayey marls. The compressibility of the Frankfurt Limestone is small compared to the Frankfurt Formation. The properties of the Frankfurt Limestone are documented by Holzhäuser (1998).

The groundwater circulates in the quaternary sand and gravel as well as in the tertiary sand and limestone bands while the tertiary clay is practically impermeable. Because the quaternary and tertiary aquifers are connected, groundwater drawdown in the tertiary layers may result in a reduction of the hydraulic head within an area with a radius of several hundred meters.

Table 4.1 Typical properties of the clay and clayey marl predominant in the Frankfurt Formation (Ganal 2024)

Parameter			Min	Max	Mean
Liquid limit	w_L	[%]	34	107	75
Plasticity index	I_P	[%]	16	79	47
Water content	w	[%]	15	57	36
Consistency index	I_C	[-]	0.48	1.20	0.85
Particles < 0.002 mm	$d_{0.002}$	[%]	21	81	49
Wet density	ρ	[to/m³]	1.55	2.10	1.82
Uniaxial compression strength	q_u	[kPa]	45	1356	323
Angle of friction	φ'	[°]	8	37	21 (20*)
Cohesion	c'	[kPa]	10	90	40 (20*)

* value after Breth (1970) usually applied in geotechnical analysis

In Frankfurt am Main (Germany), high-rise buildings have been erected on piled rafts since the early 1980s with the Torhaus der Messe (see Sections 1.3 and 2.2) being the pioneering structure. The long-term behaviour of the well-known Messeturm was discussed in Section 2.3.3, while a comparison of the results achieved with different analysis methods with measured data from the Westend 1 building was presented in Section 3.2.2.

The American Express building (Figure 4.1a) consisted of a 75-m high-rise building and adjacent low-rise sections. The piled raft foundation included

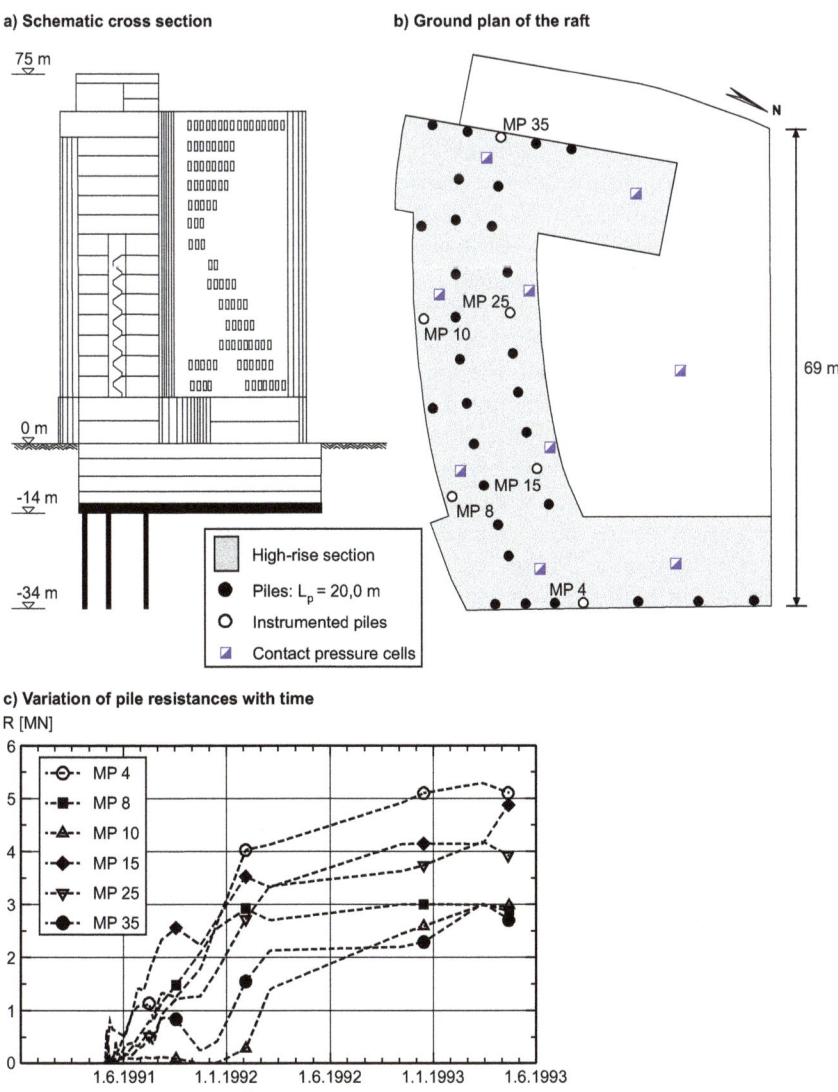

Figure 4.1 American Express building.

35 piles (d_p = 0.9 m, L_p = 20 m), arranged under the most heavily loaded area of the monolithic raft ($A \approx 3600$ m²; t_r = 2 m) as depicted in Figure 4.1b. The design of the piled raft followed a modified conventional design. As typical for the piled rafts in Frankfurt of this generation, the foundation was equipped rather extensively with measurement devices, including 6 piles instrumented with load cells at the pile head and strain gauges in 5 measuring planes. In addition, 9 contact pressure cells and 2 pore pressure cells were installed under the raft. The variation of pile resistances with time is presented in Figure 4.1c, where the pile resistances at the end of construction in May 1993 ranged between R = 2.7 MN and R = 5.1 MN. According to Ripper and El Mossallamy (1999), the structure experienced maximum settlements of s = 5.5 cm by February 1996.

Lutz et al. (1996) and Ripper and El Mossallamy (1999) described the 115-m-high Japan Center in Frankfurt am Main, which was founded on a piled raft mainly because of the high eccentricity of the structural load of 7.5 m. Under the 36.9 m × 52.7 m raft, the thickness of which reduced from 3.5 m in the centre to 2 m at the edge, there were 25 piles (d_p = 1.3 m, L_p = 22 m), six of which were equipped with load cells at the pile head. The pile base level was located only about 3 m above the rocky Frankfurt limestones, which therefore influenced the foundation response significantly. Half a year after completion of the shell, the maximum settlement relative to the time before concreting the raft amounted to approximately s_{max} = 6.5 cm (Philipp Holzmann 1996). At that time, the piles transferred 40% of the structural load to the subsoil. Two more high-rise buildings, Taunusturm (Appendix B, Table B.4) and Omniturm (Section 4.3.3), were constructed more recently in close proximity to the Japan Center. The two high-rise buildings "Kastor" (height H = 95 m) and "Pollux" (H =130 m) of the Forum building complex in Frankfurt am Main are founded on a monolithic 88 m × 120 m raft. Under the high-rise building "Pollux" with a total of 22 piles (d_p = 1.3 m, L_p = 30 m), pile resistances ranged between R = 7.4 MN and R = 11.7 MN with maximum settlements of s_{max} = 7 cm measured at the end of construction of the shell (Lutz et al. 1996). Under the high-rise building "Kastor" 26 piles (d_p = 1.3 m, L_p = 20 m to 30 m) are located. The measured pile resistances ranged between R = 5.0 MN and R = 12.6 MN at the end of construction of the shell with a maximum settlement of s_{max} = 5.5 cm (Lutz et al. 1996).

The majority of buildings on piled rafts completed since the mid-1990s have been designed according to the design philosophy of the KPP-Richtlinie (DGGT/DIBt 2001), which in the meantime has been adopted in the *ISSMGE Combined Pile-Raft Foundation Guideline* (ISSMGE TC 212 2013). The measurements on the buildings WestendDuo, Parktower, Omniturm and Neue Messehalle 3 are documented in more detail in Sections 4.3.1 to 4.3.4.

Without claiming to be complete, Appendix B, Table B.4 gives an overview on piled rafts constructed in Frankfurt over the last decades, spanning early developments such as those mentioned above and the later CPRF-designed developments.

4.3 FOUNDATIONS DESIGNED AS CPRFs

4.3.1 WestendDuo, Frankfurt am Main

4.3.1.1 Building

The building complex WestendDuo in Frankfurt, a reinforced-concrete skeleton construction consisting of two connected, 96-m-high office towers (Figure 4.2) and a low-rise section with a maximum height of approximately 20 m, was constructed from November 2004 to November 2006. The WestendDuo has replaced an existing office tower, which was founded on a raft. The demolition works included the 83-m-high tower and the 3-storey basement.

The basement area of the new building complex with a size of approximately 4100 m^2 is equivalent to the basement area of the demolished building. At the north-western corner of the site, the heritage-protected historic Miquel-Villa is situated in the immediate vicinity of the 4-storey basement of the WestendDuo (Figure 4.2b).

The load of the superstructure of the WestendDuo amounts to approximately $P_s = G + Q = 695$ MN + 220 MN = 915 MN (G: dead loads; Q: live loads) and

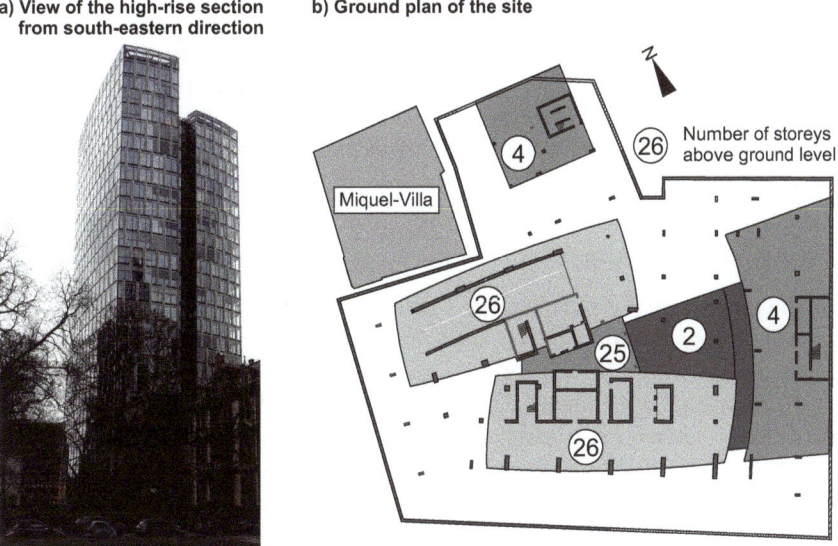

Figure 4.2 WestendDuo, Frankfurt am Main.

is transferred to the raft mainly in the approximately 17 m × 30 m core area of the two towers via walls and at the edge of the building via columns (maximum column load $G+Q$ = 23.3 MN).

For the construction of the 4-storey basement of the WestendDuo a 15-m-deep excavation pit was required after the demolition of the existing basement. The retaining wall comprised bored piles with a diameter of 0.9 m supported by four rows of pre-stressed anchors. The spacing between the bored piles was covered with shotcrete. The groundwater was lowered by means of wells equipped with pumps located outside the excavation pit. Additional borings were placed inside the excavation pit to help reduce pressures in the groundwater circulating in tertiary sand and limestone bands and to ensure the stability of the excavation pit against uplift.

The heating and cooling power of 300 kW required for the building is extracted from the groundwater. Groundwater is pumped with a rate of approximately 43 m³/h from two 140-m-deep production wells, sent through a heat pump/cooling unit and re-infiltrated via three 140-m-deep injection wells. The groundwater is pumped with a temperature of 18°C and re-infiltrated with a temperature of 10°C in winter (heating cycle) and 26°C in summer (cooling cycle), respectively.

4.3.1.2 Subsoil conditions

The subsoil condition on the project site is characterised mainly by tertiary soils and rock with artificially filled soils and quaternary sand and gravel with a thickness of approximately 6 m just below the ground surface. The tertiary soils consist of the Frankfurt Formation with a thickness of approximately 79 m at the top underlain by the Frankfurt Limestone.

The natural groundwater level is situated approximately 6.8 m below the ground surface. However, measurements showed the groundwater level at a depth of approximately 10.4 m due to neighbouring groundwater drawdowns.

4.3.1.3 Foundation design

In the course of the technical and economic design process for the foundation five design alternatives were investigated by means of 3D FEA. In a pre-design study the raft foundation F1 with a constant raft thickness of t_r = 2.4 m had been identified as the starting configuration of the design process. Furthermore, a raft foundation with a constant raft thickness of t_r = 4.0 m (F2) and piled rafts K1, K2 and K3 where the number of piles, the pile length, the pile diameter and the raft thickness were varied were investigated. The design and dimensioning followed the German "KPP-Richtlinie" (see Section 3.1.3). A detailed documentation of the numerical analyses carried out for the foundation design of the WestendDuo was given by Reul et al. (2006a). Randolph and Reul (2019) presented the results of

Table 4.2 WestendDuo: Results of the foundation analyses for the design process

Foundation configuration	t_r (hr/lr) [m]	n_p [-]	L_p [m]	d_p [m]	$n_p·L_p$ [m]	s_{max} [cm]	α [-]	α_{pr} [-]
F1 raft foundation	2.4/2.4	-	-	-	-	13.7	~1/300	-
F2 raft foundation	4.0/4.0	-	-	-	-	13.5	~1/500	-
K1 CPRF	2.4/2.4	28	30	1.3	840	6.2	~1/900	0.460
K2 CPRF	2.4/2.4	26	25/20	1.2	585	7.0	~1/700	0.377
K3 CPRF	1.8/1.2	26	25/20	1.2	585	7.2	~1/500	0.380

t_r: thickness of the raft (hr: high-rise section; lr: low-rise section); n_p: number of piles; L_p: pile length; d_p: pile diameter; $n_p·L_p$: total pile length; s_{max}: maximum settlement; α: angular distortion of the raft; α_{pr}: piled raft coefficient

a simplified analysis using an iterative approach with the program PIGLET (Randolph 2019).

Table 4.2 compares the main results, i.e. the maximum settlement, the angular distortion of the raft and the piled raft coefficient. For the raft foundations maximum settlements of s_{max} = 13.7 cm (F1) and s_{max} = 13.5 cm (F2) were calculated. For configuration F1 the angular distortion $\alpha \approx$ 1/300 was critical because the resulting cracks in the concrete structure might cause problems for the impermeable basement. For the piled rafts the maximum settlements and the angular distortions were significantly reduced, ranging between s_{max} = 6.2 cm (K1) and s_{max} = 7.2 cm (K3) and $\alpha \approx$ 1/900 (K1) and $\alpha \approx$ 1/500 (K3), respectively.

The share of the piles in the load transfer to the subsoil, as described by the piled raft coefficient α_{pr}, was only 8% higher for configuration K1 than for configurations K2 and K3. This is remarkable since the total pile length $n_p·L_p$ is approximately 44% higher for configuration K1 than for configurations K2 and K3.

For the case of a raft that is sagging; i.e. the settlements at the centre of the raft are larger than at the edge, a maximum angular distortion of α = 1/500 is generally considered to be the limit to prevent significant cracks, which could endanger the impermeability of the structure (e.g. Schultze & Horn 1995). For the piled rafts (K1, K2, K3) and the raft foundation F2 (t_r = 4.0 m), this constraint is fulfilled.

As a result of the optimisation process foundation, configuration K3 (Figure 4.3) was selected for construction, yielding tolerable deformations and the most efficient solution economically. The costs for the different foundation configurations established by Reul et al. (2006a) are shown in Table 4.3 where the costs of the single components are scaled to the total cost of foundation configuration K3. Note that only the components that are directly affected by the foundation design are considered.

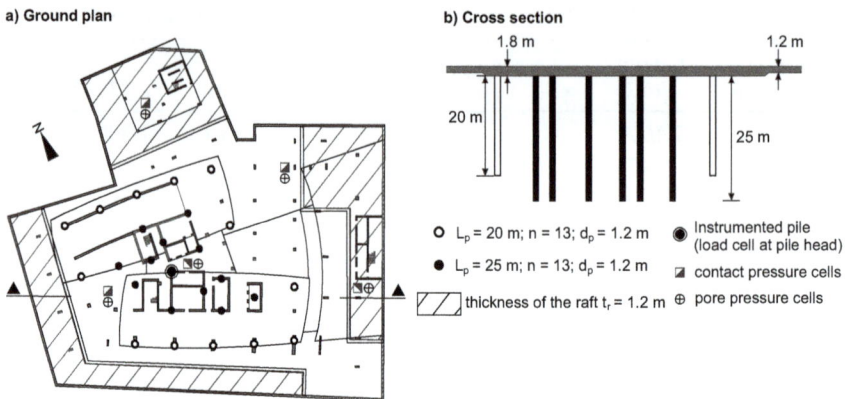

Figure 4.3 WestendDuo: Layout of the foundation.

Table 4.3 WestendDuo: Costs for the different foundation configurations (adapted from Reul et al. 2006a)

Component	Raft foundation		CPRF		
	F1 [%]	F2 [%]	K1 [%]	K2 [%]	K3 [%]
Retaining structure for the excavation pit	45.5	50.0	45.5	45.5	42.5
Excavation and disposal of soil	17.9	22.5	17.9	17.9	15.5
Dewatering system	8.1	8.3	8.1	8.1	8.1
Foundation piles	0.0	0.0	9.9	6.1	6.1
Raft	50.6	54.2	28.3	29.7	26.4
Measurement devices (foundation, dewatering, retaining structure)	0.7	0.7	1.4	1.4	1.4
Σ	122.9%	135.8%	111.2%	108.8%	100.0%

Foundation configuration K3 has approximately 9–11% lower costs than the other piled rafts K1 and K2 and approximately 23–26% lower costs than the raft foundations F1 and F2. Moreover, for foundation configuration F1 the serviceability cannot be ensured due to the high angular distortion.

The comparison shows that, for the selected foundation configuration K3, a significant cost reduction could be achieved, especially for the raft. Compared to foundation configuration F2 the selected foundation configuration K3 reduced the required reinforcement steel from approximately 2,000 tons to approximately 1,150 tons.

4.3.1.4 In situ measurements

Three foundation piles were equipped with a load cell at the pile head, and five contact pressure cells and five pore pressure cells were placed beneath the raft to establish the load transfer to the subsoil. Figure 4.3 shows the location of the measurement devices in the ground plan.

The deformations of the foundation were monitored with 23 geodetic survey points located in the basement. Figure 4.4 shows the maximum settlement measured in November 2006 when the building had just been finished, amounting to s = 4.7 cm. It can be concluded that even allowing for small time-dependent settlements due to consolidation and creep, the maximum predicted settlements (Table 4.2) will probably not be exceeded.

Figure 4.5 shows the variation of the measured mean pore pressure and the measured pile resistance with time. As observed for the Messeturm (Reul and Randolph 2003, section 2.3.3), the pile resistance is significantly influenced by the groundwater level, i.e. the uplift acting on the foundation. Two groundwater drawdowns for construction sites at a distance of approximately 300 m from the WestendDuo caused a loss of uplift and therefore an increase of the settlement inducing load of up to $\Delta P \approx 155$ MN. The maximum pile resistance was measured in November 2008 as R = 7.4 MN. The last measurement available in November 2009 showed a value of R = 5.9 MN.

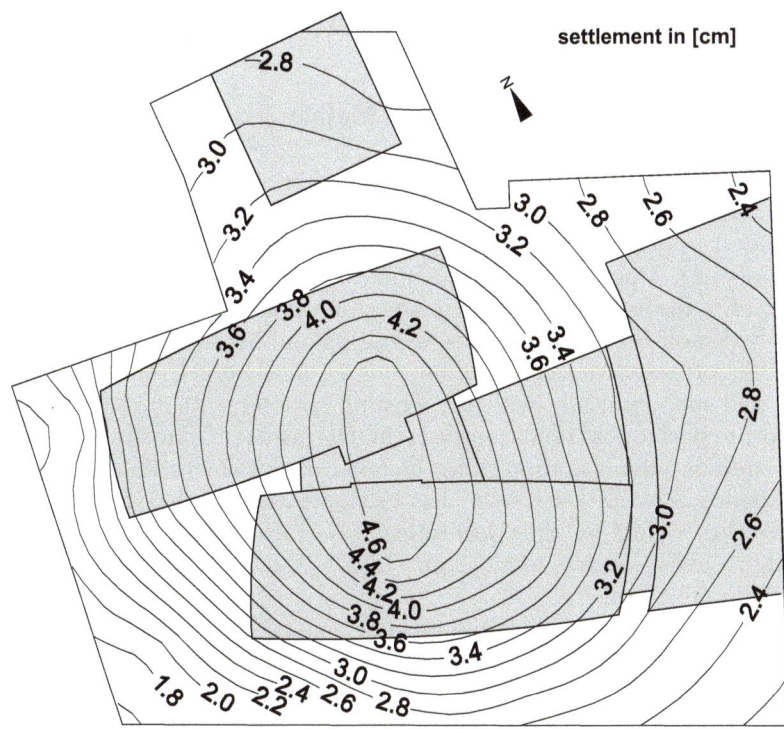

Figure 4.4 WestendDuo: Measured settlements in November 2006 after the building had been finished.

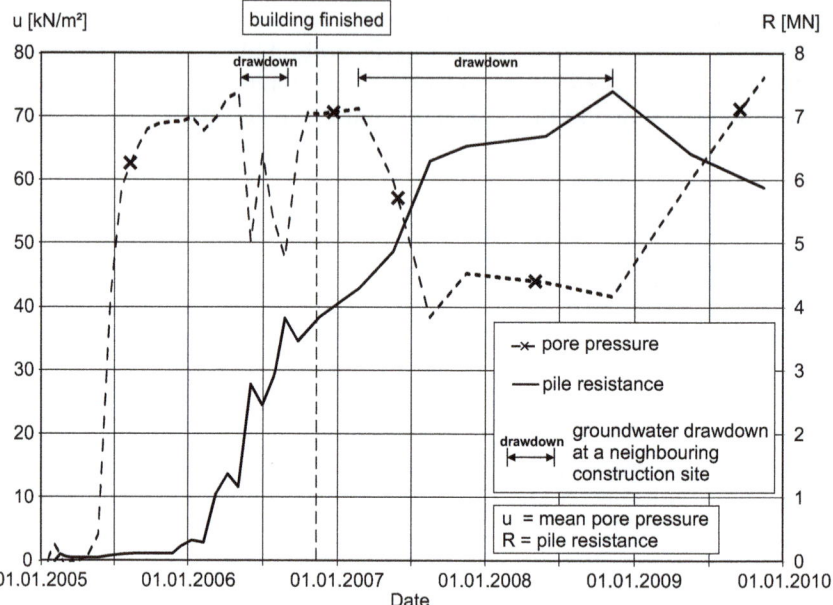

Figure 4.5 WestendDuo: Measured variation of pore pressure and pile resistance with time.

4.3.2 Park Tower, Frankfurt am Main

4.3.2.1 Construction history of the site

In Frankfurt, Germany, the existing 97-m high-rise building SGZ-Bank was increased in height and extended to become the 111-m-high Park Tower (Figure 4.6). During the construction works the existing building was stripped to the load-bearing system and storeys No. 22 to No. 24 were completely demolished between September 2005 and March 2006 (Figure 4.7). On the north-eastern side of the building 28 new storeys were constructed and tied monolithically with the existing structure. Reul and Krajewski (2010) report on a similar project in Darmstadt, Germany, where the 65.7-m-high administration office building of the Hochschule (University of Applied Sciences Darmstadt) has been extended (Annex B, Table B.5). While the original building was founded on raft and strip foundations in sand, the extension was founded on a CPRF ($t_r = 1$ m; $n_p = 18$; $d_p = 0.9$ m; $L_p = 10$ m to 13 m).

The original 24-storey high-rise building and a 2-storey building were constructed in the 1970s. The whole building complex SGZ-Bank has two basement storeys. The structure of the old high-rise building consists of a reinforced-concrete skeleton construction. The foundation comprises a raft with a thickness of $t_r = 2.7$ m with the foundation level situated at a depth of 10.8 m below ground level in the Frankfurt Formation (Figure 4.8). The core of

Case histories 153

Figure 4.6 SGZ-Bank and Park Tower.

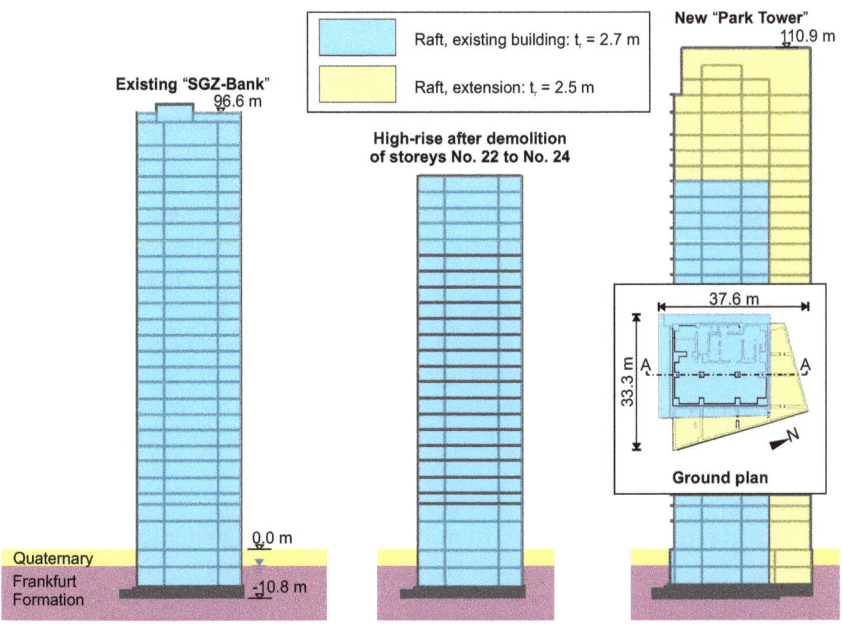

Figure 4.7 Park Tower: Construction stages.

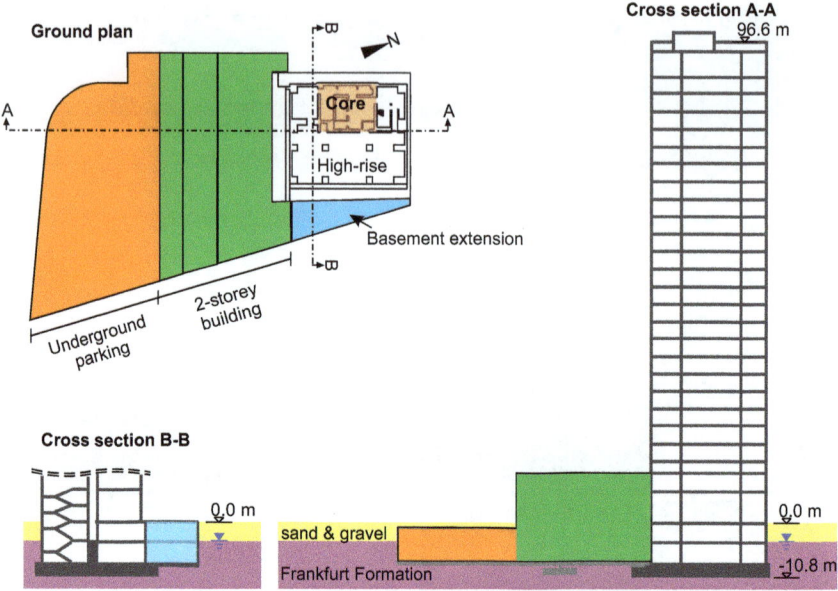

Figure 4.8 Park Tower: SGZ-Bank (constructed from 1970 to 1972).

the high-rise building, which is located at the north-western edge of the building, was built in advance using the slip-form construction method.

Due to the asymmetric position of the core, nonuniform settlements of the raft occurred from the beginning of the construction process, resulting in an increasing tilt towards the northern corner (Leonhardt 1972). As a consequence, the measures shown in Figure 4.9 were taken during the construction process to reduce the tilting, namely:

- Material was dug out along two strips at the south-western and south-eastern edge of the raft of the high rise. According to Leonhardt (1972), this undercutting at the south-western edge closed during the construction process as soon as March/April 1971.
- Application of additional dead weight on the eastern extension of the raft, which was constructed as a cantilever beam.
- The raft of the two-storey building was founded on two strip foundations, one of them located on the south-western edge of the raft of the high-rise.

The cavities under the eastern extension of the raft and under the two-storey building were created with the help of a Styrofoam layer which was dissolved by injection of an organic solvent. As a result, a CHC-pollution of the groundwater was noticeable during all construction activities that took place on the site in the meantime.

Case histories 155

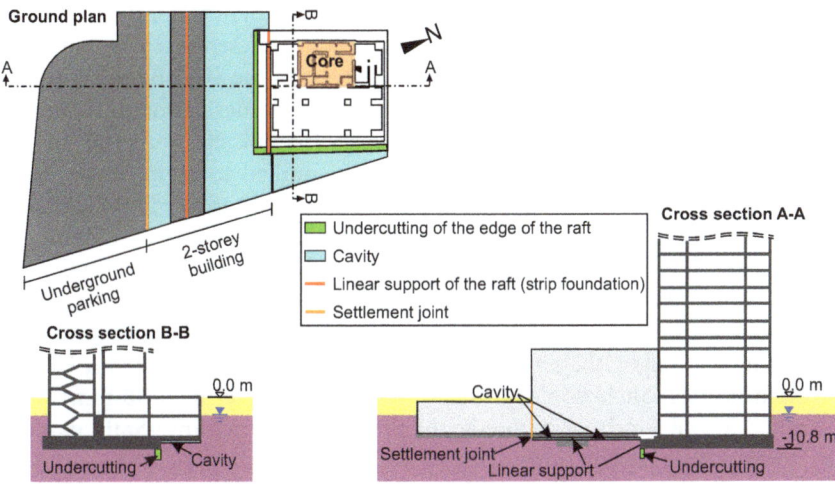

Figure 4.9 Park Tower: Measures to reduce the nonuniform settlements of the SGZ-Bank.

The various measures mentioned above showed no long-term improvement of the settlement performance of the high-rise building. For the last available settlement measurement of this construction phase in 1980, the settlements amounted to $s = 20.9$ cm at the south-eastern corner (MP2) and $s = 30.6$ cm at the north-western corner (MP4) resulting in an angular distortion of the raft of $\alpha = 1/340$ (Figure 4.10). However, there were no reports of damage to the structure of the high rise or impairments of technical

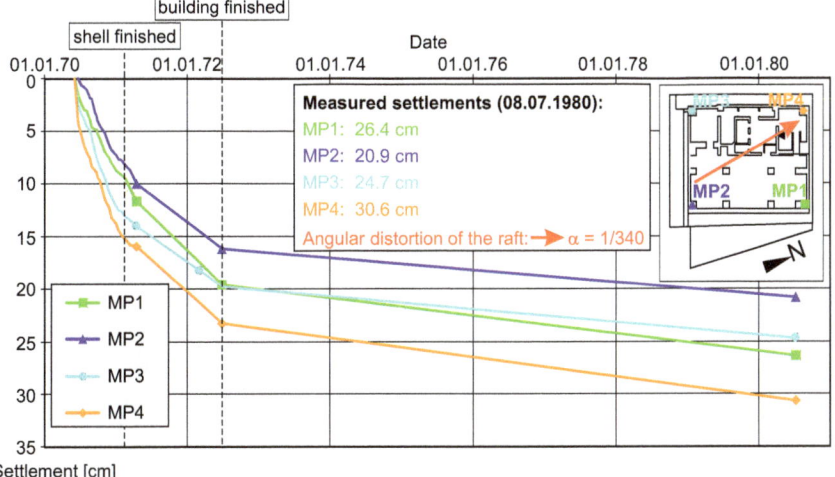

Figure 4.10 Park Tower: Measured settlements of the SGZ-Bank.

installations such as elevators caused by the significant deformations (Ripper and El Mossallamy 1999). Ganal and Reul (2023) presented the results of a numerical simulation of the construction process of the SGZ-Bank and compared them to the measurements taken over a decade. In the 3D coupled pore pressure-displacement FEA, the visco-hypoplastic material model AVISA (Tafili and Triantafyllidis 2020) was applied to capture the time-dependent material behaviour of the Frankfurt Formation. The FEA indicated that the measures taken during construction of the SGZ-Bank to reduce the differential settlements probably had some positive effects.

In 2000–2001 the two-storey building was demolished and replaced by the new six-storey Atrium building. To prevent heave, hence an increase of the tilting, caused by the removal of the strip foundation on the raft, 18 pre-stressed vertical anchors (length L_a = 46.5 m; diameter d_a = 36 mm) were installed at the south-western edge of the high-rise building (Stahlmann et al. 2001). The construction of the new Atrium building yielded only relatively small settlements of the high-rise building, i.e. s = 1.2 cm at the south-eastern corner (MP2) and s = 0.3 cm at the north-western corner (MP4) related to the beginning of this construction phase (Figure 4.11).

Figure 4.12 shows the angular distortion of the high-rise core and the high-rise columns in August 2005 before the start of the Park Tower construction project. According to this, the high-rise core exhibited an angular distortion of α = 1/340 to the north (averaged over the height of the structure), while the high-rise columns had an average angular distortion of α = 1/700 to the west. The different inclinations are caused by the advanced construction of the high-rise core and the correction carried out during construction.

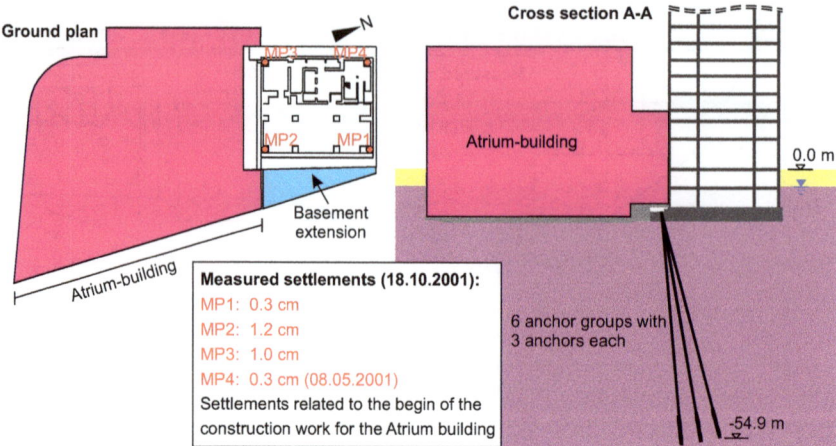

Figure 4.11 Park Tower: Atrium-building (constructed from 2000 to 2001).

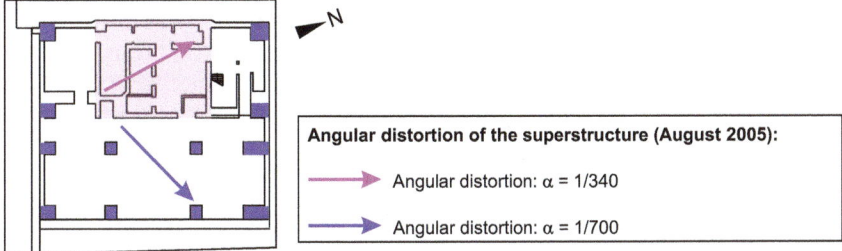

Figure 4.12 Park Tower: Angular distortion of the superstructure in August 2005 before the construction process of the Park Tower.

4.3.2.2 New building

The structural design intended to connect the extension area of approx. 2 × 168 m² per storey to the existing building without joints in order to avoid restrictions during the building's service life. Despite the footprint extension and the increase in the building's height, additional columns inside the building and reinforcement measures of the existing high-rise core were to be avoided. As a result, the design of the extension and the additional storeys had to transfer as minimal loads as possible to the existing building and itself contribute to the transfer of the horizontal loads. The following section summarises the detailed discussion of the structural aspects of the construction given by Remmel et al. (2006).

The roof structure in the core area of the existing building and the three upper floors were removed because their higher ceiling construction would not have allowed the required room height. The screed was removed from all floors in order to be able to compensate to a large extent for the higher loads caused by the planned increase in height in combination with a new, lighter ceiling structure. By using lightweight concrete for the core walls of the new floors, the additional loads on the existing structure were limited to an acceptable level.

The floor area in the extension area is composed of a 15-cm-thick reinforced concrete slab, which is stiffened in the grid of the facade axis by 28-cm-thick concrete beams. Due to the flexible rigid connection to the facade columns, the greater part of the slab loads (approx. 60%) from the extension is transferred via the facade. The remaining 40% of the slab loads are transferred via the five former exterior columns of the existing building. The monolithic connection of the new slab to the existing slab ensures that both the old building core and the facade in the area of the extension contribute to the transfer of the horizontal loads.

The existing reinforced concrete columns are initially subjected to additional normal forces due to the heightening. However, the largest increase in load is experienced by the columns in the intersection area of the existing structure and the footprint extension. Therefore, the concrete strength of the

158 Combined Pile-Raft Foundations

existing columns was determined on the basis of core sample tests as well as non-destructive testing techniques at an early stage of the project. It was shown that the strengths envisaged for the original structure were exceeded for the most part, so that reinforcements were required for a total of 11 columns.

The existing building is stiffened solely by a reinforced concrete core consisting of a staircase and elevator and utility shafts. Together with the core, the reinforced concrete facade of the extension carries the increased horizontal loads resulting from the addition of floors and the extension. The stiffness of the facade is controlled by dimensions and material qualities in such a way that the internal forces acting on the core can be accommodated without reinforcing the existing structure.

By connecting the new basement to the existing structure, it was possible to transfer the additional loads that the former exterior columns of the SGZ-bank tower experience due to the extension to the new foundation. For this purpose, it was possible to use to a large extent the connection rebars with which the former basement extension was connected to the basement of the high-rise building via the raft and walls. On the northern side of the building, the connection was achieved via adhesively bonded connection rebars and a shear connection. Shear forces from the existing structure are transferred to the foundation via bulkheads in the new basement. The connection is made by means of a force transmitting connection of the raft and the slab above the first basement, which is not shown in the structural model in Figure 4.13.

4.3.2.3 Subsoil conditions

The subsoil conditions at the project site are characterised mainly by tertiary soils and rock with artificially filled soils and quaternary sand and gravel with a thickness of approx. 5 m just below the ground surface. The tertiary soils consist of Frankfurt Formation with a thickness of 64 m at the top underlain by the rocky Frankfurt Limestone. Measurements showed the groundwater level in a depth of approx. 6.8 m at the project site.

4.3.2.4 Foundation design

The foundation of the new Park Tower comprising the existing structure and the fully connected extension was designed as a CPRF. For this reason, 16 piles with a length of L_p = 33.5 m and a diameter of d_p = 1.38 m and a raft with a thickness of t_r = 2.5 m were located in the area of the extension (Figure 4.14).

The foundation design of the new Park Tower was based on 3D FEA taking into account the construction history of the site starting with the SGZ-Bank whereby the material behaviour of the Frankfurt Formation was modelled with an elasto-plastic hardening-soil model. Table 4.4 compares

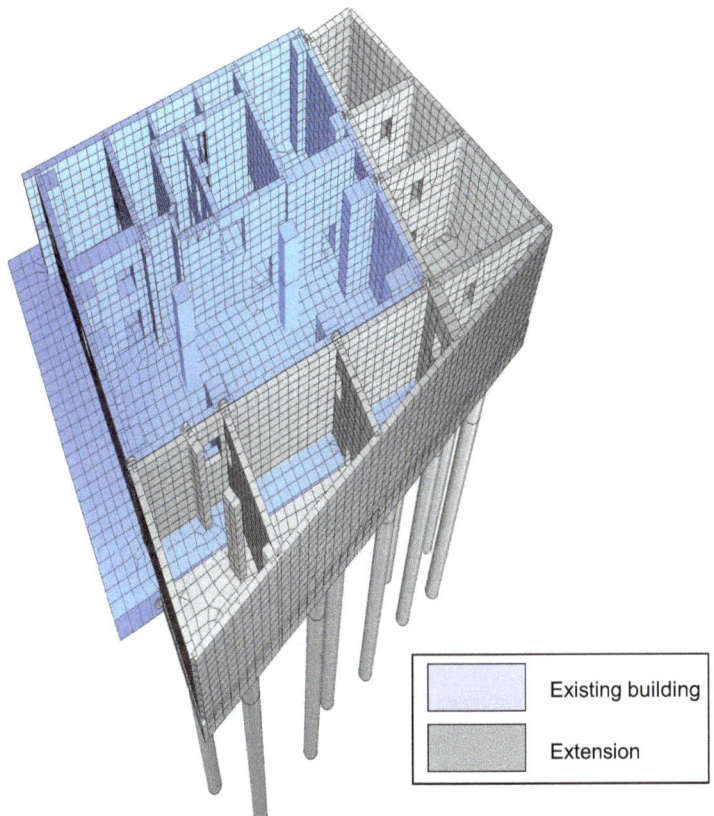

Figure 4.13 Park Tower: Finite element model for the structural analysis of the coupling of existing building and extension. (After Remmel et al. 2006)

the main results of the FEA for the CPRF and a hypothetical raft foundation. The FEA shows that a raft foundation would yield maximum angular distortions of $\alpha_B = 1/150$ (existing building) and $\alpha_N = 1/220$ (extension), respectively, which could not be tolerated. With the CPRF the deformations were significantly reduced and the serviceability of the foundation thus ensured. A detailed discussion of the FEA carried out in the scope of the design process was given by Reul et al. (2007) and Reul and Remmel (2009).

4.3.2.5 Pile installation

Due to the small footprint of the excavation pit, it was not possible to drill the 33.5-m-long foundation piles in the excavation pit from an advanced excavation level. The foundation piles were therefore installed from the ground surface after the basement extension had been demolished down to the raft and backfilled. The piles had a casing to a depth of 14.5 m below

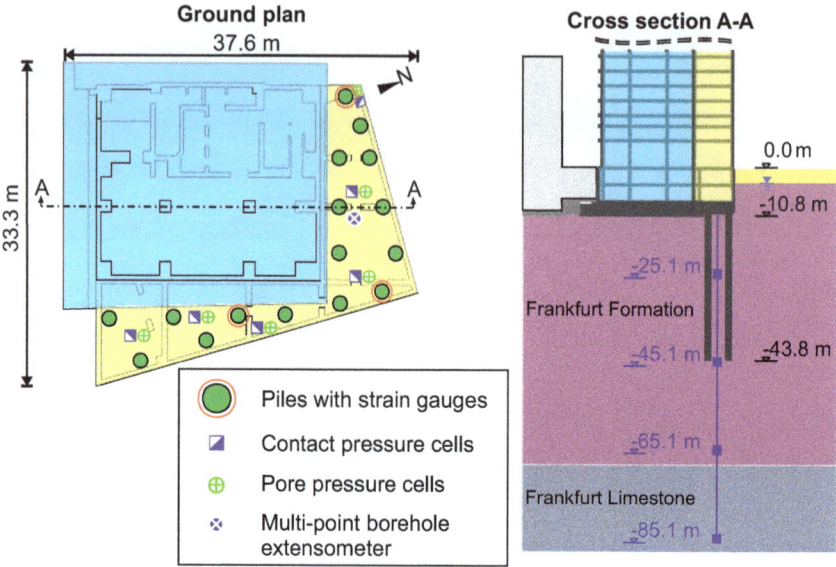

Figure 4.14 Park Tower: Layout of the foundation and measurement devices.

Table 4.4 Park Tower: Results of FEA for the new Park Tower – Comparison of the CPRF with a hypothetical raft foundation

	$s_{max,B}$ [cm]	$s_{MW,B}$ [cm]	$s_{max,N}$ [cm]	$s_{MW,N}$ [cm]	α_B [-]	α_N [-]	α_{pr} [-]
Hypothetical raft foundation	19.6	9.5	25.0	17.2	1/150	1/220	-
CPRF (n_p = 16; L_p = 33.5 m; d_p = 1.38 m)	2.7	2.1	3.6	2.7	1/2250	1/8390	0.72

$s_{max,B}$: maximum settlement in the area of the existing building; $s_{MW,B}$: mean settlement in the area of the existing building; $s_{max,N}$: maximum settlement in the area of the extension; $s_{MW,N}$: mean settlement in the area of the extension; α_B: angular distortion in the area of the existing building; α_N: angular distortion in the area of the extension; α_{pr}: piled raft coefficient

ground surface with a diameter of d_p = 1.5 m, the uppermost 10.1 m being an empty borehole. Below this, the borehole was drilled without casing to 43.6 m below ground surface with the borehole stabilised with a suspension and a pile diameter that was tapered to d_p = 1.38 m. 13 piles were designed in material strength class C30/37, the three highest-loaded piles in material strength class C35/45.

Young's modulus and uniaxial compressive strength were determined for the three instrumented piles (Figure 4.14) by means of uniaxial compression tests on nine of the samples obtained from core drillings. Table 4.5 shows the comparison of the test results with the theoretical values assigned to the

Table 4.5 Park Tower: Results of uniaxial compression test on concrete sample taken from piles by means of core drilling

Pile	Strength class	Theoretical value for the strength class		Test results	
		Characteristic compressive strength f_{ck} [MN/m²]	Young's modulus E_{cm} [MN/m²]	Uniaxial compressive strength* σ_u [MN/m²]	Young's modulus* E [MN/m²]
MP1	C30/37	30	31900	31.8	17700
MP2	C30/37	30	31900	32.3	16400
MP3	C35/45	35	33300	36.0	15100

* Mean value of all tests carried out for one pile

respective strength class. The elastic moduli determined in the unconfined compression tests are smaller than the respective theoretical values. Similar results, i.e. a reduction of the modulus of elasticity of the pile concrete compared to the theoretical value according to the strength class, have been documented by Franke and Lutz (1994), Holzhäuser (1998) and Reul (2000) for construction projects in Berlin and Frankfurt.

4.3.2.6 In situ measurements

Three foundation piles were each equipped with three strain gauges just beneath the pile head for measuring the pile resistances. The contact stresses and the pore pressures beneath the raft were determined with six contact pressure cells and six pore pressure cells. A multi-point borehole extensometer with anchorage points at depths of 25 m, 45 m, 65 m and 85 m below the ground surface was used to determine the deformation profile over the depth. The location of the geotechnical measuring devices is shown in the ground plan in Figure 4.14.

The measured settlements and resistances in October 2007 after the building was finished are documented in Figure 4.15. The measured pile loads show a significant bandwidth between $R = 1.3$ MN and $R = 8.6$ MN. The maximum settlements amount to $s = 1.6$ cm in the area of the building extension and to $s = 1.3$ cm in the area of the existing building. The initial settlements of the raft, which are not included in the values documented above, can be estimated from other case histories as approximately $s_{initial} \leq 0.5$ cm. The settlement profile derived from the extensometer after the completion of the building indicates a block deformation of pile group, raft and soil (Figure 4.16).

The variation over time of the settlements up to about nine years after completion is shown in Figure 4.17a. During the observation period, the pore pressures measured beneath the raft decreased by an average of $u \approx 23$ kN/m² during the construction phase, resulting in a corresponding loss of

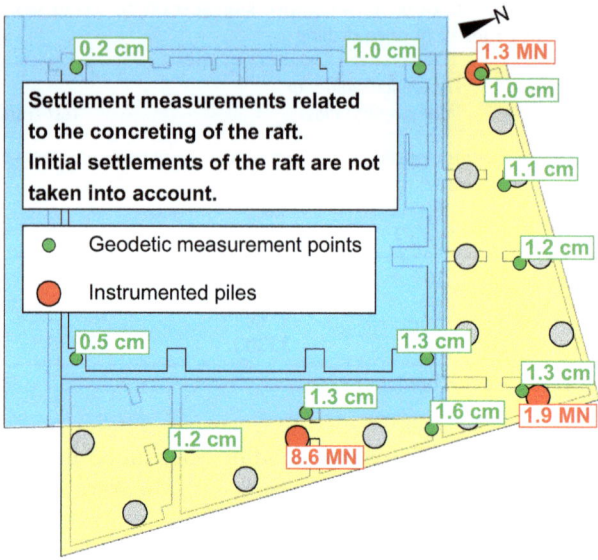

Figure 4.15 Park Tower: Settlements and pile resistances at the completion of the building (17.10.2007).

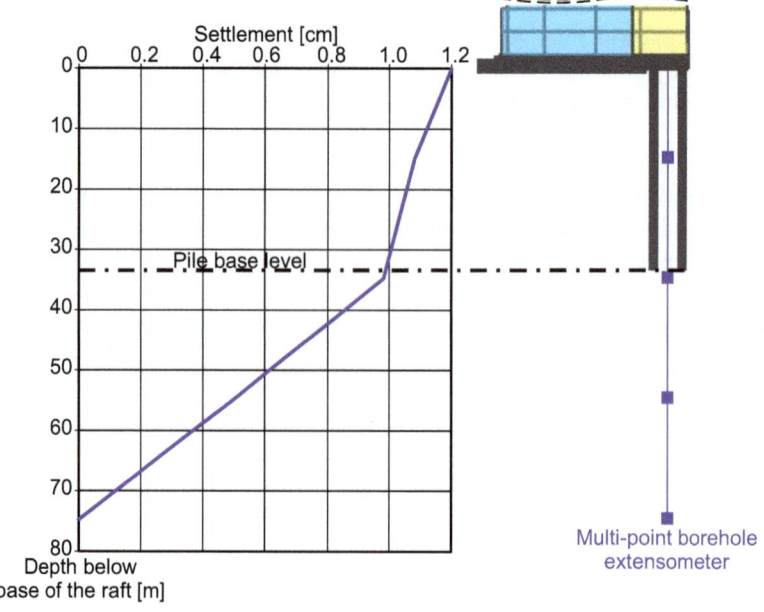

Figure 4.16 Park Tower: Settlement profile at the completion of the building (17.10.2007).

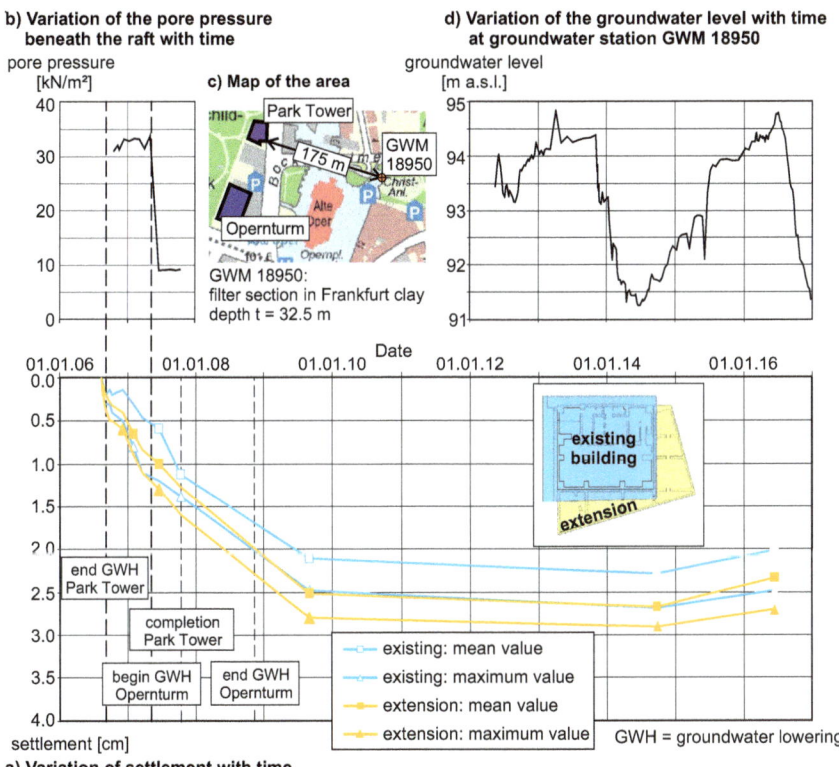

Figure 4.17 Park Tower: Variation of settlement with time.

buoyancy, which was due to the lowering of the groundwater level caused by the immediately neighbouring construction of the Opernturm (Figure 4.17b). The maximum settlements measured during the observation period decreased from $s_{max,B}$ = 2.7 cm and $s_{max,N}$ = 2.9 cm in 2014 to $s_{max,B}$ = 2.5 cm and $s_{max,N}$ = 2.7 cm in 2016. During the same period, a groundwater rise of about 3 m was recorded in a tertiary groundwater monitoring well about 175 m away (Figure 4.17d). Ganal (2024) presented the results of a 3D coupled pore pressure-displacement FEA with the visco-hypoplastic material model AVISA (Tafili and Triantafyllidis 2020) to back-analyse the long-term deformation behaviour of the Park Tower.

4.3.3 Omniturm, Frankfurt am Main

4.3.3.1 Building

Just as for the Queen Elizabeth II Conference Centre (e.g. Burland and Kalra 1986, section 4.2), the Maintower in Frankfurt (e.g. Katzenbach et al.

2000, Appendix B, Table B.4), the Amuplaza building in Kagushima, Japan (e.g. Sonoda et al. 2009, Appendix B, Table B.7) and the Abeno Harukas building in Osaka, Japan (Hirakawa et al. 2016, Appendix B, Table B.7) the Omniturm (Figure 4.18a) in the centre of Frankfurt's banking district is an example of a piled raft constructed with the top-down method. The Omniturm was built between 2016 and 2019 in the immediate vicinity of several existing high-rise buildings (Figure 4.18b), resulting in complex constraints and a large number of restrictions for the planning and execution of the excavation.

On a footprint of approximately 50 m × 50 m, the approx. 190-m-high building provides 46 stories of office and residential space as well as areas for the public. A key architectural feature of the building is that some of the floors project beyond the actual floor plan at about half the height of the structure, so that the residential apartments located there have balconies (Figure 4.18a).

For the construction of the four basement levels, which are mainly used as underground parking, excavation of an approximately 16-m-deep excavation pit was necessary. Along the north-east side of the site, the basement levels were constructed on a strip approx. 15 m wide without overbuilding. Before construction work began, an existing office building with up to seven upper floors and two basement floors was demolished.

The construction site is bordered by roads with heavy traffic (Figure 4.18b). The existing Deutsche Bank building on the north-east boundary directly adjacent to the construction site was still in use at the time the excavation pit was built. The building was since demolished, and the Four Frankfurt project (Meissner et al. 2019) comprising four high-rise buildings with heights between 100 m and 220 m was constructed on the site. Construction site set-up areas outside the building site were therefore only available to a very limited extent for the foundation engineering work.

The load of the reinforced concrete superstructure of the Omniturm amounts to approximately $P_s = G + Q = 1174$ MN + 285 MN = 1459 MN (G: dead loads; Q: live loads) and is transferred to the raft in the approximately 17 m × 20 m core area via walls and at the edge of the tower via columns (maximum column load $G + Q = 31.5$ MN).

4.3.3.2 Subsoil conditions

At the project site, the Quaternary sands and gravels are present at depths of approximately 5 m–11 m below artificial fill, remains of old buildings and residual Quaternary silty flood deposits. These are underlain by the Frankfurt Formation which has a layer thickness of approximately 36 m at the site with the bottom of the layer dipping from approximately 43 m below ground level to approximately 47 m below ground level to the west. The embedded limestone bands were evaluated with thicknesses ranging from about 0.1 m–2.0 m. The lower surface of the thickest limestone band

Case histories 165

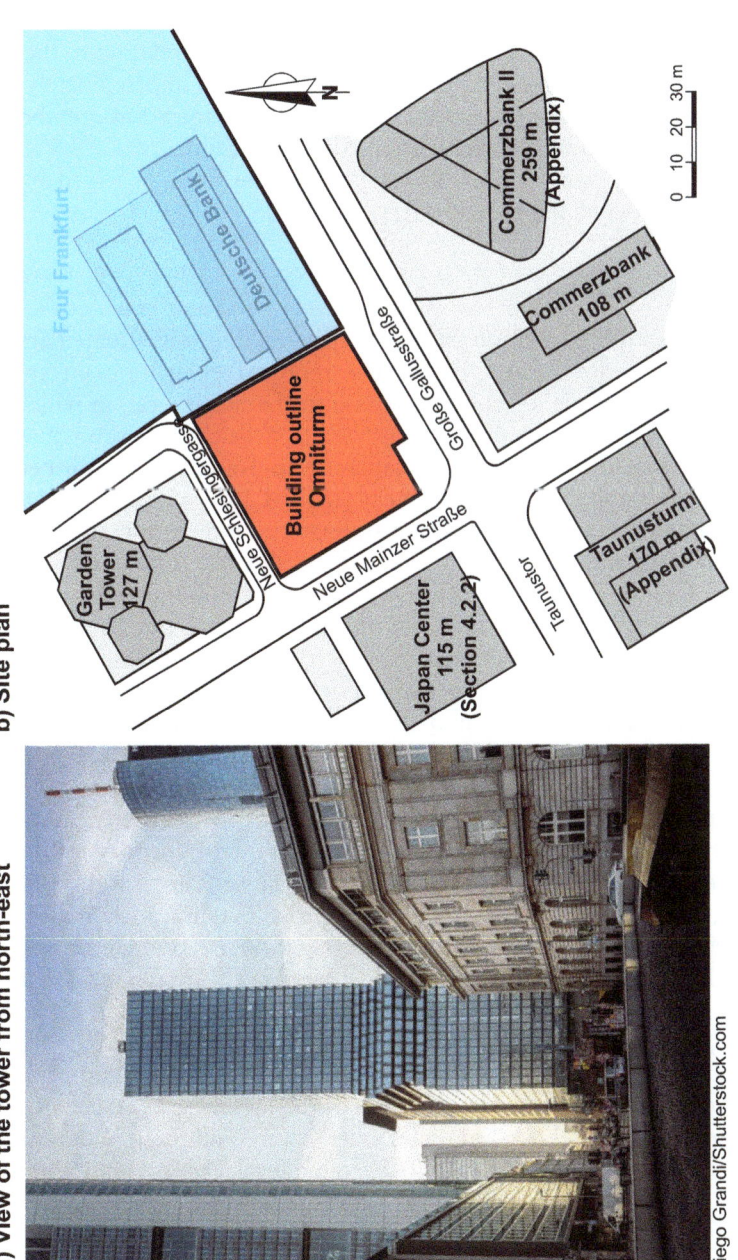

Figure 4.18 Omniturm, Frankfurt. (Adapted from Ramm et al. 2020)

investigated within the Frankfurt Formation dips from about 20 m below ground level at the eastern corner of the project site to about 24 m below ground level at the western corner with an inclination of about 7°. The proportion of limestone bands as well as sand layers in relation to the layer thickness of the Frankfurt Formation both amount to approximately 14% at the project site. Below the Frankfurter Formation, the 90-m-deep borings investigated Frankfurt Limestone up to their final depths.

During the site investigation, the groundwater level was observed approximately 5.5 m below the ground surface.

4.3.3.3 Excavation pit

The bottom of the excavation pit for the new building was situated approximately 16 m below ground level. The required excavation pit support along the adjacent Neue Schlesingergasse to the north-west, in which several power, water and media lines as well as a brick-built sewage system run, is dominated by the small distance of approximately 10 m between the Omniturm and the Garden Tower (Figure 4.18b) which is founded on a raft foundation at a depth of approximately 18 m below ground level. The supporting wall with timber sheeting and shotcrete lining of the former excavation pit of the Garden Tower was still present in the underground.

On the north-east side of the Omniturm project site, the Deutsche Bank (Figure 4.18b) building was located with a one-storey basement immediately at the property boundary and a two-storey basement further away. With up to seven above-ground floors, the maximum contact pressures caused by the Deutsche Bank building amount to $\sigma_0 = 282$ kN/m².

To the south-east of the building site along Grosse Gallusstraße, the Commerzbank complex is located at a distance of approximately 22 m–28 m to the Omniturm. The Commerzbank I is founded on a raft foundation with the bottom edge of the up to 3-m-thick raft situated at approximately 12 m below ground level. The so far tallest high-rise building in Frankfurt, the Commerzbank II, is founded on a pile foundation; i.e., the contribution of the raft in the load transfer was not considered in the design process, in the tertiary Frankfurt Limestone (Katzenbach et al. 1994, 1996) and is located outside the influence area of the excavation pit of the Omniturm.

The Japan Center is located on the opposite side of Neue Mainzer Straße, approximately 17 m away. The piled raft of the Japan Center (Section 4.2.2) was carried out with 22-m-long piles ($d_p = 1.3$ m) and the foundation base situated at approximately 15 m below ground level. The excavation pit support for the Japan Center was designed as a tied-back secant bored pile wall ($d = 0.9$ m), with the anchors remaining in the ground after construction.

Based on the experience gained from previous construction projects carried out with the top-down method, the construction consortium for the Omniturm developed a bracing system using two basement slabs (Ramm et al. 2020). For this purpose, slab sections of the second and fourth basement

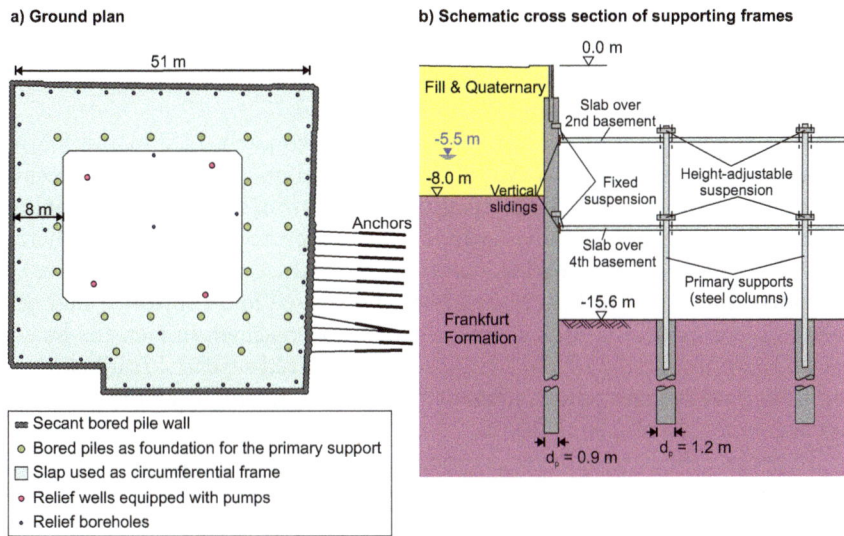

Figure 4.19 Omniturm: Retaining system. (Adapted from Ramm et al. 2020)

floors were constructed in the form of approximately 8-m-wide circumferential frames (Figure 4.19). Only in the area adjacent to the Deutsche Bank at the north-east additional tie-back anchors were required to reduce the deformation of the neighbouring building to an acceptable amount.

For the vertical support of the slab sections, a distinction was made between the edge of the excavation and the inside. In the load-bearing concept of the high-rise building, the retaining wall comprising secant bored piles (d_p = 0.9 m; d_p = 1.2 m adjacent to the Deutsche Bank) takes on the function of a permanent shielding of the earth pressure in the horizontal direction; however, it is not used for the vertical transfer of building loads and is therefore not part of the foundation. To take these boundary conditions and the predicted relative displacements in the order of several centimetres into account, the bedding of the slabs on the retaining wall basically worked as vertically sliding by means of sliding bearings. Only during the phase of the soil excavation until connection of the rising exterior walls were the slab sections temporarily suspended firmly on the retaining walls (Figure 4.19b).

In principle, the use of foundation piles is an option for supporting the slab sections inside the excavation pit. However, in this case the distances between the foundation piles were too large in some areas. In addition to some contractual aspects, the foundation design of the CPRF had not yet been completed. The planning and execution of the support in the centre of the excavation pit was therefore carried out independently of the building foundation with a total of 22 temporary steel columns (primary supports) founded on bored piles (d_p = 1.2 m; L_p = 11 m to 18 m) as shown in Figure 4.19b. In the course of the subsequent construction of the basement floors, the two slab sections were relocated onto columns and walls of the actual building. The steel sections were then demolished and the piles were covered with a compressible layer to prevent punching of the raft as a result of the settlements of the high-rise building (Ramm et al. 2020).

After construction of the slab sections, approximately 11.0 m and 5.5 m of soil, respectively, were excavated by means of mini-excavators. As a result of the stress relief due to the excavation, nonuniform heave of the retaining wall and the temporary primary supports was expected. A system was therefore developed for the bearing points of the slab sections on the primary supports that allowed them to be lowered or raised in a controlled manner before incompatible vertical displacements were reached. In fact, the heave of up to approximately 2 cm that occurred required repeated readjustment of individual bearing points, demonstrating the functionality of this system several times (Ramm et al. 2020).

To protect the base of the excavation pit against uplift, 43 relief boreholes and 4 relief wells equipped with pumps were installed inside the excavation pit, reaching down to a depth of approximately 27.5 m below ground level. The groundwater inside the excavation pit was successively lowered in advance of the excavation. In particular, as a prerequisite for using vehicles on the water-saturated clay, dewatering was required by actively pumping the relief wells by means of vacuum well points. Since the groundwater relief had to be operated beyond the time of concreting the raft, openings were installed in the raft for the relief wells, through which also all injection pipes for sealing the boreholes and wells after completion of the measures were directed. In total, the groundwater lowering was operated over a period of 13 months with an average pumping rate of approximately 34 m³/h and a maximum pumping rate of 58 m³/h (Ramm et al. 2020).

4.3.3.4 Foundation design

Based on a numerical parametric study carried out by means of 3D FEA, a CPRF with 41 piles (d_p = 1.5 m) with pile lengths between L_p = 23.4 m below the facade columns and L_p = 27.4 m below the core of the high-rise as well as a raft thickness of t_r = 2.9 m in the high-rise area was identified as the optimised foundation variant for the Omniturm. The pile lengths were established taking into account a minimum distance of 2 m between the pile base and the top edge of the Frankfurt Limestone, to avoid inadmissible high loads in the piles. In the eastern part of the building without superstructure, additionally 5 piles (L_p = 12 m to 20 m; d_p = 1.2 m) were placed under columns to reduce the stresses in the raft with a reduced thickness of only t_r = 0.5 m in this area. Figure 4.20 shows the layout of the foundation. The retaining wall and the piles for the primary columns are decoupled from the foundation and are therefore not involved in the transfer of the structural loads into the subsoil. Based on a holistic design approach of the entire foundation and the structure when assessing the safety against uplift,

Case histories 169

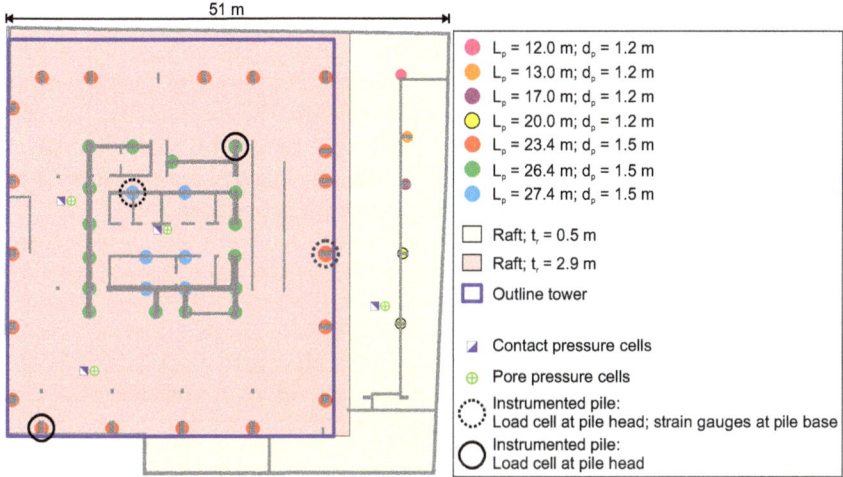

Figure 4.20 Omniturm: Ground plan of the foundation with measurement devices.

Table 4.6 Omniturm: Results of the 3D FEA

Load case	s_{max} [cm]	s_{ave} [cm]	α [-]	α_{pr} [-]
Settlement inducing load; $G_{raft} + G + Q/3 - U$	8.3	5.9	1/510	0.65
Full load for structural design; $G_{raft} + G + Q$	11.1	8.1	1/370	0.59

s_{max}: maximum settlement; s_{ave}: average settlement of the pile heads in the area of the high-rise section; α: maximum angular distortion of the raft; α_{pr}: piled raft coefficient

it was possible to omit anchoring elements such as micropiles to provide uplift resistance in the areas without superstructure.

Based on the 3D FEA, maximum settlements of s_{max} = 8.3 cm were determined for the settlement inducing load $P_{eff} = G_{raft} + G + Q/3 - U$. Since the 3D FEA did not take into account the stiffness of basement walls and superstructure, the predicted deformations represent a conservative estimate in terms of differential settlements and angular distortions. The most relevant analysis results for the settlement inducing load, which is decisive for the serviceability limit state design, as well as for the full load relevant for structural design $G_{raft} + G + Q$ are summarised in Table 4.6.

4.3.3.5 In situ measurements

Four foundation piles were equipped with load cells at the pile head to measure the pile resistances. Two of these piles were also equipped with strain gauges at the base, which can be used to determine the pile base resistances estimating the pile stiffness from pile P07 which had an additional pair of strain gauges installed just beneath the pile head. Contact pressure and pore pressure under the raft are determined with four contact pressure cells and four pore pressure cells each. The location of the geotechnical measuring devices is shown in the ground plan in Figure 4.20. Geodetic measurements were carried out during construction to monitor the settlements. For this purpose, a total of 22 measuring points were arranged on the raft, distributed over the entire footprint of the building, which were transferred to neighbouring walls and columns as construction progressed.

Figure 4.21 presents the measurements together with the development of the total building load (Figure 4.21a). Figure 4.21b shows the variation of

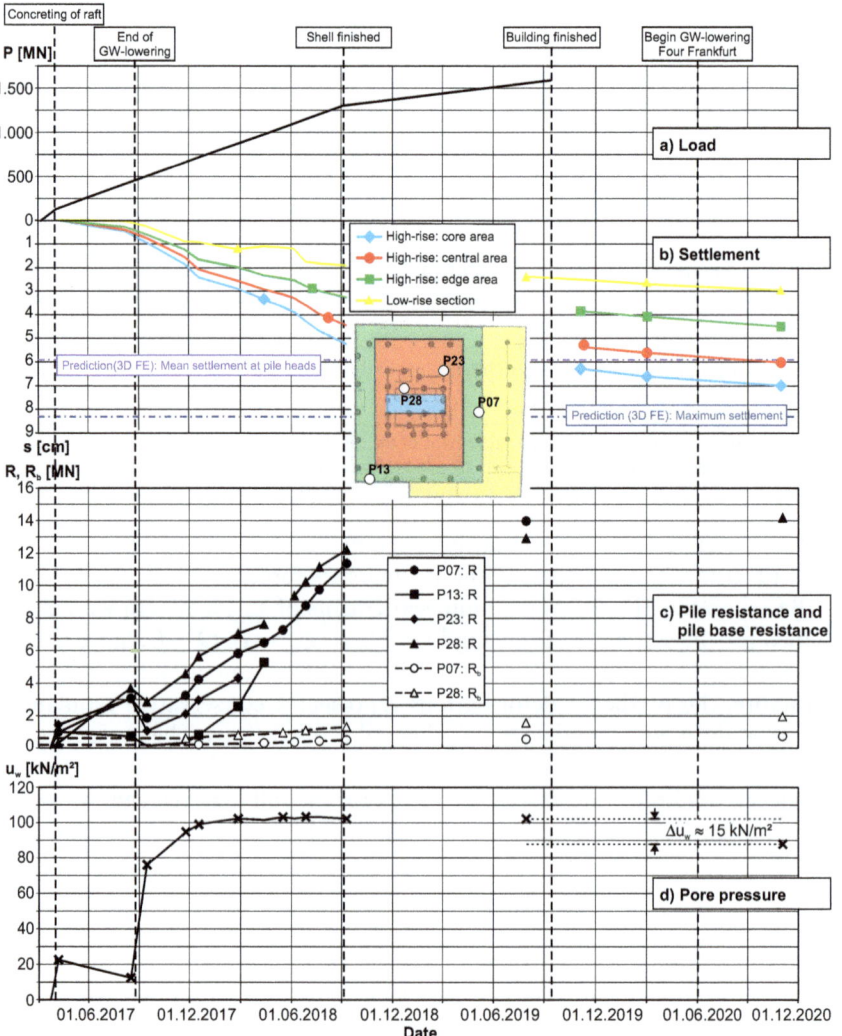

Figure 4.21 Omniturm: Variation of load, settlements, pile resistances, pile base resistances and pore pressure with time.

settlements over time in the different sections of the building complex, whereby no measured values are available for the first 10 months after completion of the shell. The immediate settlements from concreting the raft, which can be estimated at $s_{immediate} = 0.5$ cm to $s_{immediate} = 1.0$ cm based on experience with comparable foundations, are not included in the measured values. In November 2020, more than a year after commissioning of the building, the settlements ranged between $s = 3.0$ cm (area without superstructure) and $s = 7.0$ cm (core area of the high-rise section). The settlements predicted for the settlement inducing load (Table 4.6) will probably be reached in the long term, taking into account the time-dependent deformations to be expected in the Frankfurt Formation and the effect of reduced uplift due to the neighbouring construction site Four Frankfurt (Figure 4.18b).

Figure 4.18d shows the average pore pressure measured at the four pore pressure cells beneath the raft. With the start of the groundwater lowering in the Frankfurt Formation in June 2020 in the course of excavation works for the building complex Four Frankfurt, the pore pressure drops by $\Delta u_w \approx 15$ kN/m² until the last measurement in November 2020, causing a loss of uplift and a corresponding increase of the settlement inducing load of up to $\Delta P \approx 38$ MN. The concept for groundwater lowering for the building complex Four Frankfurt is discussed by Meissner et al. (2019).

The variation of pile resistances (for P07, P13, P23, P28) and pile base resistances (for P07, P28) with time are plotted in Figure 4.21c. The measurements available in August 2020 shortly before the building was finished showed pile resistances of $R = 14.0$ MN (P07; $L_p = 23.4$ m) and $R = 12.9$ MN (P28; $L_p = 27.4$ m). The significant increase of pile resistance to $R = 14.2$ MN observed for pile P28 at the last available measurement in November 2020 can probably be associated with the loss of uplift mentioned above. For both piles P07 and P28 it was possible to collect strain gauge data at the pile base until the end of the measurement campaign, indicating maximum pile base resistances of $R_b = 0.7$ MN (P07) and $R_b = 1.9$ MN (P28), respectively.

4.3.4 Neue Messehalle 3, Frankfurt am Main

4.3.4.1 Building

The exhibition hall Neue Messehalle 3 (Figure 4.22) was built from April 2000 to Juli 2001 at the trade fair centre in Frankfurt. With a length of 215 m, a width of 130 m and a maximum height of 45 m covering a total exhibition area of 39,000 m² on two levels, the Neue Messehalle 3 is one of the largest exhibition halls in Europe. Towards the east and west, the exhibition area is bounded by the east and west wings, respectively (Figure 4.23a). These wings are each divided into five freestanding, reinforced concrete structures, the so-called boxes. The roof with a free span of 165 m was designed as a double curved, three-dimensional load bearing structure consisting of five arched compression trusses and six arched tension trusses

Figure 4.22 Neue Messehalle 3, Frankfurt.

(Katzenbach and Turek 2004). On the eastern and western side of the building the roof rests on six so-called A-frames (Figure 4.22b), which are located between the boxes. The 24-m-high A-frames comprise two vertical steel tubes carrying vertical tension forces and two inclined steel tubes carrying compression forces (Figure 4.24a). The vertical loads are fully transferred to the A-frames, while the horizontal loads resulting from arch thrust in the main compression arches of the roof are split between the A-frame and the tension arches of the roof (Turek 2006). The load-bearing behaviour of the roof structure and the A-frames is governed to a large extent by the interaction between structure, foundation and subsoil. High flexibility of the foundation would lead to relief of the A-frames from the horizontal shear from the main compression arches and to a higher load on the tension arches of the roof structure. On the other hand, very stiff behaviour of the foundation would cause higher loading of the A-frames with a simultaneous relief of the tension arches (Turek 2006). A detailed discussion of the design and dimensioning of the roof structure of the Neue Messehalle 3 is provided by Meese and Ziesler (2002) and Stroetmann and Möll (2003).

The roof structure was partitioned into a total of 44 segments weighting a maximum of 130 tons, which were pre-assembled at an assembly yard. In the hall, the roof segments were then set down on temporary towers arranged at the quarter points of the roof arches. After welding the roof segments, the roof arches were lowered onto the A-frames using hydraulic presses. The installation and lowering of the roof arches took place over a period of five weeks, working from the southern edge of the hall to the north.

a) Ground plan of the building with measurement devices

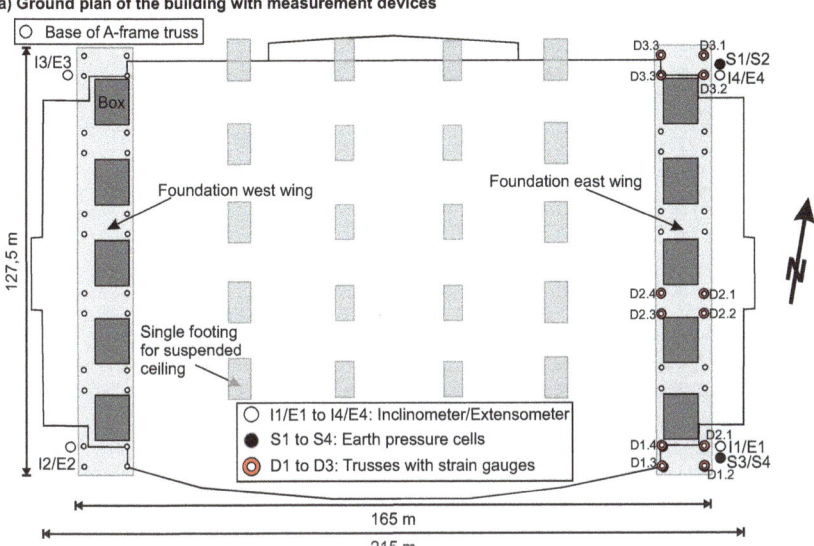

b) Front view of the building and cross section through the subsoil

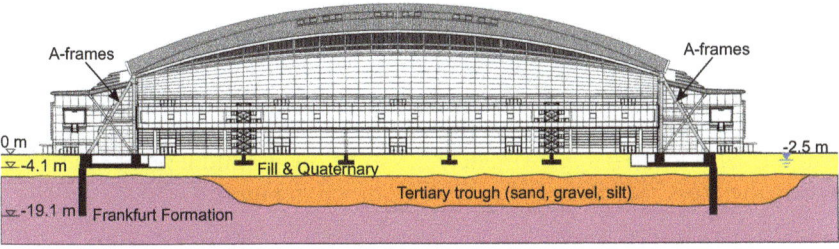

Figure 4.23 Neue Messehalle 3: Groundplan and front view of the building. (Adapted from Turek 2006.)

The lowering of the roof arches onto the A-frames mobilised a large part of the high vertical and horizontal loads on the foundations of the east and west wings (Turek 2006). The construction process of the Neue Messehalle 3 was documented in detail by Altner and Roth (2002).

4.3.4.2 Subsoil conditions

Below artificial fill with layer thickness of 0.5 m–4.0 m, Quaternary silty flood deposits of the river Main are present with layer thickness of up to 2.0 m. This is followed by Quaternary sands and gravels with layer thickness varying between 1.5 m and 5.0 m. The Quaternary layers are underlain by Tertiary sediments which at the site can be divided in to the Frankfurt Formation and in sands, gravels and silts located in a trough-like geological

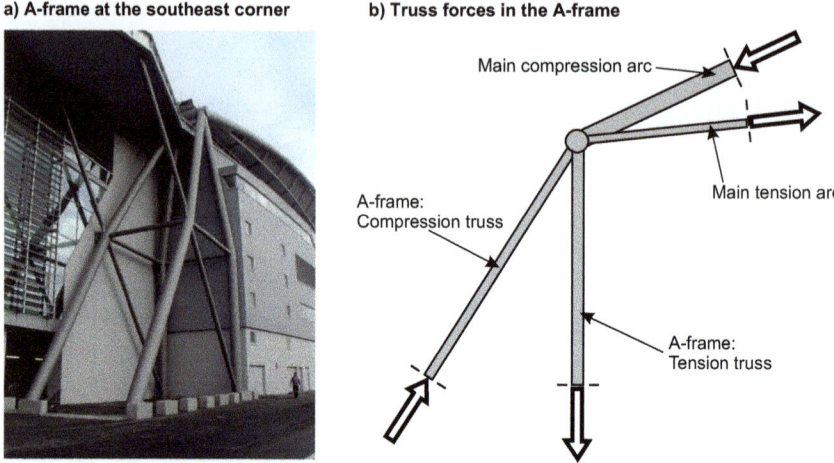

Figure 4.24 Neue Messehalle 3: A-frame as support for the roof. (Adapted from Turek 2006)

structure reaching to a depth of 16 m below ground surface. This Tertiary trough runs diagonally in a northwesterly-southeasterly direction through the building site. Especially in the area of the south-western corner of the exhibition hall, the Frankfurt Formation is already present at shallow depths exhibiting lower stiffnesses than the soil in the Tertiary trough. The groundwater level is located approximately 2.5 m below ground level at the site. A schematic cross section of the subsoil is shown in Figure 4.23b.

4.3.4.3 Foundation design

Since the settlements and horizontal deformations had to be within a range of a few centimetres due to their influence on the load-bearing behaviour of the roof structure and the A-frames, the east and west wings of the building were founded on a CPRF. The design of the CPRF was based on 3D FEA employing an elasto-plastic cap model for the Frankfurt Formation as well as for the Quaternary and Tertiary sands. The material parameters taken into account for the Frankfurt Formation and the Quaternary and Tertiary sand are documented, e.g. by Reul and Randolph (2003). In addition to distributed loads resulting from the boxes and the dead weight of the raft, concentrated loads resulting from the supports of the A-frames were applied in the analysis (Figure 4.25a). As a result of an extensive parametric study for the east and the west wing, two CPRFs each comprising a 127.5 m × 22.3 m raft (t_r = 1.4 m) and 14 bored piles (L_p = 15 m, d_p = 1.5 m) were chosen for the foundation. For the proof of the transfer of the horizontal loads to the subsoil, the ultimate lateral capacity of the piles did not need to be considered; i.e. the lateral capacity of the raft alone was sufficient.

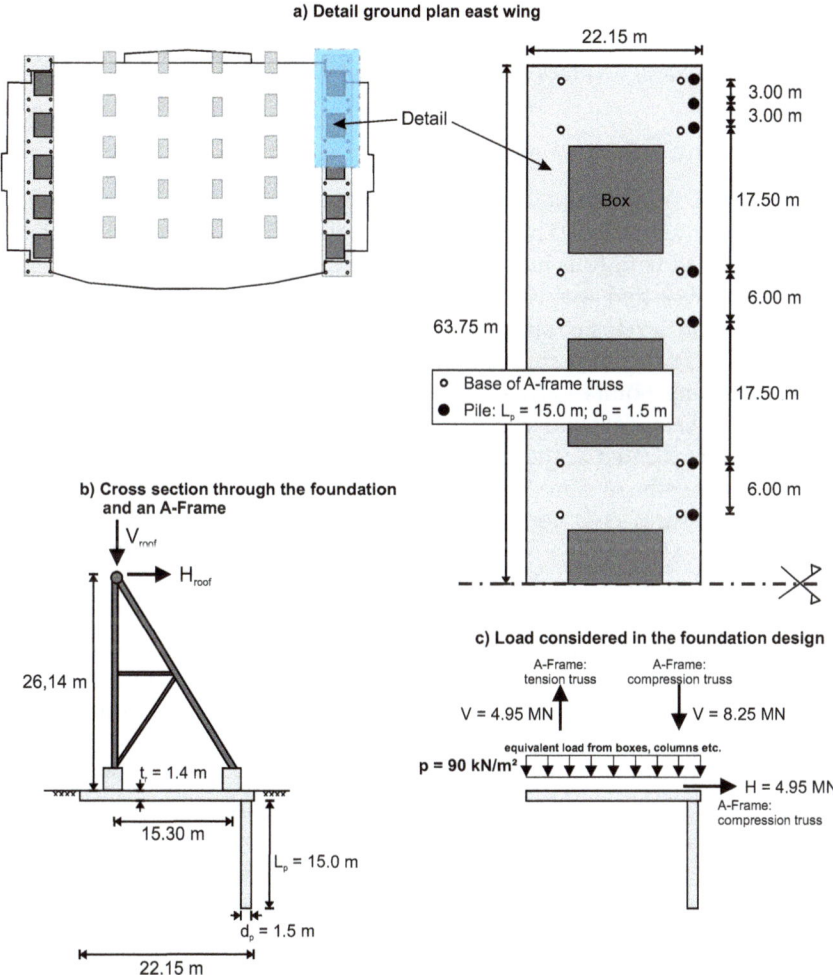

Figure 4.25 Neue Messehalle 3: Foundation for the A-frames. (Adapted from Turek 2006)

Figure 4.25b shows the ground plan of the raft of the east wing with the piles located in a line at the edge of the raft. According to Katzenbach et al. (2002), the predicted settlements and horizontal displacements of the CPRF amounted to $s_{max} = 35$ mm and $u_{max} = 30$ mm for this configuration. However, the inhomogeneities of the subsoil at the site imposed significant challenges on the accurate prediction of deformations. Therefore, the joints between the column base of the vertical tension trusses of the A-frames and the raft were designed such that a correction of vertical displacements was possible by means of hydraulic jacks (Katzenbach and Turek 2004). After the construction of the roof and the deformations due to consolidation processes were assumed to be completed, a correction of vertical displacements of approximately 2 cm–3 cm was carried out, leading to forces in the structure similar to the design values (Katzenbach and Turek 2004).

4.3.4.4 In situ measurements

Figure 4.23a shows the locations of measurement devices in the ground plan of the foundation. The horizontal displacements of the subsoil have been measured with four inclinometers I1, I2, I3 and I4 located outside the rafts of the east and west wing, respectively, reaching to a depth of 50 m below ground level, i.e. approximately 30 m below pile base. The inclinometers were installed together with extensometers E1, E2, E3 and E4, with anchoring points located in depth of 10 m, 20 m, 35 m and 50 m, respectively, below ground level. Close to the boreholes for the Inclinometer/Extensometer located at the east wing four sets of earth pressure cells were installed at depths of 5 m (S2, S3) and 10 m (S1, S4) respectively. Each set of earth pressures cells comprises three hydraulic pressure pads installed vertically to measure the horizontal stresses in three sections with area normals pivoted at 45° to each other. Strain gauges were applied to a total of twelve trusses of three A-frames of the east wing (D1.1 to D3.4) to derive the changes in the normal forces acting in the trusses from the axial strains.

Figure 4.26 shows the last available extensometer and inclinometer measurements for each position. The significantly larger vertical and horizontal displacements of extensometer E2 (Figure 4.26a) and inclinometer I2 (Figure 4.26b, at least at the inclinometer head) are due to the fact that the Tertiary trough, with its stiffer sands and gravels compared to the Frankfurt Formation, is essentially absent there.

With reference to the beginning of the measurements, absolute values of the horizontal displacements of a maximum of $u = 8$ mm (I2) and $u = 5$ mm (I1, I4) were recorded (Figure 4.26b). The value of $u = 3$ mm for inclinometer I3 was determined only three months after the roof segments were

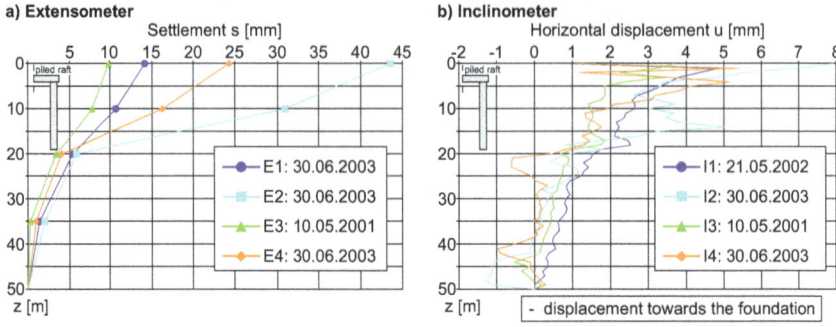

Figure 4.26 Neue Messehalle 3: Extensometer and inclinometer measurements. (Adapted from Turek 2006)

lowered onto the A-frames and is therefore not representative. Turek (2006) reports that the displacement increments that were reached from the beginning of the lowering of the roof segments to the last measurement date as a result of arc forces in the structure were between $\Delta u = 5$ mm (I3) and $\Delta u = 10$ mm (I2) and thus at the lower end of the predicted range.

The variation of truss forces, settlements, horizontal displacement and changes in earth pressure with time is shown in Figure 4.27 for the measurement devices located in the northern part of the east wing. The influence of the lowering of the roof segment on the A-frame in late January 2001 is

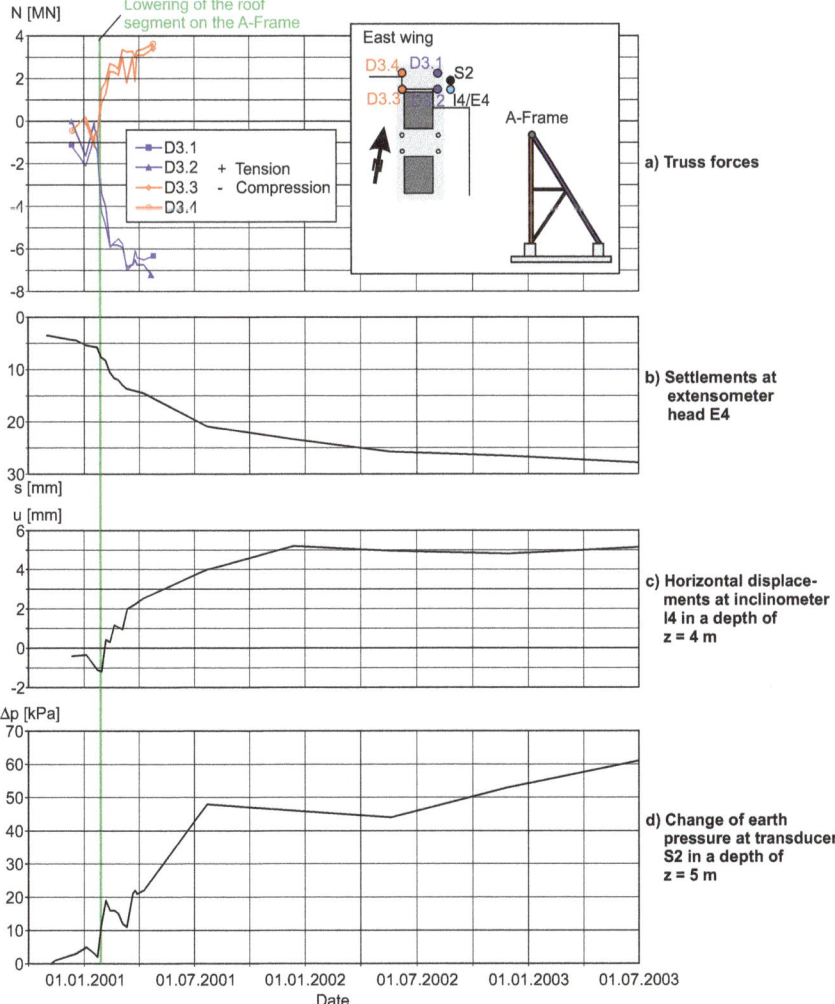

Figure 4.27 Neue Messehalle 3: Variation of truss forces, settlements, horizontal displacement and changes in earth pressure with time. (Adapted from Turek 2006)

clearly visible. Due to the significant increase of truss forces that are transferred to the subsoil by the foundation, the settlements, horizontal displacements and the earth pressures also markedly increase. A detailed summary and discussion of the in situ measurements carried out on the foundation of the Neue Messehalle 3 is provided by Turek (2006).

4.3.5 Haus der Wirtschaft, Offenbach

4.3.5.1 Building

The building complex Haus der Wirtschaft constructed from 1997 to 1999 in Offenbach, Germany, is made up of several building sections with maximum heights of 68 m, which are positioned on an approximately 5100 m² monolithic raft. The northern 18-storey high-rise tower (Building 4) overlaps the public street space from the second floor onwards, so that part of the high-rise loads there are transferred via three columns in the sidewalk area (Figures 4.28a and 4.29b). The southern 16-storey high-rise tower (Building 2) and the northern high-rise tower (Building 4) are separated from each other by an inner courtyard and surrounded by a 2- to 6-storey peripheral development (Building 1, Building 3 and Building 5) (Figure 4.28b). The structure has a two-storey basement across its entire footprint. The load of the reinforced concrete superstructure of the Haus der Wirtschaft amounts to approximately $P_s = G + Q = 589$ MN $+ 291$ MN $= 880$ MN (G: dead loads; Q: live loads).

The railway tunnel of the Offenbach city line runs beneath the Berliner Strasse, in the immediate vicinity of the northern high-rise building (Figure 4.28b). This settlement-sensitive underground construction, which is permanently and heavily frequented by the double-track train operation, was built in the years 1990 to 1993 at a depth of 14 m below ground level, using the cut-and-cover method (Schultz 1994, Zabel et al. 1994).

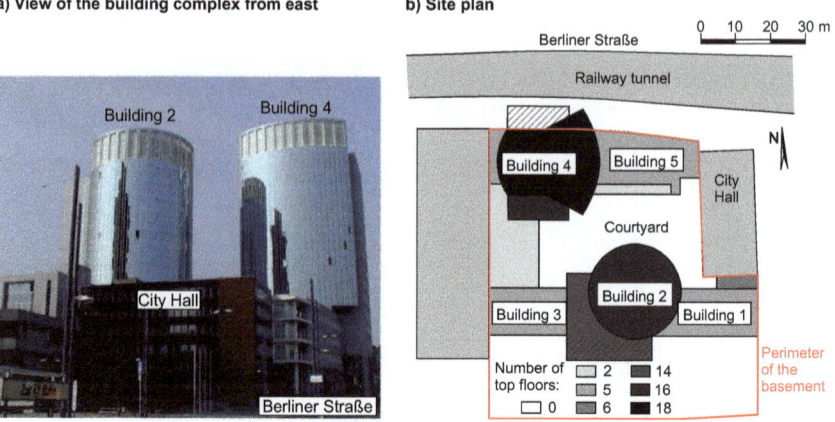

Figure 4.28 Haus der Wirtschaft, Offenbach.

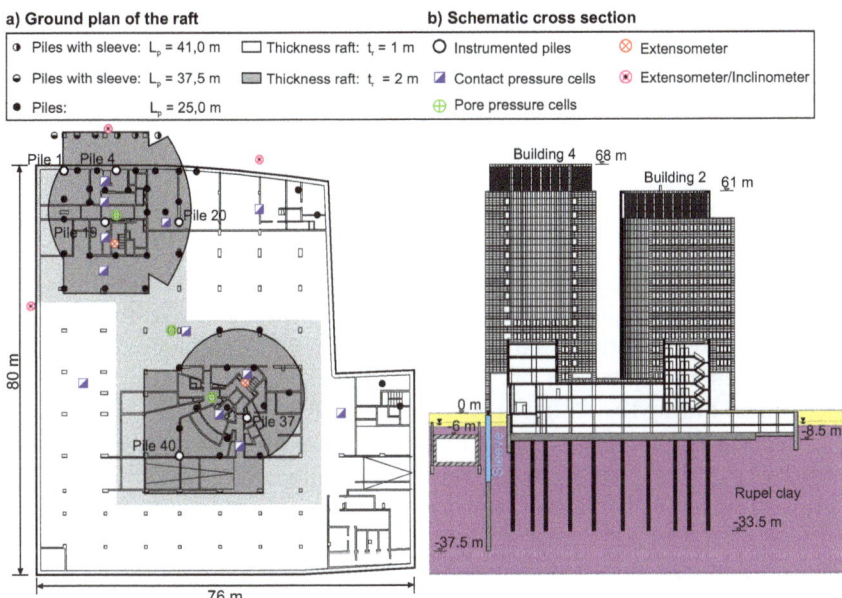

Figure 4.29 Haus der Wirtschaft: Foundation.

4.3.5.2 Subsoil conditions

In the area of the Haus der Wirtschaft, beneath up to 1.4-m-thick artificial fill, Quaternary sands and gravels (also in part Quaternary) there are alluvial deposits present down to a depth of 3.5 m–4.0 m below ground level. Below this, the Tertiary Rupel clay follows to at least 70 m below ground level and is underlain by the Rotliegend Group.

The Rupel clay can be classified as a stiff, highly plastic clay. Table 4.7 summarises typical properties derived from samples that were obtained from depths of up to 60 m below ground level as part of the subsoil investigation for various construction projects in Offenbach. Only very thin layers of sand are embedded in the Rupel clay, which is therefore much more

Table 4.7 Properties of the Rupel clay

Parameter			Min	Max	Mean
Liquid limit	w_L	[%]	47	73	58
Plasticity index	I_P	[%]	27	46	35
Water content	w	[%]	18	30	24
Consistency index	I_C	[-]	0.87	1.15	0.97
Unit weight	γ	[kN/m³]	19.7	21.3	20.5
Undrained shear strength	s_u	[kPa]	99	988	400

uniform than the Frankfurt Formation. The mean undrained shear strength of $s_u = 400$ kN/m² is significantly higher than the value of $s_u = 150$ kN/m² given by Breth (1970) for the Frankfurt Formation.

The groundwater is located approximately 3 m below ground level and flows in the Quaternary from south to north towards the river Main.

4.3.5.3 Foundation design

Particularly demanding requirements resulted for the design of the high-rise foundation due to the close proximity to the settlement-sensitive railway tunnel, whose operational safety must be ensured by complying with a permissible settlement of $s = 3$ cm. Katzenbach et al. (1998b) report on the FEA performed during the design process of the CPRF.

The monolithic raft ($t_r = 1.0$ m to $t_r = 2.0$ m), with its bottom located at a depth of 7.5 m–8.5 m below ground level, is subjected to significant eccentric loading by the rising structure. By means of 47 bored piles with a length of $L_p = 25$ m and a diameter of $d_p = 1.2$ m, which are mainly located under the two high-rise towers (Figure 4.29a), the reaction forces of the CPRF are centred under the resulting building load. The part of the northern high-rise building (Building 4) projecting into the sidewalk area of Berliner Straße is founded with a conventional pile foundation on six large-diameter drilled piles with lengths of $L_p = 37.5$ m and $L_p = 41.0$ m, respectively, and a diameter of $d_p = 1.2$ m to minimise the impact on the neighbouring railway tunnel. Load transfer from the piles into the ground to a depth of 4 m below the tunnel base is prevented by means of a sleeve construction.

4.3.5.4 In situ measurements

To measure the pile resistances, four foundation piles in the area of Building 4 and two foundation piles in the area of Building 2 were equipped with load cells at the pile head. Contact and pore pressure under the raft were determined with twelve contact pressure cells and three pore pressure cells, which were installed under the two high-rise buildings as well as under the low-rise sections and in the area of the courtyard. A total of five multi-point borehole extensometers were used to measure the settlement distribution over the depth, with three located either outside the excavation pit or in the retaining wall. The three extensometers located outside the building outline were combined with inclinometers to also measure horizontal displacements. The extensometers installed beneath Building 2 and Building 4 comprised anchoring points located at depth of 12 m, 25 m, 40 m and 60 m (Building 4 only) below the foundation base. Additionally, geodetic measurements were carried out during construction to monitor the settlements. For this purpose, measuring points were arranged on the raft, distributed over the entire footprint of the building complex as well as in the neighbouring railway tunnel. The location of the geotechnical measuring devices is shown in Figure 4.29a.

Figure 4.30 shows the variation of load and settlements over time. The geodetic settlement measurements were taken at the extensometer heads located at the centre of Building 2 and Building 4 and in the railway tunnel (maximum settlements in vicinity of the Haus der Wirtschaft). The measurements are related to the point in time prior to concreting the raft. A detailed description of the construction process and the development of building loads is given by Reul (2000). The development of the settlements under Building 2 and Building 4 over time show a similar course, which is characterised by the largest increase in settlement up to the end of construction of the building shell. The settlement of Building 2 follows the settlement of Building 4 with a time lag due to the later concreting of the raft. In November 1999, after completion of the building complex, the settlements amounted

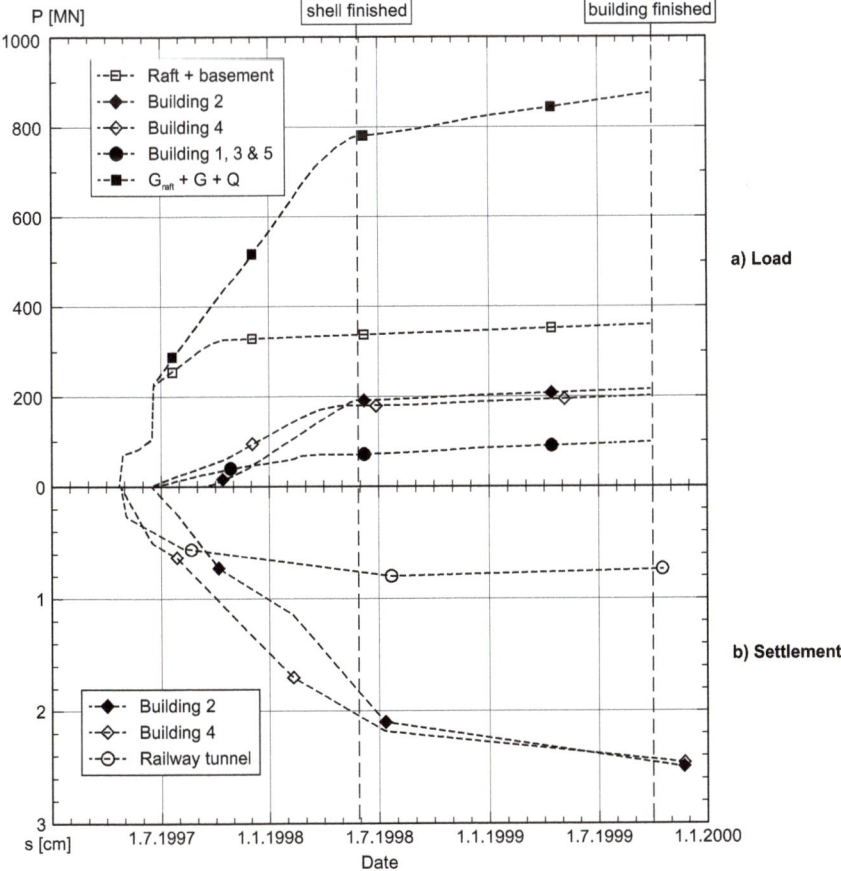

Figure 4.30 Haus der Wirtschaft: Variation of load and settlement with time.

to $s = 2.5$ cm in both Building 2 and Building 4. At $s = 0.8$ cm, the maximum settlement of the railway tunnel remains well below the permissible value of $s = 3.0$ cm for maintaining the operational safety of rail traffic. After completion of the shell, a slight decrease in the maximum tunnel settlement can be identified.

Contour lines of settlements are shown in Figure 4.31 related to the time after casting the raft; i.e. the immediate settlements resulting from the self-weight of the raft are not considered. The maximum settlements can be observed under the two high-rise sections both when the shell is finished (13.07.1998) and after the completion of the structure (24.11.1999). While the settlements under the two high-rise buildings increase between the end of the shell construction and the completion of the structure, significant heave occurs under the foundation areas subjected to lower loads. In November 1999, the heave at the south-east corner of the raft where no upper floors are present amounted to 3 cm.

The assessment of the groundwater level in the vicinity of the building complex showed no significant change in the groundwater level. The hydrostatic pore water pressure on the raft ($t_r = 1.0$ m in the south-east corner of the building complex) amounted to $u_w = 48$ kN/m². The dead weight of the raft and the two basement floors results in a total contact pressure of approximately $\sigma = 51$ kN/m², so that the measured heave cannot be attributed to the effect of buoyancy.

The maximum angular distortion of the raft resulting from the settlement trough shown in Figure 4.31 amounts to approximately $\alpha = 1/1000$ and does not compromise the serviceability of the structure. The significant heave under the low-loaded areas in the south-east corner can possibly be

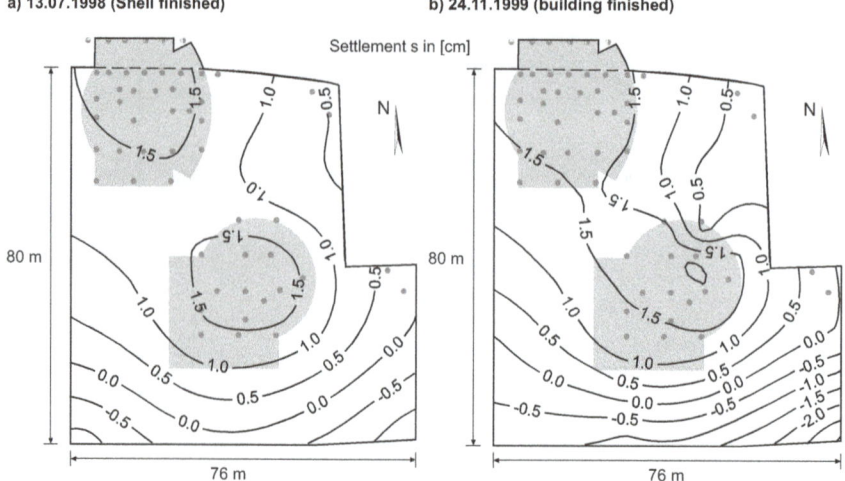

Figure 4.31 Haus der Wirtschaft: Contour lines of settlement (Settlements related to the installation of the raft).

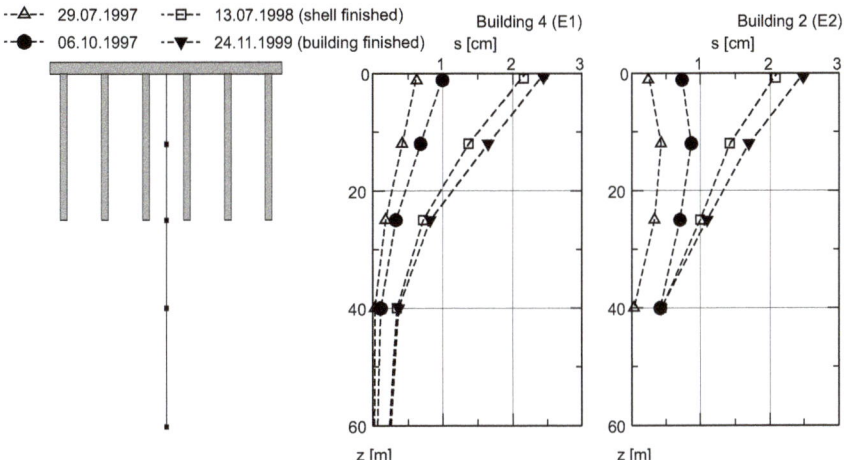

Figure 4.32 Haus der Wirtschaft: Settlement profile.

attributed to the development of swelling pressures in the Offenbach Rupel clay. Mader (1989), for example, reported swelling pressures of $\sigma = 17$ kN/m² to $\sigma = 109$ kN/m² for clay samples from the Frankfurt Formation.

The settlement profile over the depth is shown in Figure 4.32 for the extensometers located underneath Building 2 and Building 4. For the measurements after completion of the shell (13.07.1998) and after completion of the building (24.11.1999), the two extensometers show comparable settlement distributions. Forty metres below the raft, 14–18% of the settlements at the foundation base are observed at both extensometers. At the extensometer beneath Building 4, the settlements at a depth of 60 m are still 9% and 12% of the settlements at the foundation base for these construction stages. At the time of the first two measurements on July 29, 1997 and on October 6, 1997, the settlements at the extensometer beneath Building 2 are larger at a depth of 12 m than the settlements at the foundation base. This is due to the stress changes in the subsoil caused by the preceding construction of Building 4 and the resulting superposition of the stresses caused by the construction of Building 4 (loading) and the excavation in the area of Building 2 (unloading).

The variations of pile resistances with time are plotted in Figure 4.33a. The concreting of the raft (Building 4: 28. & 29.04.1997; Building 2: 18. & 19.06.1997) caused pile resistances that on average correspond to the self-weight of the concrete above the pile head.

The maximum pile resistance during the observation period amounted to $R = 4.1$ MN (Pile 1, Building 4) in April 1999. At the time of the last measurements available in November 1999, the maximum and minimum pile resistances amounted to $R = 3.1$ MN (Pile 1 and Pile 4, Building 4) and $R = 1.4$ MN (Pile 40, Building 2).

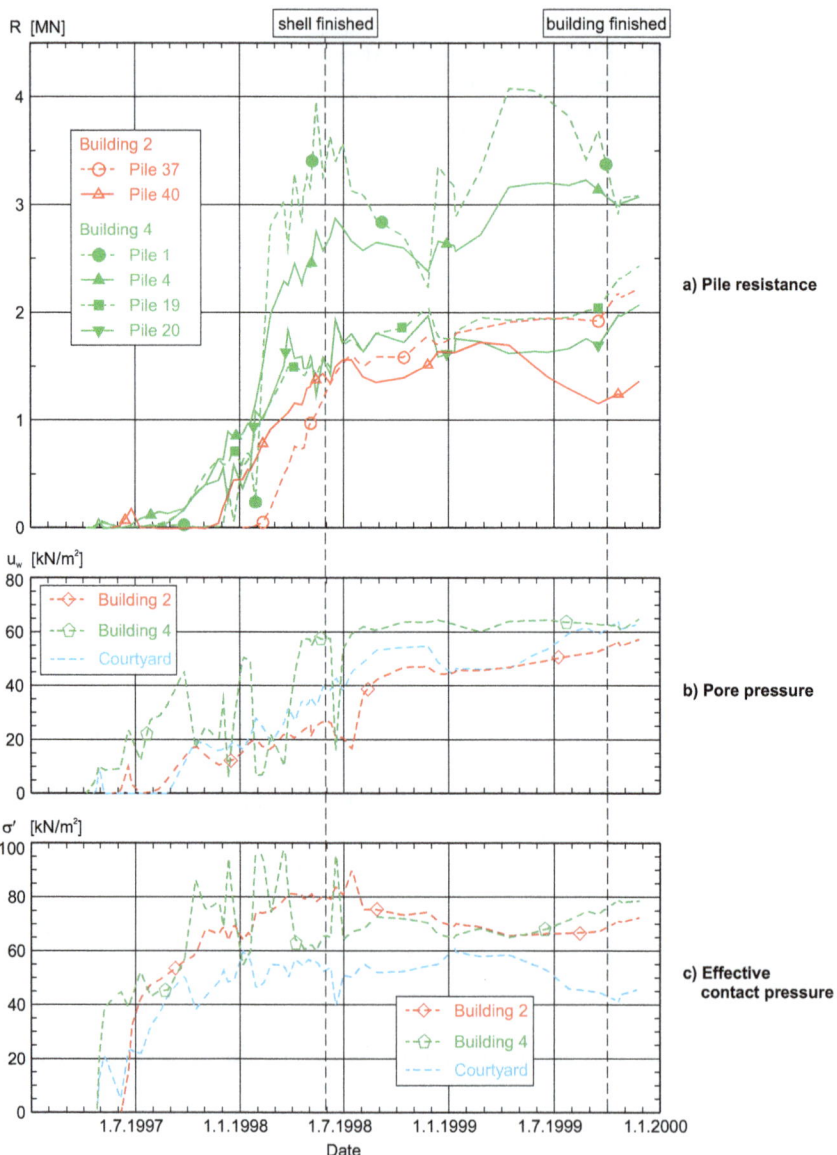

Figure 4.33 Haus der Wirtschaft: Variation of pile resistance, pore pressure and effective contact pressure with time.

For the instrumented piles under Building 4, there are several redistributions of the pile resistances during the observation period, as can be recognised distinctly, for example, in November 1998 and in October 1999. With a decrease of the pile resistances of Pile 1 and Pile 4, which are located at the border to the Berliner Straße, an increase of the pile resistances of Pile 19 and Pile 20 can be observed. Holzhäuser (1998) observed similar redistributions of the pile resistances at the foundation of the Commerzbank II high-rise building and suggested that the cause for the redistribution was the change in load transfer in the contact zone between the pile and the soil. According to this hypothesis, the development of shaft friction, the relative displacement between pile and soil, and the deformation of the soil do not represent a continuous process, but occur in stages.

In November 1999, at the time of the last available measurement, the pore pressures were u_w = 57 kN/m² (Building 2), u_w = 65 kN/m² (Building 4) and u_w = 63 kN/m² under the courtyard (Figure 4.33b). Since the pore pressure cells are located in the area with a 2-m-thick raft approximately 5.8 m below the groundwater level, a hydrostatic pore pressure of $u_w \approx 58$ kN/m² can be expected. The development of pore pressure under Building 4 shows significant fluctuations of up to Δu_w = 42 kN/m² until the shell is finished. However, a correlation with the changes in the pile resistances (Figure 4.33a) is not evident from the available data.

The effective contact stresses σ' in Figure 4.33c may be determined from the measured total contact stresses σ and the pore water pressures u_w using the principle of effective stresses:

$$\sigma' = \sigma - u_w \tag{4.1}$$

In November 1999, the effective contact stresses were σ' = 73 kN/m² (Building 2), σ' = 79 kN/m² (Building 4) and σ' = 46 kN/m² (Courtyard). The main increase in effective contact stresses occurred beneath all building parts essentially by January 1998. The fluctuations of the effective contact stresses under Building 4 are due to the already mentioned fluctuations of the pore pressure.

4.3.6 Bahn Tower, Berlin

4.3.6.1 Building

The 103-m-high BahnTower (Figure 4.34a) is located at the eastern periphery of the multifunctional building complex Sony Center at the Potsdamer Platz in Berlin and was constructed from 1997 to 1999. The reinforced concrete high-rise building is directly connected to a neighbouring regional train station via its two underground levels. The office tower houses the headquarters of the German national railway company Deutsche Bahn AG within its 26 storeys (floor space of 22,000 m²).

4.3.6.2 Subsoil conditions

Berlin is located in the upper moraine landscape of the North German-Polish Basin, whose deposits were formed during the Weichselian glaciation.

186 Combined Pile-Raft Foundations

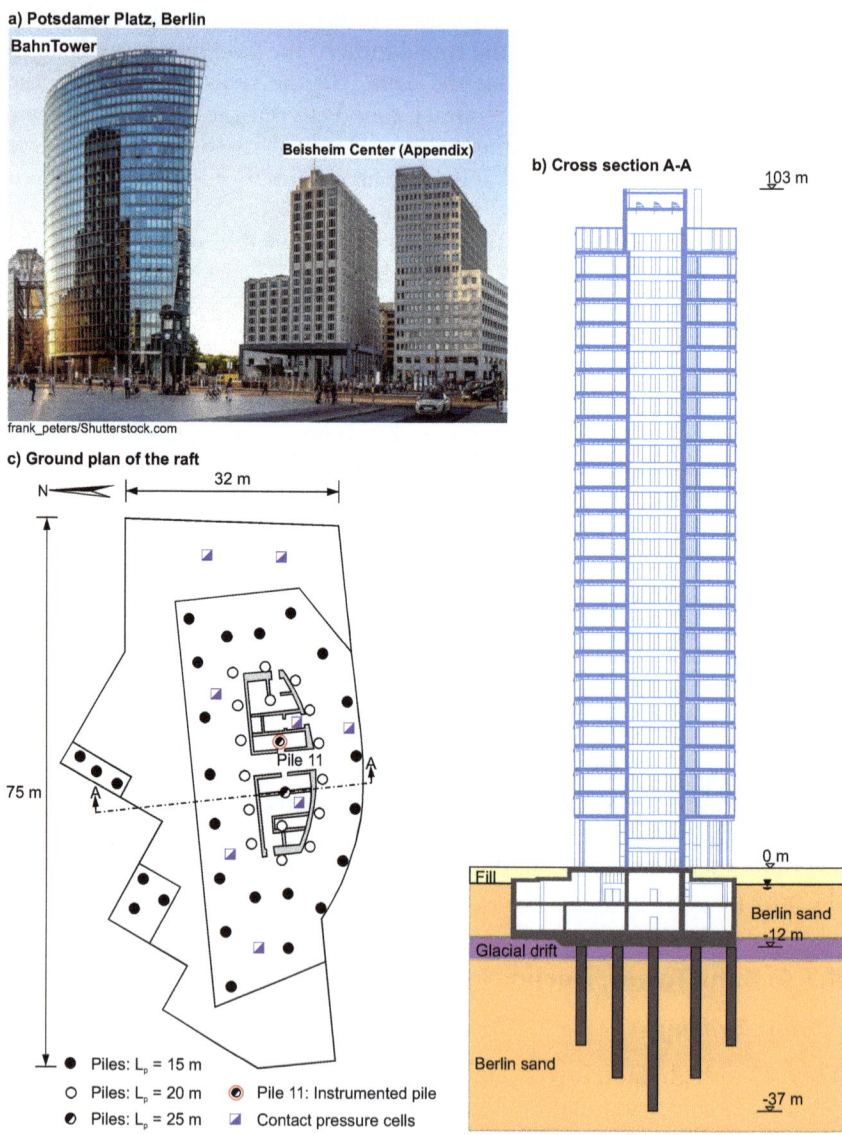

Figure 4.34 BahnTower, Berlin: Cross section of the building and ground plan of the raft.

Underneath lie further glacial deposits of the Saale and Elster glaciation, generally compressed by subsequent ice advances and also eroded in places. An overview of the material behaviour of soil layers relevant to construction practice in Berlin, i.e. mainly sand, gravel and glacial drift, can be found, for example, in Weiß (1978).

Underneath fill with a mean thickness of approximately 3 m, medium dense sands are present at the site down to a depth of approximately 11 m. These are underlain by a 2-m- to 5-m-thick layer of glacial drift, which has a stiff to semi-solid, locally solid consistency. Beneath the layer of glacial drift, medium-dense to dense sands are present. The groundwater level is located approximately 3 m below ground level.

4.3.6.3 Foundation design

In addition to the proximity to the regional train station, the determining factor for the design of the high-rise building's foundation with its ground area of 2600 m² as a CPRF was the glacial drift layer located directly beneath the raft (Figure 4.34a). A total of 44 bored piles with a diameter of $d_p = 1.5$ m and a length between $L_p = 15$ m and $L_p = 25$ m are arranged under the 1.5 m to 2.5 m-thick raft (Figure 4.34b). During the construction phase, the bored piles were also used to limit any adverse effects from uplift of the structure. The location of the foundation base up to 9 m below groundwater level resulted in an uplift of approximately $U \approx 225$ MN. The total load of the BahnTower amounts to approximately $P = G_{raft} + G + Q \approx 880$ MN.

At an early stage of the project a load test was carried out on a bored pile at the construction site. The test pile ($L_p = 23$ m; $d_p = 1.5$ m) was provided with a sleeve excluding shaft friction on the upper 11.6 m (Katzenbach and Moormann 1997). Figure 4.35 shows the pile configuration and the resistance settlement curves derived from the load test. The foundation design

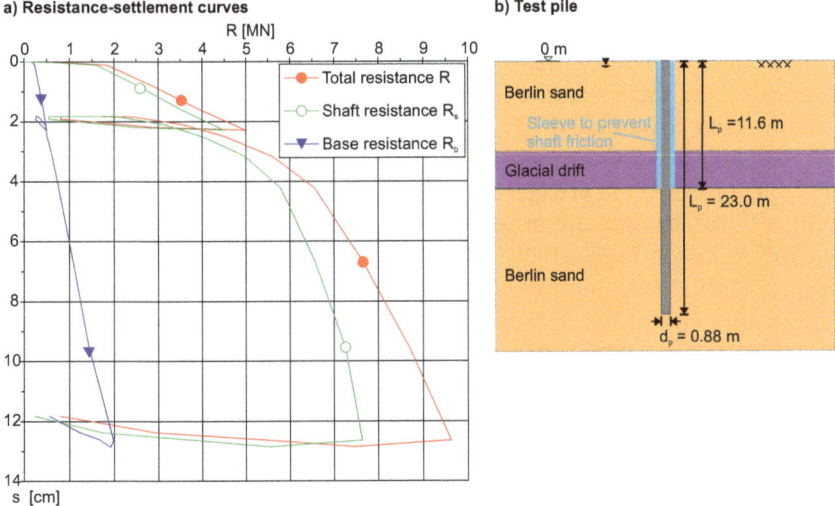

Figure 4.35 BahnTower: Pile load test. (Adapted from Katzenbach and Moormann 1997)

188 Combined Pile-Raft Foundations

was based on 3D FEA with the model calibrated by means of the back-analysis of the pile load tests (Richter et al. 1998).

4.3.6.4 In situ measurements

The vertical displacements of the structure were measured at several measurement points starting after concreting of the raft; i.e. the weight of the raft was not included. Figure 4.34b shows the location of the geotechnical measuring devices in the ground plan. The total contact stresses under the raft were determined from eight contact pressure cells. Measuring devices for determining the pile resistance were initially installed in four piles.

Since the piles were also required to secure safety against uplift of the structure during construction, the pile reinforcement also had to be passed into the raft from the instrumented piles. For this reason, the use of load cells at the pile head was not possible. It was therefore planned to determine the pile resistances from the measurement of axial strains with the aid of displacement transducers inserted in boreholes in the pile heads. If the Young's modulus of the pile concrete and the reinforcement ratio are known, the axial pile force can be determined neglecting the stiffness of the displacement transducer. However, since three of the four installed displacement transducers failed immediately after installation, only the measurements on Pile 11 are available. The Young's modulus of the concrete of Pile 11 amounts to $E_{concrete}$ = 20,500 MN/m². With a reinforcement ratio of approximately 0.7%, the modulus of elasticity of the reinforced concrete cross section at the pile head of Pile 11 can be established as E_{pile} = 21,800 MN/m². The average Young's modulus of all four piles was $E_{concrete}$ = 19,800 MN/m² and is thus well below the theoretical value for a concrete of the now replaced strength class B25 of $E_{concrete}$ = 30,000 MN/m².

The measurements are presented in Figure 4.36 together with the development of the total building load (Figure 4.36a). When the structure was completed in November 1999, the settlement in the core area of the highrise building was s = 2.8 cm (Figure 4.36b) and thus remained below the settlement of s = 5 cm predicted in the course of the design process (Richter et al. 1996). The decrease in settlement in July and August 1998 and from December 1998 to July 1999 is particularly noticeable. This may be due to an increase in the groundwater level, the development of which is not known to the authors, and the resulting decrease in the effective load of the structure. Also, some deterioration of the height benchmarks cannot be ruled out because of the intensive construction activity in the area of Potsdamer Platz during that time.

Measurements of pile resistance and total contact pressure are available until the shell of the building was finished in October 1998. The resistance of the 25 m-long Pile 11 amounts to R = 15 MN at this construction stage (Figure 4.36d). No change in the curve can be seen in July and August 1998, during the first decrease in settlement. The same is true for the total contact

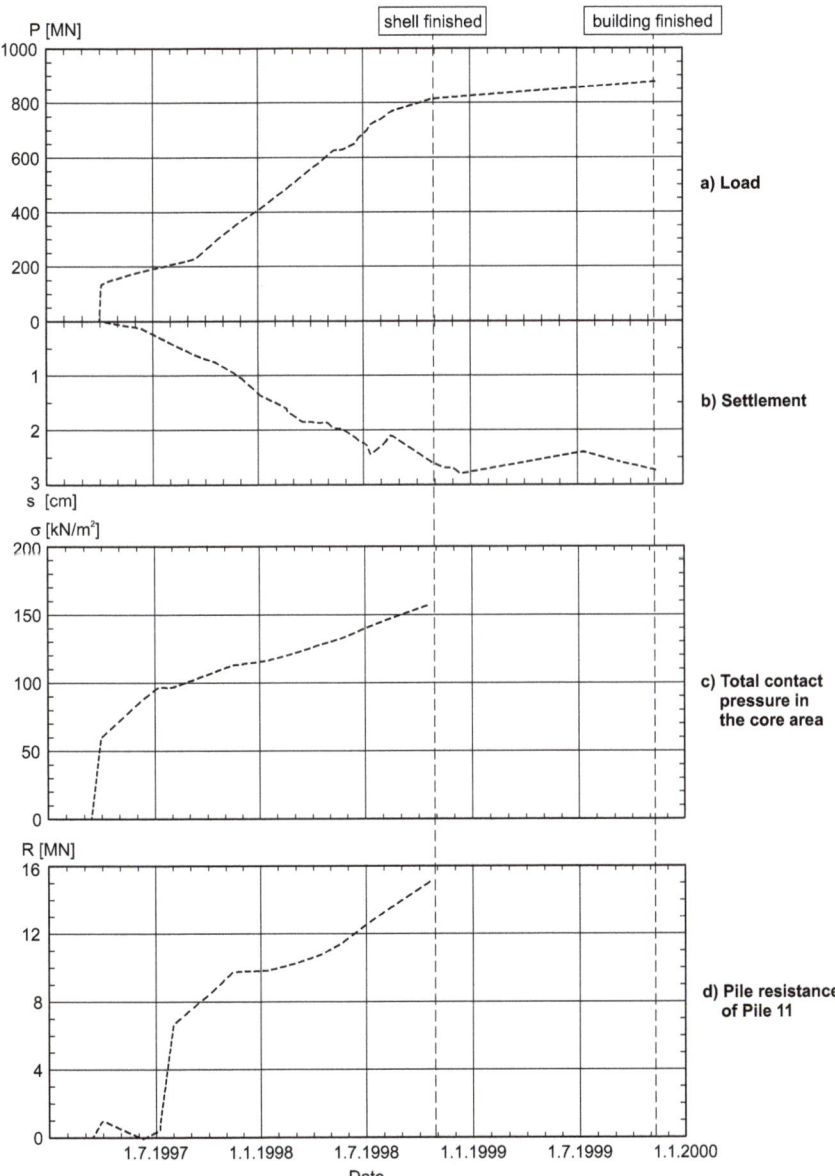

Figure 4.36 Bahn Tower: Variation of load, settlement, total contact pressure and pile resistance with time.

stresses in the core area of the office tower (Figure 4.36c), which was determined from mean values of the two contact pressure cells located at the central part of the building. When the shell of the building was finished, the total contact stress amounted to $\sigma = 157$ kN/m².

4.3.7 Treptowers, Berlin

4.3.7.1 Building

The 121-m-high Treptowers high-rise building with 32 floors, erected on the site of a former electrical appliance factory of the state-owned VEB Kombinat Elektro-Anlagen-Werke (EAW) in Berlin-Treptow, is located in the immediate vicinity of the river Spree (Figure 4.37). The high-rise building and the adjacent 10-storey building complex, with a total of approximately 165,000 m² of office space, were built on behalf of the Allianz-Versicherungs AG for the company's Berlin headquarters.

The load of the superstructure of the Treptowers amounts to approximately P_s = 662 MN and is transferred to the raft mainly in the approximately 17 m × 17 m core area via walls and at the edge of the building via columns.

4.3.7.2 Subsoil conditions

The subsoil conditions in the vicinity of the Treptowers are characterised mainly by quaternary sediments. On the first 4 m below ground level, the subsoil consists of fill and organic soils which are followed up to a depth of 35 m below ground level by loose sand and medium dense to dense sands of various grain size distributions (Figure 4.38). The loose sand can be found down to a maximum depth of 10 m below ground level, while the dense sands are located from a depth of 22 m below ground level onwards. A layer

Sinuswelle/Shutterstock.com

Figure 4.37 Building complex Treptowers.

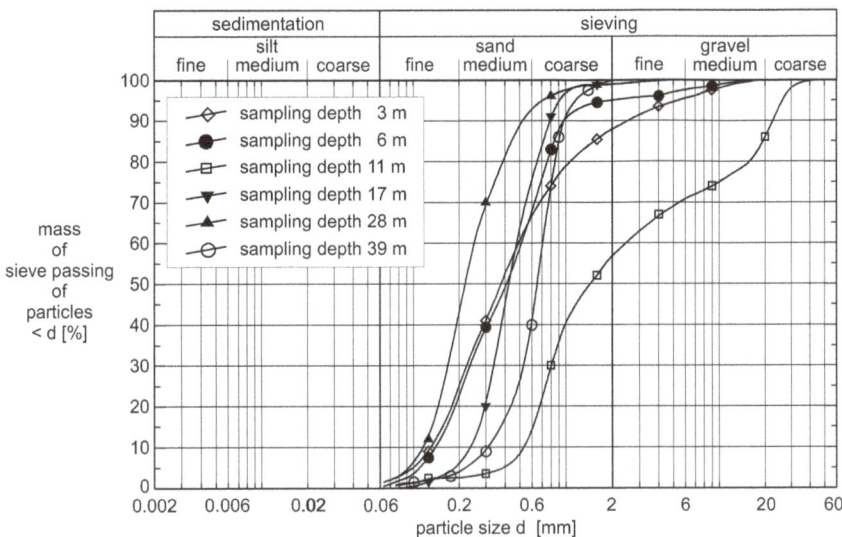

Figure 4.38 Treptowers: Grain size distribution of Berlin sand at the site.

of glacial drift, with a thickness of up to 1.5 m, is situated at a depth of 35 m. The glacial drift has stiff consistency and medium plasticity and is underlain again by dense sand. The groundwater level is located approximately 3 m–4 m below ground level with the groundwater flowing with a small hydraulic gradient in the direction of the Spree.

4.3.7.3 Foundation design

Arslan et al. (1997) and Katzenbach et al. (1998a) reported on the design process of the CPRF based on 3D FEA. With the CPRF a part of the load of the high-rise building is transferred to the medium dense to dense sands. A total of 54 piles with a diameter of $d_p = 0.88$ m and length between $L_p = 12.5$ and $L_p = 16$ m are located under the raft with a thickness of $t_r = 2$ m to $t_r = 3$ m (Figure 4.39a and b). To improve shaft friction between the pile shaft and the surrounding sand, shaft grouting was performed. The foundation level lies 8 m below ground level (5 m below groundwater level) in the area of the elevator pit and 4.5 m below ground level in the remaining part of the building.

4.3.7.4 In situ measurements

The settlements of the structure were registered by geodetic measurements at four measuring points at the corners of the structure. Pile load measurements are available for two out of the three instrumented piles, No. 1 ($L_p = 16.0$ m) and No. 17 ($L_p = 16.0$ m), both of which were equipped with load cells at the

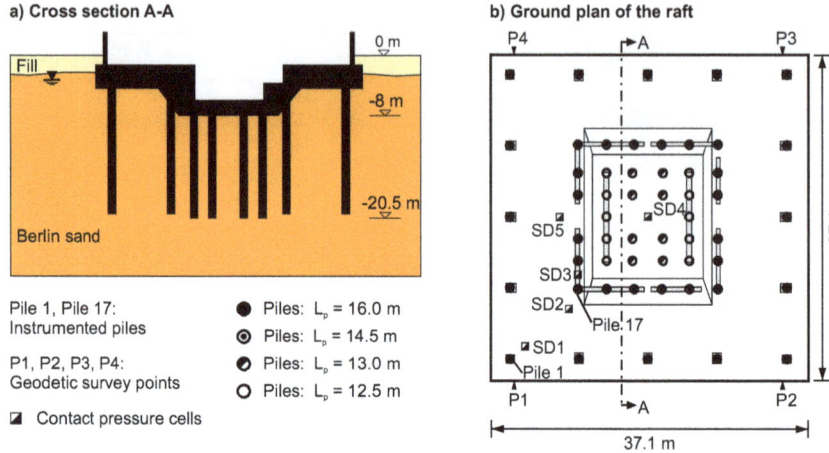

Figure 4.39 **Treptowers: Foundation.**

pile head and strain gauges along the pile shaft which were arranged in pairs at five measuring planes. Moreover, five contact pressure cells were installed beneath the raft. Figure 4.39b shows the location of geotechnical measuring devices and geodetic measuring points in the ground plan of the raft.

Figure 4.40 summarises the variation of load, settlement and pile resistance over time. The settlements of the high-rise building were measured regularly after completion of the second basement in January 1996 (Figure 4.40b). The largest increases in settlement were recorded from July 1996 to January 1997, analogous to the load development (Figure 4.40a). The maximum settlement measured at the completion of the structure in March 1998 was $s_{P4} = 7.3$ cm in the north-west corner of the structure. The minimum settlement of the structure was measured in the south-east corner of the structure at $s_{P2} = 5.0$ cm. The resulting maximum angular distortion of $\alpha < 1/2000$ means that the serviceability of the structure was not affected. The mean settlement after completion of the construction was $s_{mean} = 6.3$ cm.

The development of the measured pile resistances over time in Figure 4.40c shows the largest increase for Pile 1 as well as for Pile 17 from July 1996 to January 1997. The significantly larger pile resistance of Pile 17 compared to Pile 1 after concreting the raft is due to the location of the two piles within the pile group and the preliminary construction of the core area of the superstructure. While Pile 1 had a pile resistance of only $R_1 = 0.05$ MN, corresponding to the weight of the concrete in the area of the pile cross section due to the concreting process, the pile resistance of Pile 17 at that time was ten times greater at $R_{17} = 0.5$ MN. In July 1996, the pile resistances were $R_1 = 1.0$ MN and $R_{17} = 2.4$ MN. However, as construction progressed, the pile resistances became increasingly similar. At the last available measurement in October 1998, the pile resistances were $R_1 = 7.0$ MN and $R_{17} = 6.7$ MN,

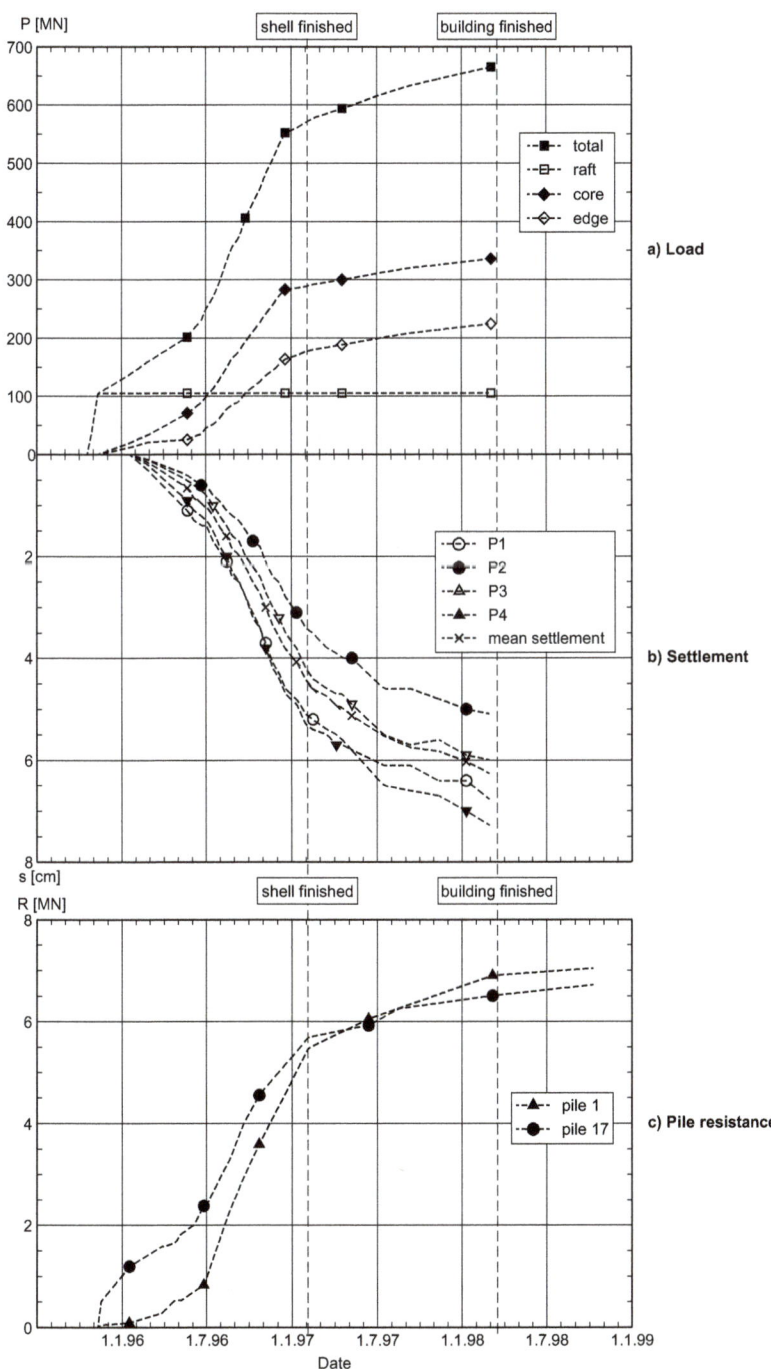

Figure 4.40 Treptowers: Variation of load, settlement and pile resistance with time.

with an increase in pile resistances of $\Delta R_1 = 0.1$ MN and $\Delta R_{17} = 0.2$ MN after completion of the structure in March 1998.

Figure 4.41 compares the distribution of pile load and shaft friction along the pile shaft for piles No. 1 and No. 17 for the construction stages after the

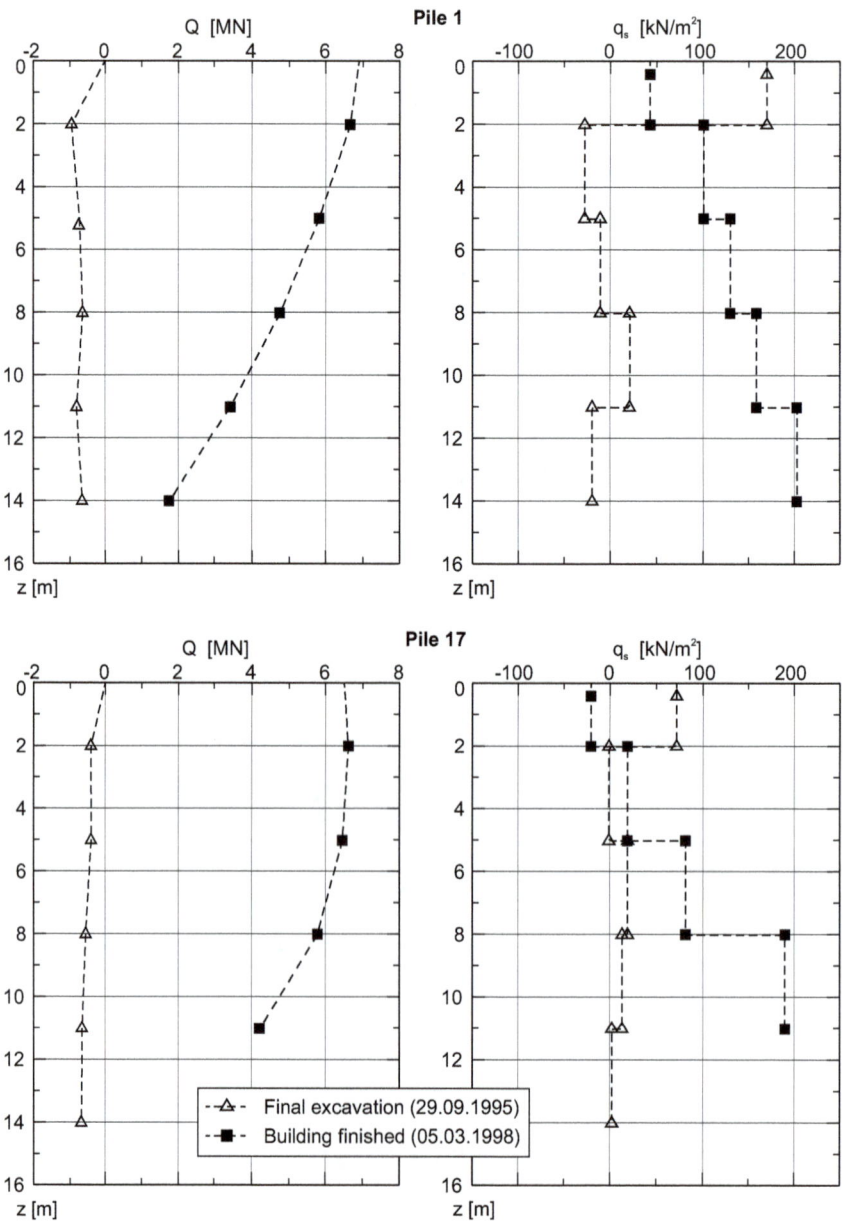

Figure 4.41 Treptowers: Distribution of pile load and shaft friction along the pile shaft.

final excavation (approximately 2 m below the installation level of the piles and 4 m below ground level) and after completion of the building. The pattern is similar in both cases. Initially, after the final excavation, negative friction was induced in the lower part of each pile with resulting tensile loads in the piles. These reduced gradually as the external loading caused by the weight of the raft and the superstructure was applied.

A full documentation of the measurements carried out was provided by Reul (2000). Reul (2000, 2005) presented the results of a back-analysis of the bearing behaviour of the foundation by means of 3D FEA with an elasto-plastic cap model for Berlin sand. Based on those analyses the settlement of a hypothetical raft foundation was estimated as $s_{raft,ave}$ = 11.9 cm; i.e. the CPRF reduced settlements to approximately 53%. Kudella and Reul (2002) presented the results of the numerical analysis of the Treptowers foundation with a hypoplastic material model for the Berlin sand. Richter et al. (1998) compared the measurements with the results of an analysis where the raft was placed on Winkler springs representing the soil and the piles were simplified as elastic springs. That study indicated that the interaction between raft, piles and soil must be taken into account in estimating appropriate spring stiffnesses to achieve realistic results.

4.3.8 Weser Tower, Bremen

4.3.8.1 Building

The 82 m-high Weser Tower was constructed as a reinforced concrete frame from 2007 to 2010 in Bremen, Germany (Figure 4.42). The 22-storey office

Figure 4.42 **Weser Tower.**

tower, with a total of approximately 18,000 m² of office space, is located in close proximity to the river Weser. The load of the superstructure of the Weser Tower amounts to approximately $P_s = G + Q = 216$ MN + 54 MN = 270 MN (G: dead loads; Q: live loads) and is transferred to the raft mainly in the core area via walls and at the edge of the building via columns (maximum column load $G + Q = 17.1$ MN).

4.3.8.2 Subsoil conditions

At the site quaternary soil layers with a mean thickness of approximately 2 m are present beneath heterogeneous fill with a mean thickness of approximately 4 m. At the foundation level the quaternary deposits mainly comprise very soft, silty soils that are not suitable for transferring loads. These are underlain by the medium dense to dense "Weser Sands" with a layer thickness of approximately 11 m at the site. The so-called "Lauenburg strata" follows below, consisting of medium dense to dense, partly silty fine sands, in which locally clay layers are embedded. The clay layers that have been investigated have a maximum thickness of 0.5 m at the site and a stiff consistency. The soils of the Lauenburg strata were not penetrated by the boreholes, up to 40 m deep, carried out during site investigation.

The groundwater level varied between 5.2 m and 6.5 m below ground level and was clearly influenced by the tide-dependent water levels of the adjacent river Weser.

4.3.8.3 Foundation design

The design of the CPRF, comprising 31 piles ($d_p = 0.9$ m; $L_p = 16$ m to 22 m) and a raft with uniform thickness of $t_r = 1.8$ m, was based on 3D FEA. The very soft quaternary soils below foundation level had been replaced prior to the concreting of the raft by fine to medium sand compacted to medium density. Figure 4.43 shows a ground plan of the raft and a cross section through the foundation. It should be noted that some pile positions had to be moved slightly out of the load axes beneath walls and columns because there were piles from previous buildings in the ground that could not be removed without unreasonable effort.

4.3.8.4 In situ measurements

The location of all the geotechnical measuring devices is shown in the ground plan in Figure 4.43b. Initially, six foundation piles were equipped with load cells at the pile head to measure the pile resistances. However, the cables from one of the load cells were damaged during installation of the raft. Contact and pore pressures under the raft were determined from six contact pressure cells and two pore pressure cells. Geodetic measurements were carried out during construction to monitor the settlements. For

Case histories 197

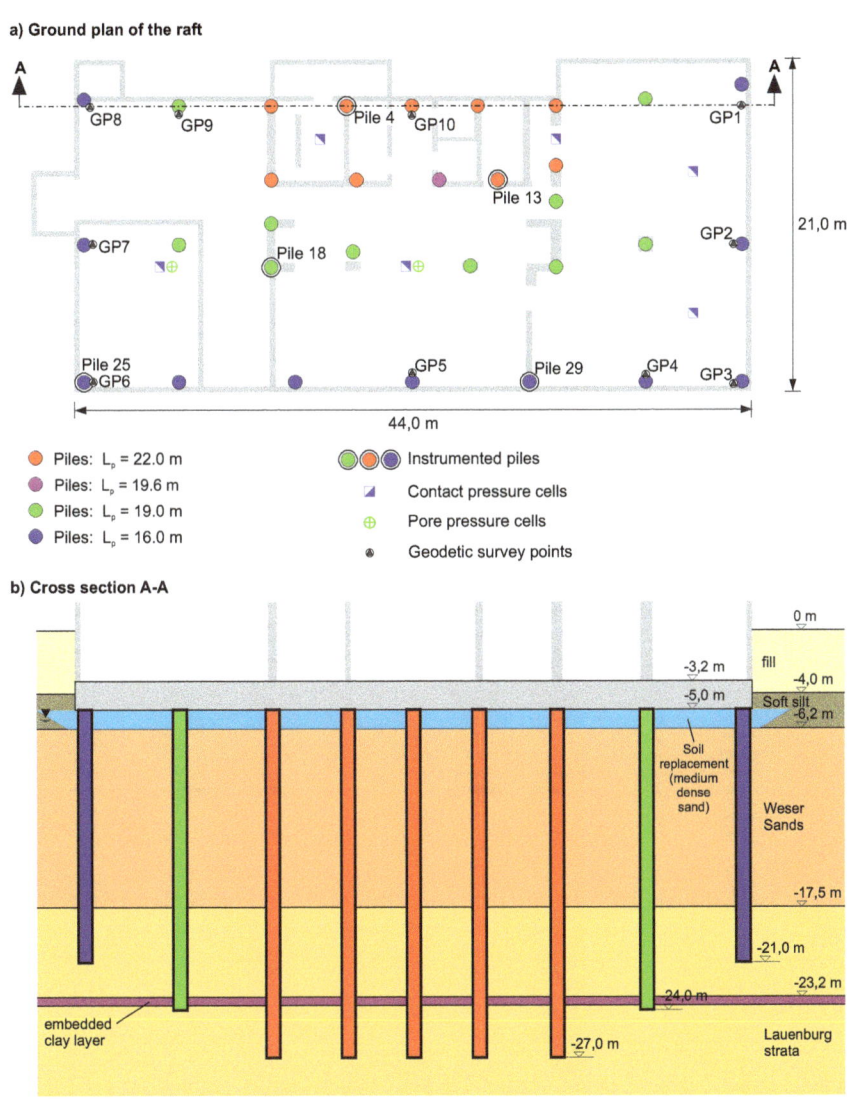

Figure 4.43 Weser Tower: Foundation.

this purpose, a total of ten measuring points were arranged on the raft and basement walls, mainly at the building edges.

Figure 4.44 shows the development of load, settlement and pile resistance over time. The measurements include the settlements due to concreting of the raft in May 2008 (Figure 4.44b). As for the Treptowers (Section 4.3.7), the largest increases in settlement were recorded at the time the shell of the building was finished in June 2009. At the last measurement, approximately three months after completion of the building, in September 2010, the mean

198 Combined Pile-Raft Foundations

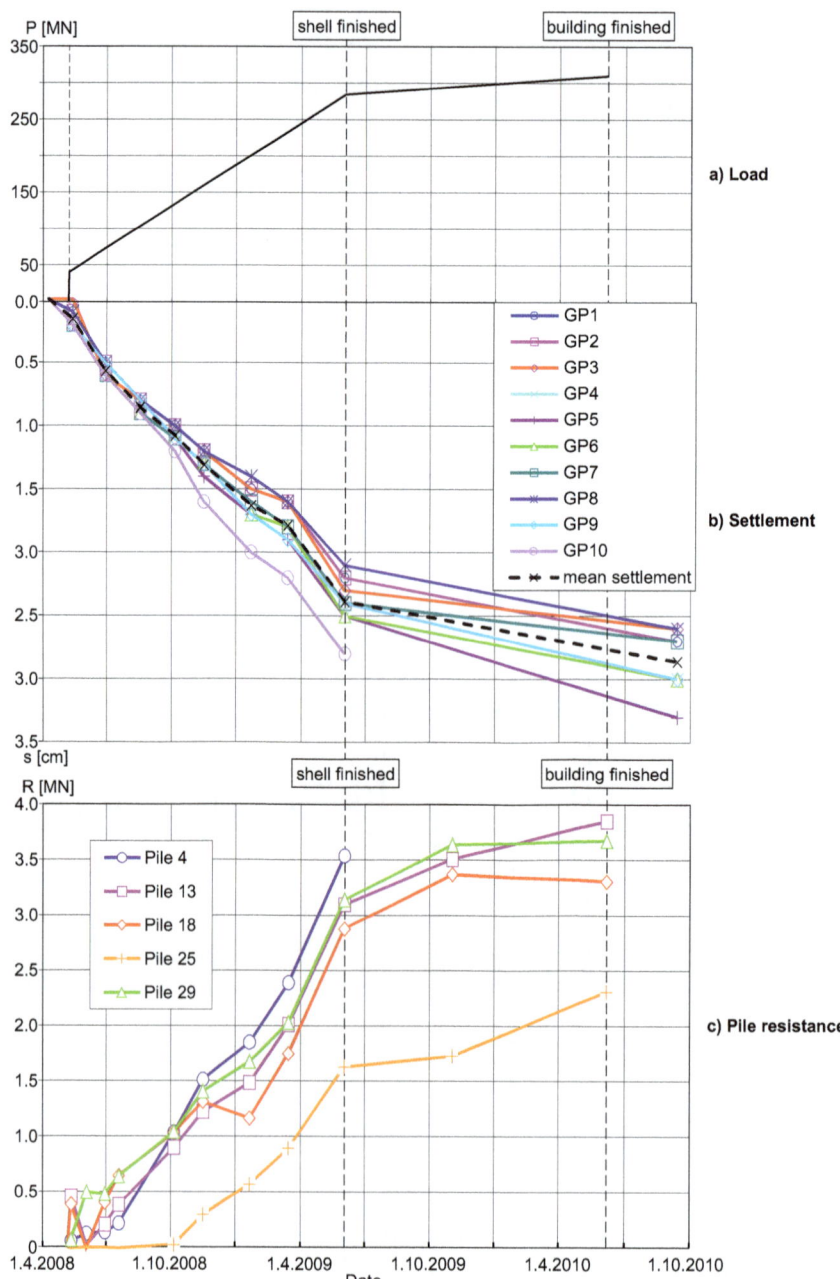

Figure 4.44 Weser Tower: Variation of load, settlement and pile resistance with time.

settlement amounted to s_{mean} = 2.9 cm. The variation of pile resistances over time in Figure 4.40c shows the largest increase for all piles from May 2008 to June 2009 with the maximum resistance measured for Pile 4 with R_4 = 3.5 MN, although unfortunately that pile delivered no reliable measurement signal after that date. In June 2010, at completion of the building, the pile resistances ranged between R_{25} = 2.3 MN and R_{13} = 3.8 MN.

4.3.9 Hegau-Tower, Singen

4.3.9.1 Building

The Hegau-Tower building complex comprises a 67.5 m-high office tower (Figure 4.45), an 18.6 m-high low-rise section and an underground car park. The Hegau-Tower, with 18 storeys and a floor area of around 720 m², and the low-rise section, with 5 storeys and a floor area of around 780 m², were constructed as a reinforced concrete frame from 2006 to 2007. The load of the superstructure of the Hegau-Tower amounts to approximately P_s = G ı Q = 299 MN + 154 MN – 454 MN (G: dead loads; Q: live loads).

4.3.9.2 Subsoil conditions

Fill and top layers, some of which are organic, are located at the surface of the site down to a maximum depth of 1.6 m. Below that, there are sediments of the Upper Gravel layer (layer 1) with loose to medium-dense sandy and silty gravel.

The Upper Gravel layer is underlain by moraine sediments, which are initially sandy silts (Layer 2) with soft to stiff consistency down to depths of 6.4 m to a maximum of 11.2 m. Below that, silts and clays (layer 3) with a soft to semi-solid consistency follow to a depth of 12.1 m to a maximum of 13.8 m, changing to gravelly, sandy silts (layer 4) with a stiff to semi-solid consistency towards the bottom of the layer.

The transition of the moraine sediments to the medium-dense to dense sands and gravels of the Lower Gravel layer (Layer 5) takes place at a depth of 17.6 m down to 20.3 m, marked by an increase of the sandy and gravelly constituents with depth.

The boundary between the Upper Gravel layer (Layer 1) and the moraine sediments (Layers 2 to 4) dips more than 5 m to the north, while the top of the Lower Gravel layer (Layer 5) is almost horizontal. The individual soil layers within the moraine sediments (Layers 2 to 4) dip at shallow inclinations in different directions and have strongly varying thicknesses.

The dipping of the Upper Gravel Layer coincides with the perimeter of the high-rise building and is of particular relevance for the foundation due to the partially soft consistency of layers 2 and 3. Geologically, a fault zone, an inclined stratification formed during sedimentation or a glacial "dead-ice" trough is conceivable as the origin of the layer submergence at the site.

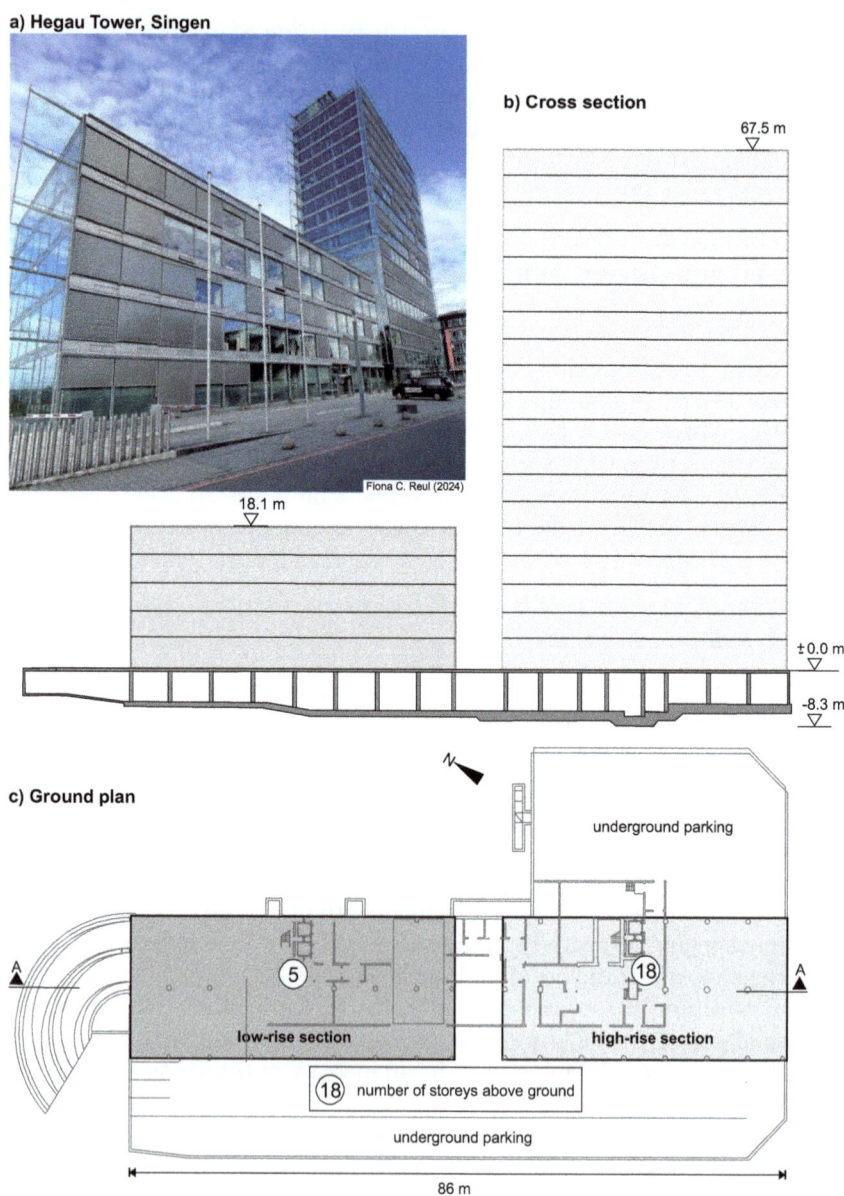

Figure 4.45 Hegau-Tower: Ground plan and cross section of the building complex.

The low-permeable cohesive moraine sediments act as an aquitard and separate the groundwater into an upper and a lower aquifer, comprising the sands and gravels of the Upper and Lower Gravel layers respectively. The groundwater level at the project site was encountered approximately 5.0 m (upper aquifer) and 19.7 m (lower aquifer) below ground surface.

The project area is located in a water protection zone, upstream of the freshwater intake where the lower aquifer in the Lower Gravel layer is used by the city of Singen for freshwater supply.

4.3.9.3 Foundation design

In order to ensure the contribution of the raft in the load transfer, soil improvement was carried out by means of rigid inclusions instead of replacement of the soft moraine sediments of layer 2. The rigid inclusions comprised 154 non-reinforced concrete columns with length between L_{cc} = 2.0 m to L_{cc} = 2.7 m distributed in a relatively uniform grid beneath a 0.3-m-thick load transfer layer (LTL) (Figure 4.46). The diameter of the concrete columns, with d_{cc} = 0.9 m, is significantly larger than usual rigid inclusions (e.g. Bohn 2015).

The 28 bored piles of the pile raft were designed with lengths of L_p = 7.5 m to L_p = 11.0 m and diameters of d_p = 0.9 m and d_p = 1.2 m, respectively, and are located beneath core walls and highly loaded columns (Figure 4.46). As a means to protect the aquifer relevant for the freshwater supply, the piles do not extend into the Lower Gravel layer (Layer 5); i.e. the pile bases are

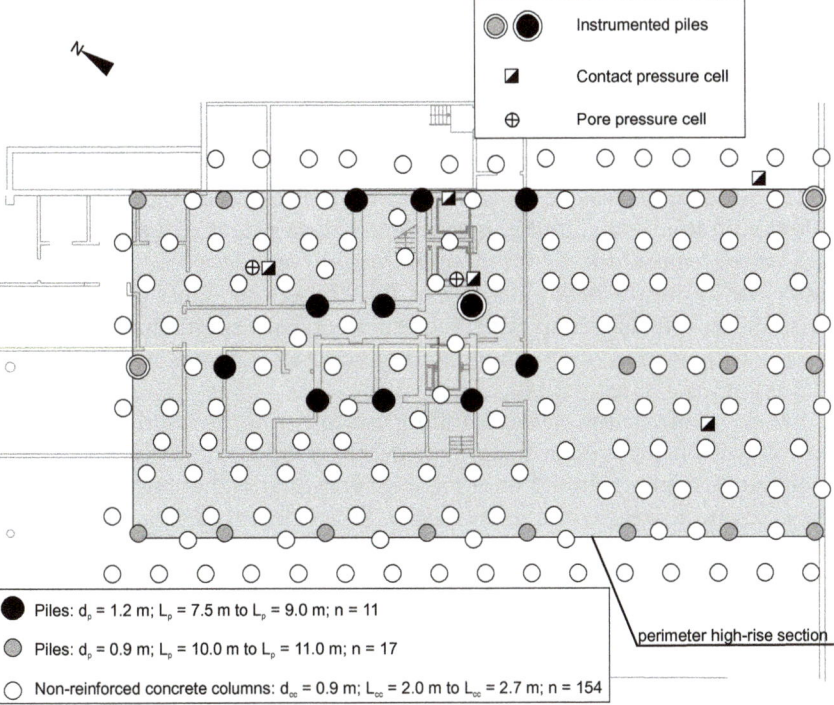

Figure 4.46 Hegau-Tower: Ground plan of the high-rise section with pile layout and measurement devices.

202 Combined Pile-Raft Foundations

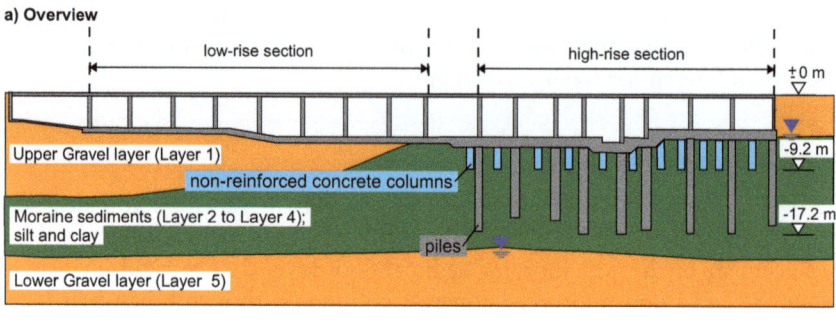
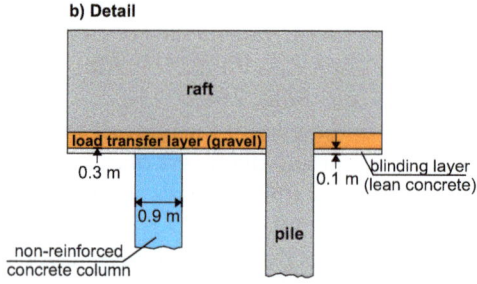

Figure 4.47 Hegau-Tower: Cross section of the foundation.

located in layer 4 (Figure 4.47). In the regular areas of the high-rise building and the low-rise section, the raft has a thickness of $t_r = 2.0$ m and $t_r = 1.0$ m, respectively. In the area of the underground parking with no superstructure, the raft thickness is $t_r = 0.5$ m to $t_r = 0.8$ m.

The foundation was designed based on 3D FEA (Reul and Wawrzyniak 2008) with Reul et al. (2006b) presenting analysis results of a pre-study. The soil improvement due to the non-reinforced concrete columns piles was modelled by means of an increase of the Young's modulus of the subsoil beneath the raft. The equivalent Young's modulus was established comparing the deformation of an infinite grid of concrete columns with the deformation of a soil layer with increased stiffness.

For the construction of the 3,400 m² basement of the building complex, an excavation pit with a maximum depth of approximately 8.3 m was required. A single anchored sheet pile wall was used as the shoring system in the excavation areas with the excavation base below the groundwater level. In the other areas, a single anchored soldier pile wall was used. For the construction of the excavation pit, the groundwater was lowered by means of wells located inside the excavation pit.

4.3.9.4 In situ measurements

Three foundation piles were equipped with strain gauges at the pile head to measure the pile strains, from which the pile loads could then be derived.

The contact pressure and the pore water pressure under the raft were determined from five contact pressure transducers and two pore water pressure transducers. The location of the geotechnical measuring devices is shown in Figure 4.45 in the ground plan. Geodetic measurements were carried out during construction to monitor the settlements. For this purpose, a total of 21 measuring points were distributed across the entire structure in the basement.

Measurements of pile resistances when 9 of 18 storeys of the high-rise section had been completed showed values between $R = 1.7$ MN ($d_p = 0.9$ m) and $R = 3.9$ MN ($d_p = 1.2$ m). The last available geodetic measurements when the shell of the building was finished showed settlements between $s_{measure} = 1.0$ cm and $s_{measure} = 1.7$ cm; these can be extrapolated to final settlements between $s_{prediction} = 1.5$ cm and $s_{prediction} = 3.4$ cm, considering a slight increase of dead and live loads and the effect of consolidation.

REFERENCES

Altner, J., Roth, E. (2002). Der Bau der neuen Messehalle 3 in Frankfurt am Main. Bauingenieur 77, Heft 9, 400–405.

Amann, P., Best, G., Schneider, W. (1976). *Bodenmechanische und geologisch-sedimentpetrographische Ergebnisse einer 100 m tiefen Kernbohrung im Untermiozän von Frankfurt am Main*. Geologisches Jahrbuch, Hannover, 23–68.

Arslan, U., Katzenbach, R., Reul, O. (1997). A numerical study of the soil-structure-interaction of a combined pile-raft foundation on loose sand in Berlin. *Proceedings of the 7th International Conference on Computing in Civil and Building Engineering*, Seoul, 2035–2040.

Bohn, C. (2015). Serviceability and safety in the design of rigid inclusions and combined pile-raft foundations. Mitteilungen des Institutes und der Versuchsanstalt für Geotechnik der Technischen Universität Darmstadt, Heft 96.

Breth, H. (1970). Das Tragverhalten des Frankfurter Tons bei im Tiefbau auftretenden Beanspruchungen. Mitteilung der Versuchsanstalt für Bodenmechanik und Grundbau der Technischen Hochschule Darmstadt, Heft 4.

Burland, J. B., Kalra, J. C. (1986). Queen Elizabeth II Conference Centre: geotechnical aspects. *Proceedings of the ICE*, Part 1, 80, 1479–1503.

Butcher, A.P., Skinner, H.D., Powell, J.J.M. (2006). Stonebridge Park – a demolition case study. *Proceedings of the International Conference on Reuse of Foundations for Urban Sites (RuFUS)*, Watford, 321–329.

Butler, F. G. (1975). Heavily over-consolidated clays - Review paper: Session III. *Proceedings of the Conference Settlements of Structures*, Cambridge, 531–578, London: Pentech.

Cooke, R. W., Bryden-Smith, D. W., Gooch, M. N., Sillett, D. F. (1981). Some observations of the foundation loading and settlement of a multi-storey building on a piled raft foundation in London Clay. *Proceedings of the ICE*, No. 70, Part 1, 433–460.

Deutsche Gesellschaft für Geotechnik (DGGT)/Deutsches Institut für Bautechnik (DIBt) (2001). Richtlinie für den Entwurf, die Bemessung und den Bau von Kombinierten Pfahl-Plattengründungen (KPP) – KPP-Richtlinie.

Franke, E., Lutz, B. (1994). Pfahl-Platten-Gründungs-Messungen. Forschungsabschlußbericht zum Forschungsauftrag Fr 600 - 11/1.

Ganal, A. (2024). Time dependent bearing behaviour of foundations subjected to alternate loading in overconsolidated clay (Ph.D. thesis). Schriftenreihe Geotechnik, Universität Kassel, Heft 30.

Ganal, A., Reul, O. (2023). Back analysis of long-term measurements of a high-rise building founded on a raft foundation in over-consolidated clay. *Proceedings of the 10th European Conference on Numerical Methods in Geotechnical Engineering*, London, https://doi.org/10.53243/NUMGE2023-116

Ganal, A., Reul, O., Tafili, M. (2024). Visko-hypoplastisches Materialmodell für tertiäre Böden in Frankfurt am Main. *Geotechnik*, 47, 84–97. https://doi.org/10.1002/gete.202300025

Green, P. A., Hight, D. W. (1976). The instrumentation of Dashwood House London. *CIRIA Report*, No. 78.

Hight, D. W., Green, P. A. (1976). The performance of a piled-raft foundation for a tall building in London. *Proceedings of the 4th ECSMFE*, 1.2, 467–472.

Hirakawa, K., Hamada, J., Yamashita, K. (2016). Settlement behavior of piled raft foundation supporting a 300 m tall building in Japan constructed by top-down method. *Proceedings of the 15th Asian Regional Conference on Soil Mechanics and Geotechnical Engineering*, 166–169.

Holzhäuser, J. (1998). Experimentelle und numerische Untersuchungen zum Tragverhalten von Pfahlgründungen im Fels. *Mitteilungen des Institutes und der Versuchsanstalt für Geotechnik der TU Darmstadt*, Heft 42.

Hooper, J. A. (1973). Observations on the behaviour of a piled-raft foundation on London Clay. *Proceedings of the ICE*, 55, Part 2, 855–877.

Hooper, J. A. (1979). Review of behaviour of piled raft foundations. *CIRIA Report*, No. 83.

ISSMGE Technical Committee TC 212 (2013). ISSMGE Combined Pile-Raft Foundation Guideline. Report of the ISSMGE Technical Committee TC 212 – Deep Foundations, ed. Katzenbach, R., Choudhury D, Technische Universität Darmstadt.

Kalra, J. C., Willows, K. R. (1986). Queen Elizabeth II Conference Centre: design and construction. *Proceedings of the ICE*, Part 1, No. 80, 1451–1477.

Katzenbach, R., Moormann, C. (1997). Design of axially loaded piles and pile groups - German practice. *Proceedings of the ERTC3 Seminar Design of Axially Loaded Piles European Practice*, Brussels, 177–201, Rotterdam: Balkema.

Katzenbach, R., Turek, J. (2004). New Exhibition Hall 3 in Frankfurt – Case History of a Combined Pile-Raft Foundation Subjected to Horizontal Load. *Proceedings of the 5th International Conference on Case Histories in Geotechnical Engineering*, Paper No. 1.33.

Katzenbach, R., Arslan, U., Gutwald, J. (1994). A numerical study on pile foundation on the 300 m high Commerzbank Tower in Frankfurt am Main. *Proceedings of the 3rd European Conference on Numerical Methods in Geomechanics*, Manchester, 271–277, Rotterdam: Balkema.

Katzenbach, R., Quick, H., Arslan, U. (1996). Commerzbank-Hochhaus Frankfurt am Main: Kostenoptimierte und setzungsarme Gründung. *Bauingenieur*, 71, Heft 9, 345–354.

Katzenbach, R., Arslan, U., Reul, O. (1998a). Soil-structure interaction of a piled raft foundation of a 121 m high office building on loose sand in Berlin. *Proceedings of the Deep Foundations on Bored and Auger Piles 1998*, 215–221, Rotterdam: Balkema.

Katzenbach, R., Arslan, U., Reul, O. (1998b). Entwurf von Hochhausgründungen auf der Basis von numerischen Computersimulationen mit der Finite-Element-Methode. *Finite Elemente in der Baupraxis - FEM'98*, 381–390, Berlin: Ernst und Sohn.

Katzenbach, R., Arslan, U., Moormann, C. (2000). Piled raft foundation projects in Germany. In *Design Applications of Raft Foundations*, ed. Hemsley, J.A., 323–391. London: Thomas Telford.

Katzenbach, R., Turek, J., Vogler, M. (2002). Entwicklung einer horizontal belasteten KPP am Beispiel der neuen Messehalle 3 in Frankfurt am Main. Bauingenieur, Band 77, Heft 9, 393–398.

Kudella, P., Reul, O. (2002). Analysis of piled raft foundation behaviour in granular soil using hypoplasticity. *Proceedings of the 5th European Conference on Numerical Methods in Geotechnical Engineering*, Paris, Presses de l'École Nationale des Ponts et Chaussées, Paris, 389–395.

Kümmerle, E., Seidenschwann, G. (2009). Erläuterungen zur geologischen Karte von Hessen – Blatt Nr. 5817 Frankfurt a.M. West. ed.: Hessisches Landesamt für Umwelt und Geologie, 3rd edition, Wiesbaden.

Leonhardt, G. (1972). Setzungskorrekturen an einem im Frankfurter Ton gegründeten Hochhaus. *Vorträge der Baugrundtagung in Stuttgart*, 211–218.

Lutz, B., Wittmann, P., El Mossallamy, Y., Katzenbach, R. (1996). Die Anwendung von Pfahl-Plattengründungen - Entwurfspraxis, Dimensionierung und Erfahrungen mit Gründungen in überkonsolidierten Tonen auf der Grundlage von Messungen. *Vorträge der Baugrundtagung 1996 in Berlin*, 153–164, Essen: DGGT.

Mader, H. (1989). Untersuchungen über den Primärspannungszustand in bindigen überkonsolidierten Böden am Beispiel des Frankfurter Untergrundes. *Mitteilungen des Institutes für Grundbau, Boden- und Felsmechanik der TH Darmstadt*, Heft 29.

Meese, L., Ziesler, D. (2002). Neue Messehalle 3 in Frankfurt/Main – Planung und Ausführung der Überdachung. Stahlbau 71, Heft 1, 1–12.

Meissner, S., Quick, H., Katzenbach, R., Werner, A. (2019). An innovative dewatering system to reduce the environmental impact. *Proceedings of the 17th European Conference on Soil Mechanics and Geotechnical Engineering*, https://doi.org/10.3 2075/17ECSMGE-2019-0441

Moormann, C. (2002). Trag- und Verformungsverhalten tiefer Baugruben in bindigen Böden unter besonderer Berücksichtigung der Baugrund-Tragwerk- und Baugrund-Grundwasser-Interaktion. *Mitteilungen des Institutes und der Versuchsanstalt für Geotechnik der TU Darmstadt*, Heft 59.

Philipp Holzmann, A.G. (1996). Hochhaus Taunustor Japan Center in Frankfurt am Main. Technical Report.

Price, G., Wardle, I. F. (1986). Queen Elizabeth II Conference Centre: monitoring of load sharing between piles and raft. *Proceedings of the ICE*, Part 1, No. 80, 1505–1518.

Ramm, H., Kissel, W., Toker, E., Ruiken, A., Reul, O. (2020). Hochhaus Omniturm – Baugrube und Gründung unter komplexen innerstädtischen. *Bautechnik*, 97, Heft 9, 656–663, https://doi.org/10.1002/bate.202000059

Randolph, M. F. (2019). PIGLET: Analysis and design of pile groups. Users' Manual, Version 6-1, Perth.

Randolph, M. F., Reul, O. (2019). Practical approaches for design of pile groups and piled rafts. *Proceedings of the 4th Bolivian International Conference on Deep Foundations*, Santa Cruz de la Sierra, Bolivia, 1–27.

Remmel, G., Sattler, F., Klug, U. (2006). Parktower, Bockenheimer Anlage 46 in Frankfurt am Main, Modernisierung der ehemaligen SGZ-Bank. *Gespräche mit Wissenschaft und Praxis 2006*, Wayss & Freytag, Frankfurt am Main, 19–24.

Reul, O. (2000). In-situ-Messungen und numerische Studien zum Tragverhalten der Kombinierten Pfahl-Plattengründung. *Mitteilungen des Institutes und der Versuchsanstalt für Geotechnik der Technischen Universität Darmstadt*, Heft 53.

Reul, O. (2005). Piled raft foundations. *Proceedings of the BSCES/ASCE Geo-Institute Seminar on Recent Advances in Geotechnical Engineering*, Waltham Massachusetts, Session III, 1–27.

Reul, O., Krajewski, W. (2010). Geotechnische Aspekte bei Umbau und Modernisierung von Hochhäusern. Vorträge des 7. Kolloquiums Bauen in Boden und Fels; Technische Akademie Esslingen, 73–81.

Reul, O., Randolph, M.F. (2003). Piled rafts in overconsolidated clay – Comparison of in-situ measurements and numerical analyses. *Géotechnique*, 53, 3, 301–315.

Reul, O., Remmel, G. (2009). Foundation design for the extension of an existing high-rise building. *Proceedings of the 17th International Conference on Soil Mechanics and Geotechnical Engineering*, Alexandria, Vol. 2, 2072–2075.

Reul, O., Wawrzyniak, C. (2008). Innovative Gründung für ein Hochhaus in setzungsempfindlichem Baugrund. Vorträge des 6. Kolloquiums Bauen in Boden und Fels; Technische Akademie Esslingen, 545–550.

Reul, O., Ehrhardt, G., Rummel, B. (2006a). Entwurfsoptimierung einer Hochhausgründung im Frankfurter Ton. *Vorträge des 5. Kolloquiums Bauen in Boden und Fels*, Technische Akademie Esslingen, 309–318.

Reul, O., Krajewski, W., Ripper, P. (2006b). Numerical Analysis of Foundations for High-Rise Buildings and Deep Excavations. *Felsbau*, 24, 2, 22–30.

Reul, O., Haebler, H., Remmel, G., Stürzl, M. (2007). Vom SGZ-Bank Hochhaus zum Parktower - Gründungstechnische Aspekte eines Bauwerks im Wandel. *Pfahl-Symposium 2007; Mitteilungen des Institutes für Grundbau und Bodenmechanik der TU Braunschweig*, Heft 84, 371–390.

Richter, T., Savidis, S., Katzenbach R., Quick, H. (1996). Wirtschaftlich optimierte Hochhausgründungen im Berliner Sand. *Vorträge der Baugrundtagung 1996 in Berlin*, 129–146, Essen: DGGT.

Richter, T., Reul, O., Arslan, U. (1998). Setzungen hoch belasteter Gründungen in Berliner Böden - Vergleich von Tief- und Flachgründungen in Berechnung und Messung. *Vorträge der Baugrundtagung 1998 in Stuttgart*, 601–613, Essen: DGGT.

Ripper, P., El Mossallamy, Y. (1999). Entwicklungen der Hochhausgründungen in Frankfurt. *Hochhäuser - Darmstädter Statik-Seminar 1999*, Bericht Nr. 16 - Technische Universität Darmstadt, Institut für Statik.

Schultz, E. W. (1994). Eine lange Baugrube im Grundwasser. Erfahrungen beim Bau des 4 km langen S-Bahn- Tunnels in Offenbach. *Mitteilungen des Institutes und der Versuchsanstalt der TH Darmstadt*, Heft Nr. 33, 137–161.

Schultze, E, Horn, A. (1995). Setzungsberechnung. In *Grundbautaschenbuch Teil 1*, 5. Auflage, ed Smoltzcyk, U., 225–254, Ernst & Sohn, Berlin.

Skempton, A. W., DeLory, F. A. (1957). Stability of Natural Slopes in London Clay. *Proceedings of the 4th ICSMFE*, London, 378–381.

Skempton, A. W., Henkel, D. J. (1957). Tests on London Clay from Deep Borings at Paddington, Victoria and the South Bank. *Proceedings of the 4th ICSMFE*, London, 100–106.

Sommer, H., Hoffmann, H. (1991). Load-settlement behaviour of the fairtower (Messeturm) in Frankfurt/Main. In: *Proceedings of the 4th International Conference on Ground Movements and Structures*. Pentech Press, London, 612–627.

Sommer, H., Tamaro, G., DeBeneditis, C. (1991). Messeturm, foundations for the tallest building in Europe. In: *Proceedings of the 4th International Conference on Piling and Deep Foundations*, Rotterdam: Balkema. 139–145.

Sonoda, R., Matsumoto, T., Kitiyodom Pastsakorn, Moritaka, H., Ono, T. (2009). Case study of a piled raft foundation constructed using a reverse construction method and its post-analysis. *Canadian Geotechnical Journal*, 46, 2, 142–159.

Stahlmann, J., El-Mossallamy, Y., Leinenbach, J., Ittershagen, M. (2001). Sicherung gegen Schiefstellung eines (Hochhaus-) Turms – nicht nur eine historische Aufgabenstellung. *Mitteilungen des Institutes und der Versuchsanstalt für Geotechnik der Technischen Universität Darmstadt*, Heft 55.

Stroetmann, R., Möll, R. (2003). Die bautechnische Prüfung und schweißtechnische Überwachung des Dachtragwerks der neuen Messehalle 3 in Frankfurt am Main. *Bautechnik*, 80, Heft 5, 285–296.

Tafili, M., Triantafyllidis, T. (2020). AVISA: Anisotropic visco-ISA model and its performance at cyclic loading. *Acta Geotechnica*, 15, 2395–2413, http://doi.org/10.1007/s11440-020-00925-9

Turek, J. (2006). Beitrag zur Klärung des Tragverhaltens horizontal belasteter Kombinierter Pfahl-Plattengründungen. *Mitteilungen des Institutes und der Versuchsanstalt für Geotechnik der TU Darmstadt*, Heft 72.

Ward, W. H., Samuels, S. G., Butler, M. E. (1958). Further studies of the properties of London clay. *Géotechnique*, VIII, 33–58.

Ward, W. H., Marsland, A., Samuels, S. G. (1965). Properties of the London Clay at the Ashford Common Shaft: In-situ and undrained strength tests. *Géotechnique*, XV, 321–344.

Weiß, K. (1978). Die Hauptbodenarten im Raum Berlin als Baugrund. *Vorträge der Baugrundtagung 1978 in Berlin*, 503–528, Essen: DGEG.

Zabel, J., Amann, P., Krajewski, W., Weiß, J. (1994.). Geotechnische Erfahrungen beim S-Bahn-Bau im Rhein-Main-Gebiet. *Vorträge der Baugrundtagung in Köln*, 271–293, Essen: DGGT.

Chapter 5

Design example

5.1 GENERAL REMARKS

In the design example the foundation for a multi-storey office building founded in non-cohesive soil is investigated with three different foundation configurations compared, namely raft foundation, pile foundation (piled raft with a conventional design approach) and CPRF. In the design example 3D FEA is carried out to estimate deformations and design parameters for structural engineering, i.e. Winkler modulus and spring stiffness. Based on these parameters the preliminary design of the raft to withstand bending moments and shear forces is carried out. The costs and performance (deformations) are compared for CPRF, raft foundation and pile foundation.

5.2 BUILDING

The load of the superstructure of the multi-storey office building amounts to $P_{s,k} = G_k + Q_k$ = 312.5 MN + 58.2 MN = 370.7 MN (G_k: dead loads; Q_k: live loads) and is transferred to the rectangular, 44 m × 19 m raft in the core area via walls and at the edge of the building via columns (maximum column load $G_k + Q_k$ = 16.9 MN). Figure 5.1 shows the ground plan of the raft together with the permanent and live loads caused by the superstructure. Due to environmental restrictions (limitation of groundwater extraction), the maximum depth of the foundation level is limited to 4 m below groundwater level, i.e. at 6.0 m below ground level resulting in a one-storey basement. With the top of the raft located at 3.0 m below ground level, a maximum thickness of the raft of $t_{r,max}$ = 3.0 m is therefore possible.

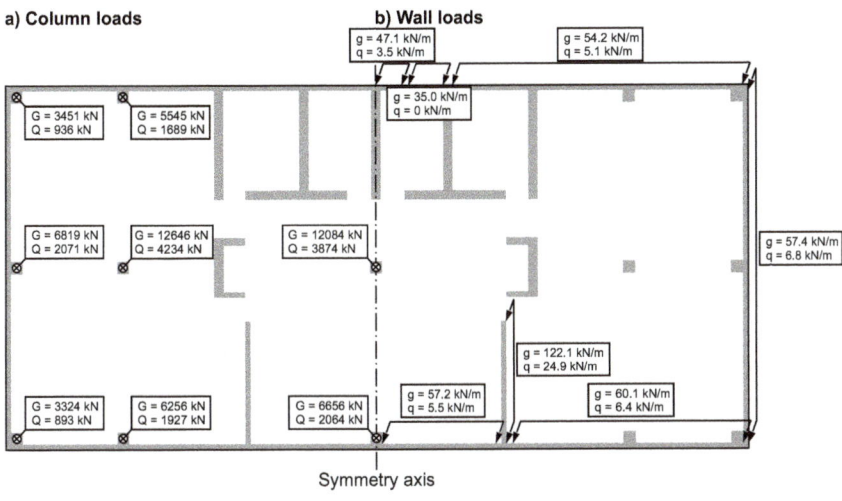
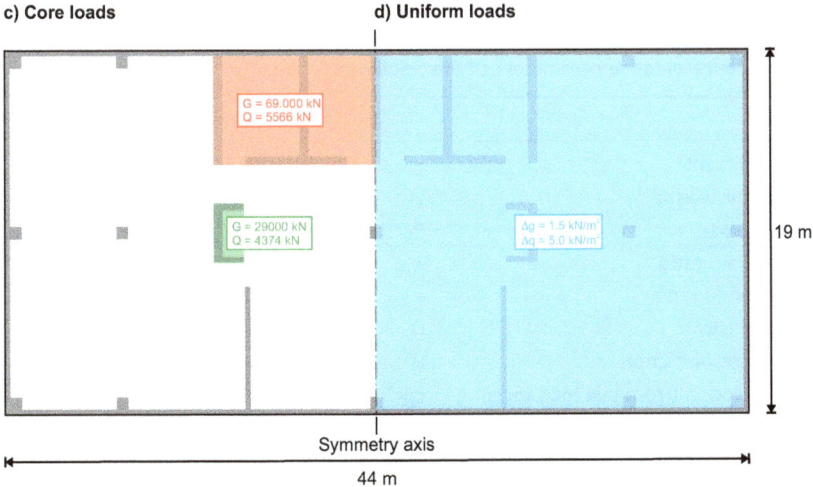

Figure 5.1 Building loads acting on the raft.

5.3 SUBSOIL CONDITIONS

The subsoil is characterised by non-cohesive soils down to large depth, namely the layers Sand 1, Sand 2 and Sand 3 with densities increasing form loose ($D_r = 0.35$) at foundation level to dense in Sand 3 ($D_r = 0.80$). Figure 5.2 shows a cross section through the basement together with the soil profile. Representative soil parameters are summarised in Table 5.1 together with characteristic pile shaft friction and unit base resistance derived from pile load tests on the site.

210 Combined Pile-Raft Foundations

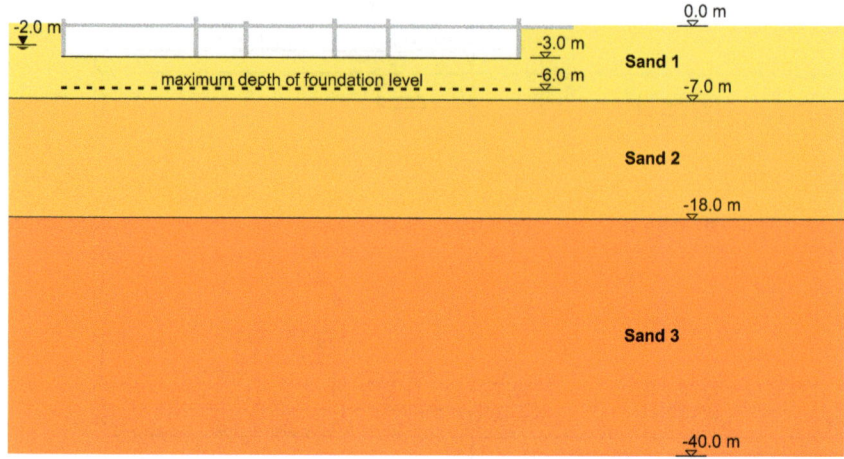

Figure 5.2 Subsoil conditions

Table 5.1 Representative parameters of the soil layers

Parameter			Sand 1	Sand 2	Sand 3
Total unit weight	γ	kN/m³	19.0	19.5	19.0
Buoyant unit weight	γ'	kN/m³	9.5	10.0	9.2
Mean grain size	d_{50}	mm	0.439	0.439	0.158
Minimum void ratio	e_{min}	-	0.460	0460	0.645
Maximum void ratio	e_{max}	-	0.776	0.776	1.059
Relative density	D_r	-	0.35	0.60	0.80
Critical angle of friction	φ'_{crit}	°	30	30	33
Pile shaft friction (from pile load tests)	$q_{s,k}$	kN/m²	32	112	375
Pile unit base resistance (from pile load tests)	$q_{b,k}$	kN/m²	-	4375	9342

5.4 FOUNDATION CONFIGURATIONS

5.4.1 General remarks

For the design process the following constraints were considered:

- Maximum depth of foundation level $z = -6.0$ m, i.e. maximum thickness of the raft $t_r = 3.0$ m.
- Maximum pile length $L_p = 30$ m and pile diameter $d_p = 0.9$ m to not restrict the number of potential piling contractors too much.
- Minimise the angular distortion which should preferably be $\alpha \leq 1/500$.
- Minimise bending moments in the raft.
- Minimise shear forces in the raft.

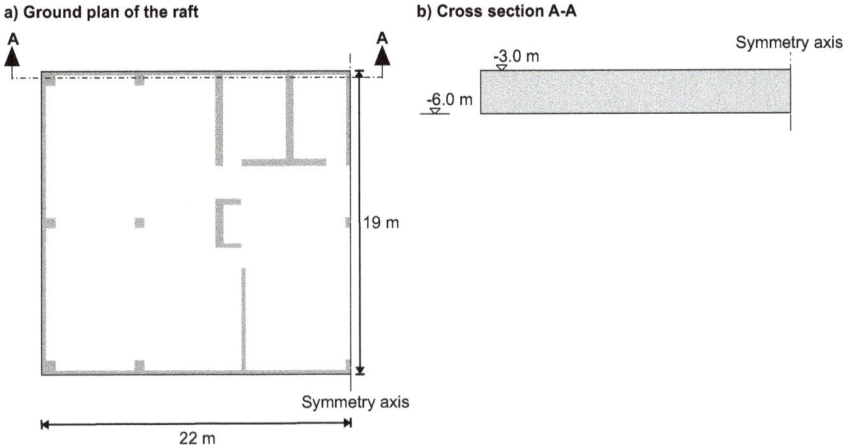

Figure 5.3 Configuration of the raft foundation.

5.4.2 Raft foundation

With the aim to minimise the angular distortion, the maximum raft thickness of $t_r = 3.0$ m was chosen (Figure 5.3).

5.4.3 Pile foundation (piled raft – conventional design)

For the pile foundation a raft thickness of $t_r = 1.6$ m was chosen. The pile configuration was established based on design code EC7-1 showing for each pile that

$$E_d = G_k\gamma_G + Q_k\gamma_Q \leq R_d = \frac{1}{\gamma_t}R_k = \frac{1}{\gamma_t}(R_{b,k} + R_{s,k}) = \frac{1}{\gamma_t}\left(q_{b,k}A_b + \sum_i q_{s,k,i}A_{s,i}\right)$$

(5.1)

where E_d = design value for the effect of actions; G_k = characteristic permanent load; Q_k = characteristic live (variable) load; R_d = design value for the pile resistance; R_k = characteristic value for the pile resistance; $R_{b,k}$ = characteristic base resistance; $R_{s,k}$ = characteristic shaft resistance; A_b = pile base area and $A_{s,i}$ = pile shaft area in layer i; $q_{b,k}$ = unit base resistance; $q_{s,k,i}$ = shaft friction in layer i; and $\gamma_G, \gamma_Q, \gamma_t$ = partial factors of safety.

The factors of safety were taken from the German code DIN 1054 as γ_G = 1.35, γ_Q = 1.50 and γ_t = 1.10 for a persistent design situation. Note that, in principle, the pile group is designed as if it were a freestanding pile group with the cap clear of the ground. However, the 3D FEA presented later allow for load transferred directly from the pile cap to the soil in order to illustrate the effect this has in practice regarding settlement and load sharing. Moreover, in the structural analysis even for the pile foundation an elastically supported raft was considered with the Winkler moduli derived from the 3D FEA. This approach was chosen to avoid a too conservative design of the raft with an excessive amount of reinforcement steel required.

Generally, the piles will be placed beneath highly loaded columns and walls to reduce the stresses and bending moments in the raft. To select an appropriate pile configuration the ground area of the foundation was divided into 11 influence areas for the loads which were assigned to different pile positions (Figure 5.4). When assigning the loads to the piles it was assumed that the weight of the raft is carried by the raft. As an example, for load areas I and II highlighted in Figure 5.4 the proof of the selected pile length is given below:

Load area I: $E_d = 46691$ kN (Figure 5.4); $n_p = 2$; $L_p = 29$ m ($L_{p,1} = 2.4$ m; $L_{p,2} = 11.0$ m; $L_{p,3} = 15.6$ m)

$$R_{b,k} = 9342 \text{kN/m}^2 \cdot \pi \cdot \left(\frac{0.9\text{m}}{2}\right)^2 = 5943 \text{kN}$$

$$R_{s,k} = \pi \cdot 0.9 \text{ m} \cdot (32 \text{ kN/m}^2 \cdot 2.4 \text{ m} + 112 \text{ kN/m}^2 \cdot 11.0 \text{ m} + 375 \text{ kN/m}^2 \cdot 15.6 \text{ m}) = 20241 \text{ kN}$$

$$R_d = \frac{1}{1.10}(5943 \text{kN} + 20241 \text{kN}) = 23804 \text{kN per pile}$$

$$E_d = 46691 \text{ kN} < n_p \cdot R_d = 2 \cdot 23804 \text{ kN} = 47608 \text{ kN}$$

Load area II: $E_d = 13486$ kN (Figure 5.4); $n_p = 1$; $L_p = 19$ m ($L_{p,1} = 2.4$ m; $L_{p,2} = 11.0$ m; $L_{p,3} = 5.6$ m)

$$R_{b,k} = 9342 \text{kN/m}^2 \cdot \pi \cdot \left(\frac{0.9\text{m}}{2}\right)^2 = 5943 \text{ kN}$$

$$R_{s,k} = \pi \cdot 0.9 \text{m} \cdot (32 \text{ kN/m}^2 \cdot 2.4 \text{m} + 112 \text{ kN/m}^2 \cdot 11.0 \text{m} + 3375 \text{ kN/m}^2 \cdot 5.6 \text{m}) = 9638 \text{ kN}$$

$$R_d = \frac{1}{1.10}(5943 \text{kN} + 9638 \text{kN}) = 14165 \text{kN}$$

$$E_d = 13486 \text{kN} < 14165 \text{kN} = R_d$$

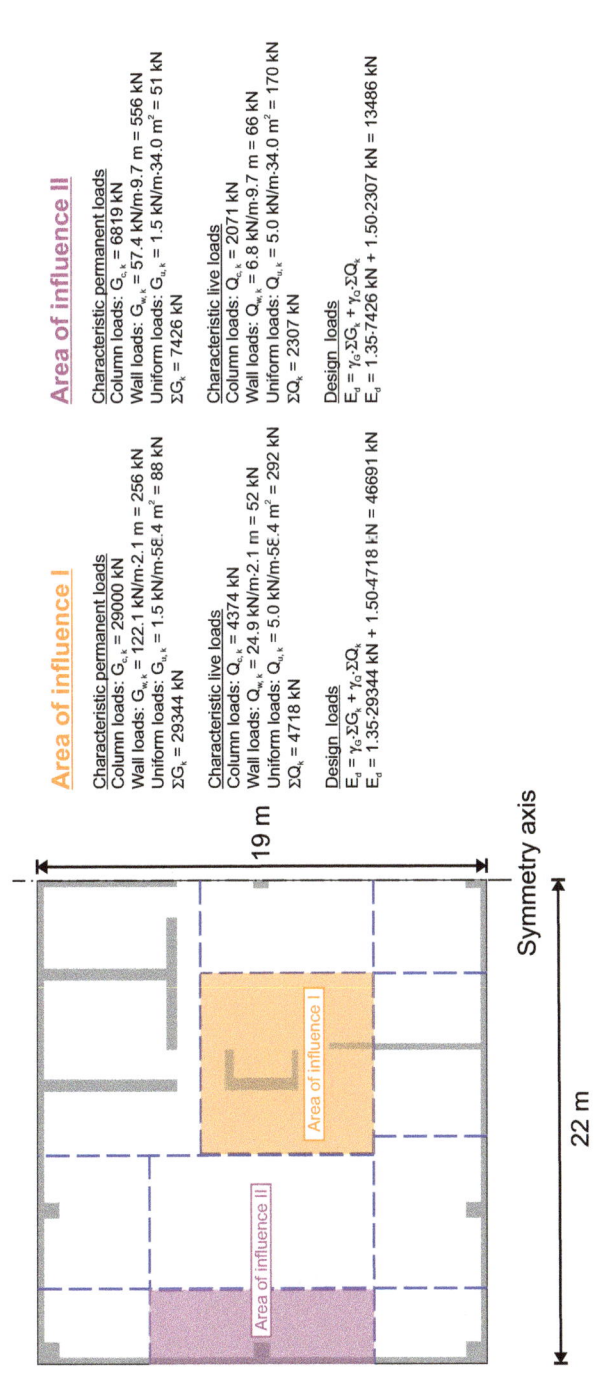

Figure 5.4 Load areas.

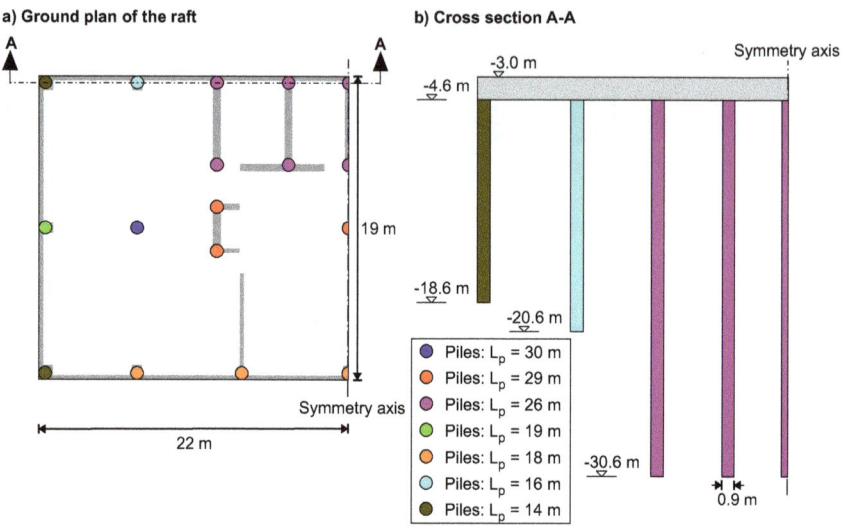

Figure 5.5 Configuration of the pile foundation.

Based on this procedure the pile configuration (n_p = 30; L_p = 14 m to 30 m; d_p = 0.9 m) shown in Figure 5.5 was established. For the complete pile foundation one then gets a total pile resistance of $\Sigma R_{pile,k}$ =585 MN and $\Sigma R_{pile,d}$ =532 MN, respectively. Note, however, that the capacity of individual piles will be significantly higher within a group, especially in sand, because of the increased effective stress level as a result of load transferred to the soil from the surrounding piles and (for a piled raft) from the raft. Moreover, the characteristic pile shaft friction $q_{s,k}$ and pile unit base resistance $q_{b,k}$ represent the values derived from pile load tests for a settlement of s = 0.1 d_p = 0.09 m; i.e. for larger settlements, larger resistances are to be expected.

5.4.4 CPRF

As for the pile foundation the raft thickness of the CPRF is kept at t_r = 1.6 m. The CPRF configuration (n_p = 30; L_p = 10 m to 26 m; d_p = 0.9 m) was derived from the configuration of the pile foundation by keeping the pile positions but reducing the pile length, especially at the edge of the raft, to reduce the angular distortion. The design of the CPRF follows the CPRF guideline (ISSMGE TC 212 2013) with the proof of external ULS and SLS documented in Sections 5.6.2 and 5.6.3.

Considering the constraints mentioned above an optimised pile configuration and raft thickness were found based on 3D FEA described below (Figure 5.6).

Design example 215

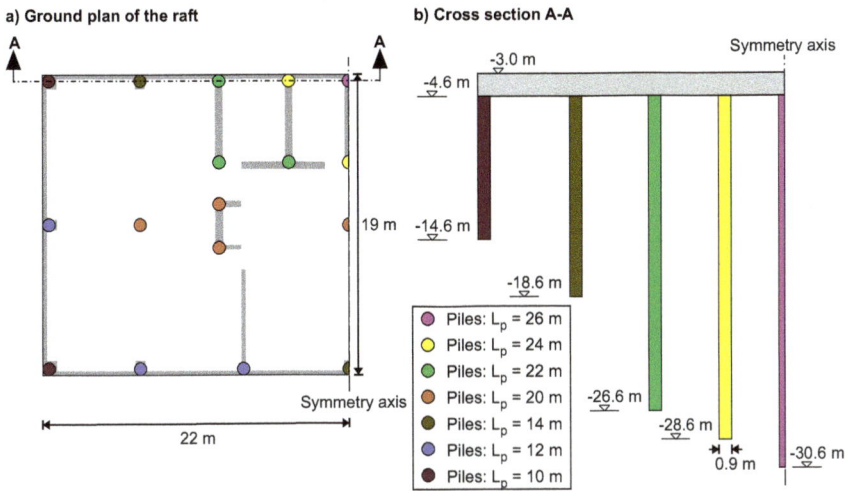

Figure 5.6 Configuration of the CPRF.

5.5 3D FEA

5.5.1 3D FEA – model

In the 3D FEA the finite element models of the soil, the piles and the raft are represented by first-order solid finite elements of hexahedra (brick) shape. The finite element mesh is shown in Figure 5.7 for the example of the CPRF foundation. The circular piles were replaced by square piles with the same shaft circumference. Thin solid continuum elements were applied for modelling the contact zone between soil and raft and soil and the large diameter bored piles. The contact between structure and soil was described as perfectly rough. This means that no relative motion takes place between the nodes of the finite elements that represent the structure and those of the finite elements that represent the uppermost layer of soil. The material behaviour in the contact area is therefore simulated by the material behaviour of the soil.

The material behaviour of Sand 1, Sand 2 and Sand 3 are modelled with the hypoplastic soil model by von Wolffersdorff (1996) and the intergranular strain extension by Niemunis and Herle (1997) which allows to consider the small strain stiffness behaviour of the sand. The material parameters are summarised in Table 5.2. As pointed out in Section 3.2, in the case of predominantly monotonic load paths and appropriately calibrated material parameters, certain analysis results, such as settlements and piled raft coefficient, can also be established with reasonable accuracy

216　Combined Pile-Raft Foundations

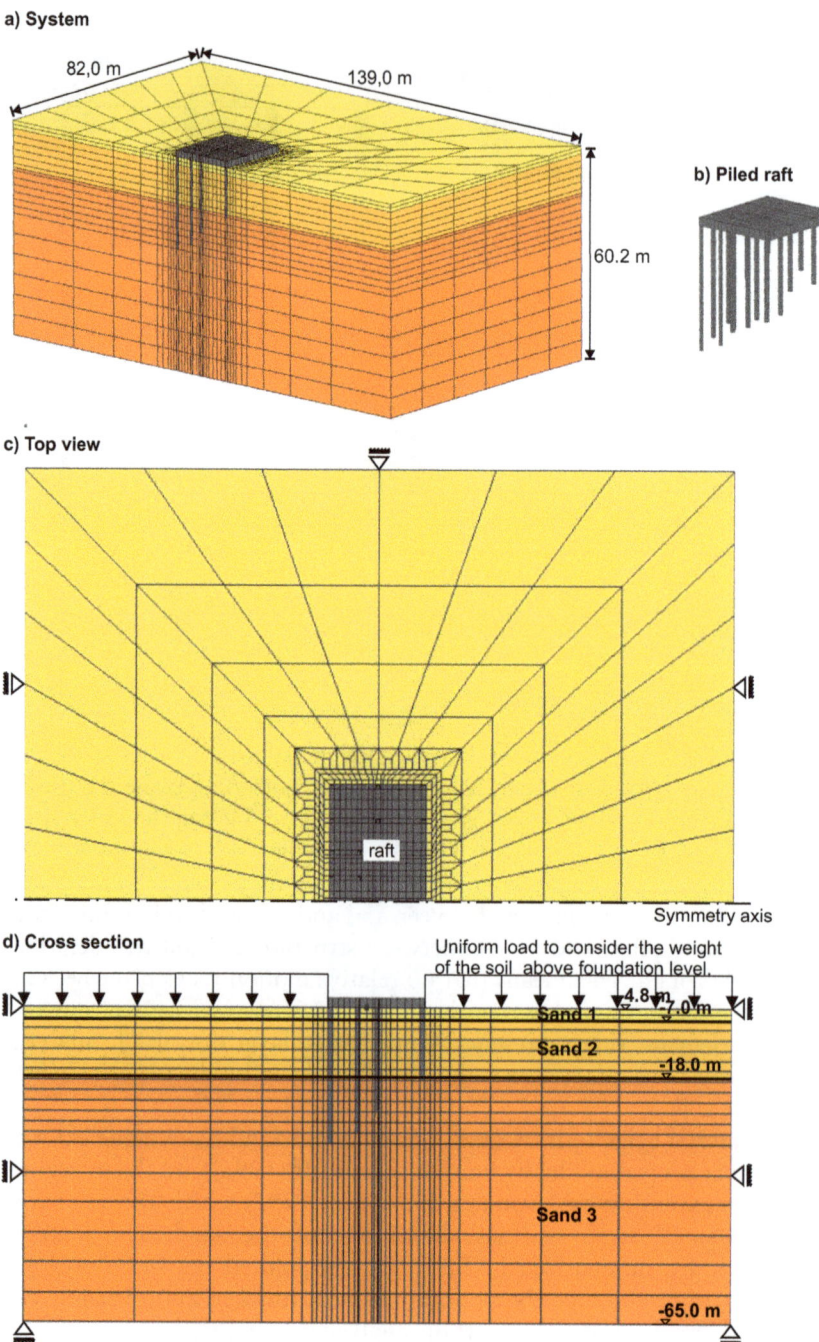

Figure 5.7 Finite element model for the CPRF.

Table 5.2 3D FEA: Material parameters for the hypoplastic model for the soil

Parameter			Sand 1	Sand 2	Sand 3
Total unit weight	γ	kN/m³	19.0	19.5	19.0
Buoyant unit weight	γ'	kN/m³	9.5	10.0	9.2
Critical angle of friction	φ'_{crit}	°	30	30	33
Granular hardness	h_s	MPa	5800	5800	5000
Exponent	n	-	0.54	0.54	0.35
Void ratio at critical state	e_{c0}	-	0.776	0.776	1.059
Void ratio at a state of maximum density	e_{d0}	-	0.460	0460	0.645
Maximum void ratio[a]	e_{i0}	-	0.892	0.892	1.218
Exponent	α	-	0.13	0.13	0.25
Exponent	β	-	1.00	1.00	1.00
Initial void ratio	$e_{initial}$	-	0.665	0.586	0.728
	R	-	$1 \cdot 10^{-4}$	$1 \cdot 10^{-4}$	$1 \cdot 10^{-4}$
	m_R	-	2.2	2.2	2.2
Parameters of the intergranular strain concept	m_T	-	1.1	1.1	1.1
	β_R	-	0.1	0.1	0.1
	χ	-	5.5	5.5	5.5
	ϑ	-	5.5	5.5	5.5

[a] representing the theoretical isotropic normal compression line of a loose soil skeleton in a gravity-free space ($e_{i0} = 1.15 e_{max}$)

using linear-elastic-ideal-plastic models. In this study the hypoplastic model was chosen simply for convenience because

- the dependence of soil stiffness on the stress level and the soil density is incorporated,
- the intergranular strain extension defines a small strain stiffness which makes the definition of the bottom boundary of the model less crucial and
- for different sandy soils material parameter sets are easily available (e.g. Mašín 2019).

However, the initial elastic stiffness profile for the soil model may be approximated by a shear modulus profile of $G = 15 + 1.54z'$ MPa, where z' is the depth below the underside of the raft. Allowing for bedrock at a depth of 65 m (60.2 m below the raft), the raft stiffness (total applied load divided by central settlement) assuming a fully flexible raft may be estimated as about 3740 MN/m (Brown and Gibson 1979). A similar calculation for the pile group, using the PIGLET program (Randolph 2021) assuming a rigid pile cap, gives a stiffness of 5310 MN/m.

The material behaviour of raft and piles has been modelled linear-elastic with the material parameters summarised in Table 5.3. Since the raft in all

218 Combined Pile-Raft Foundations

Table 5.3 3D FEA: Material parameters for raft and piles

Parameter		Unit	Raft	Piles
Young's modulus	E	MPa	34000 (23900[a]/157000[b])	34000
Poisson's ratio	ν	—	0.2	0.2
Total unit weight	γ	kN/m³	25	25
Buoyant unit weight	γ'	kN/m³	15	15

[a] equivalent Young's modulus for the pile foundation and the CPRF (t_r = 1.6 m)
[b] equivalent Young's modulus for the raft foundation (t_r = 3.0 m)

Table 5.4 Step-by-step analysis of the construction process in the 3D FEA

Step	Raft foundation P_k [MN]	Pile foundation, CPRF P_k [MN]
1 In situ stress state	-	-
2 Excavation to foundation level	-	-
3 Pile installation	-	-
4 Application of the weight of raft	62.7	33.4
5 Application of the raft stiffness	62.7	33.4
6 $P_k = G_{raft,k} + 0.5 G_k$	218.9	189.7
7 $P_k = G_{raft,k} + G_k + Q_k/3 - U$	363.1	345.5
8 $P_k = G_{raft,k} + G_k + Q_k$	439.2	410.0
9 to 14 Successive loading until $P_k = G_{raft,k} + 4 \cdot (G_k + Q_k)$	1568.9	1539.6

$G_{raft,k}$: weight of the raft; G_k: dead load; Q_k: live load; U_k: uplift

configurations was modelled with a thickness of t_r = 1.8 m, an equivalent Young's modulus was applied to achieve the appropriate bending stiffness for rafts with a thickness of t_r = 1.6 m (pile foundation, CPRF) and t_r = 3.0 m (raft foundation). In a real project, it is suggested to consider the stiffness of the superstructure by an equivalent bending stiffness of the raft. This equivalent bending stiffness should be provided by the structural engineer based on preliminary investigations.

Table 5.4 summarises the step-by-step analysis of the construction process. The piles were modelled as "wished-in-place", i.e. ignoring any changes in the soil surrounding the pile caused by the installation process. The weight of the raft minus the uplift was applied over the whole raft area before the raft stiffness was included in the model. The load of the superstructure was applied on top of the raft according to Figure 5.1.

5.5.2 3D FEA – results

All results presented here relate to analysis step 5 (Table 5.4); i.e. settlements and resistances caused by the weight of the raft are not considered. Figure 5.8 documents the load settlement curves of the raft foundation, the

Design example 219

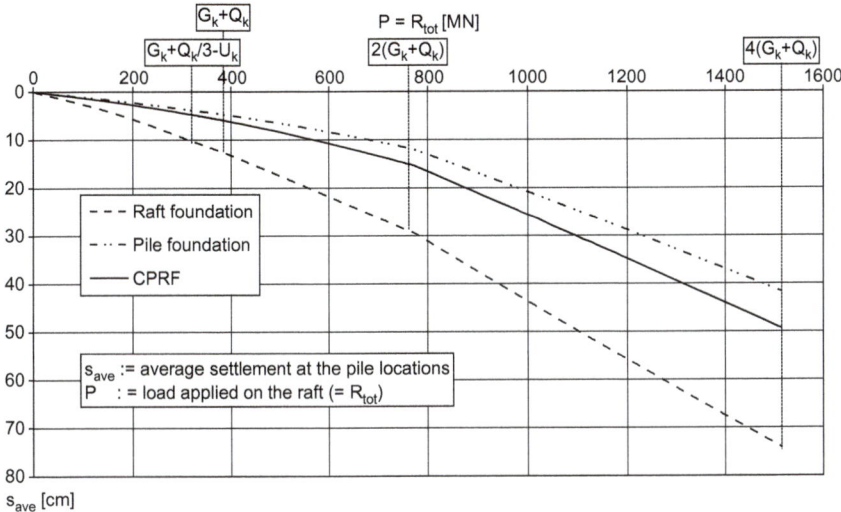

Figure 5.8 3D FEA: Load-settlement curves for the different foundation configurations.

pile foundation and the CPRF each loaded up to $P_k = 4 \cdot (G_k + Q_k)$, with none of the foundations showing excessive non-linear behaviour. This indicates that the ultimate capacity is not yet reached, which is not surprising since the ultimate capacity of the raft foundation estimated by means of a Finite Element Limit Analysis (FELA) with the program Optum gives an ultimate capacity of $R_{raft,ult} = 2487$ MN. For the FELA which is based on associated flow, equivalent angles of friction of $\varphi_1^* = \varphi_2^* = 27.5°$ and $\varphi_3^* = 29.7°$ were applied using the approach by Davis (1968) adopting an angle of dilatancy of $\psi_1^* = \psi_2^* = \psi_3^* = 5°$. The FELA was carried out with so-called mixed elements which supposedly produce a solution that falls between the lower and upper bounds.

Table 5.5 summarises main analysis results, namely average settlement at the pile positions s_{ave}, maximum angular distortion α_{max} and piled raft

Table 5.5 Results of the 3D FEA

Foundation configuration	$P_k = G_k + Q_k/3 - U_k$			$P_k = G_k + Q_k$		
	s_{ave} [cm]	α_{max} [-]	α_{pr} [-]	s_{ave} [cm]	α_{max} [-]	α_{pr} [-]
Raft foundation	9.8	1/174	0.000	12.7	1/168	0.000
Pile foundation (piled raft with conventional design approach)	3.8	1/569	0.806	4.7	1/567	0.793
CPRF	4.8	1/605	0.712	6.0	1/625	0.693

G_k: dead load; Q_k: live load; U_k: uplift; s_{ave}: average settlement at the pile locations; α_{max}: maximum angular distortion; α_{pr}: piled raft coefficient

coefficient α_{pr}, for the load cases $P_k = G_k + Q_k/3 - U_k$ and $P_k = G_k + Q_k$ which in design practice are frequently taken as relevant for evaluating the serviceability and for which the design parameters Winkler modulus and equivalent pile spring stiffness are evaluated for the corresponding structural analysis, respectively. It has to be noted that the piled raft coefficient, α_{pr}, describes the ratio of the sum of all pile mobilised resistances to the total resistance mobilised by the foundation which is equivalent to the total load on the foundation (Sections 1.2 and 2.2).

Contour plots of settlements for all three foundation configurations are plotted in Figure 5.9 for the load case $P_k = G_k + Q_k/3 - U_k$ which is assumed to give some sort of best estimate for the real behaviour of the building. The CPRF shows the least significant variation of settlements which consequently results in small angular distortion values (Table 5.5).

Figure 5.10 shows the typical decrease of the piled raft coefficient with increasing load level. As can be expected (Section 2.2), this decrease is more pronounced for the CPRF.

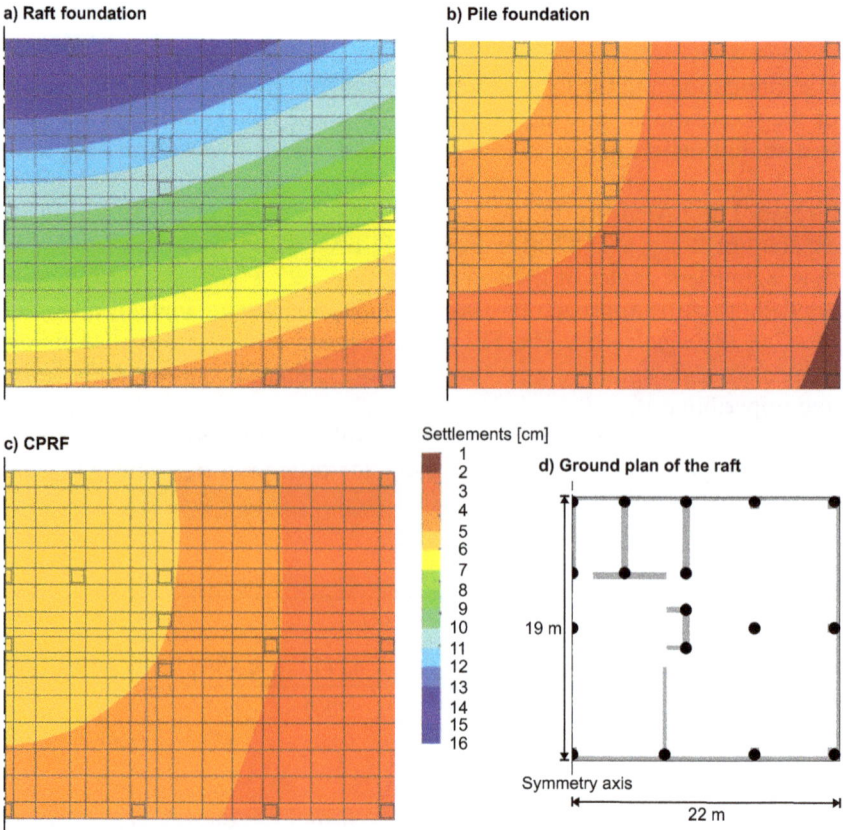

Figure 5.9 3D FEA: Contour plot of settlements ($P_k = G_k + Q_k/3 - U_k$).

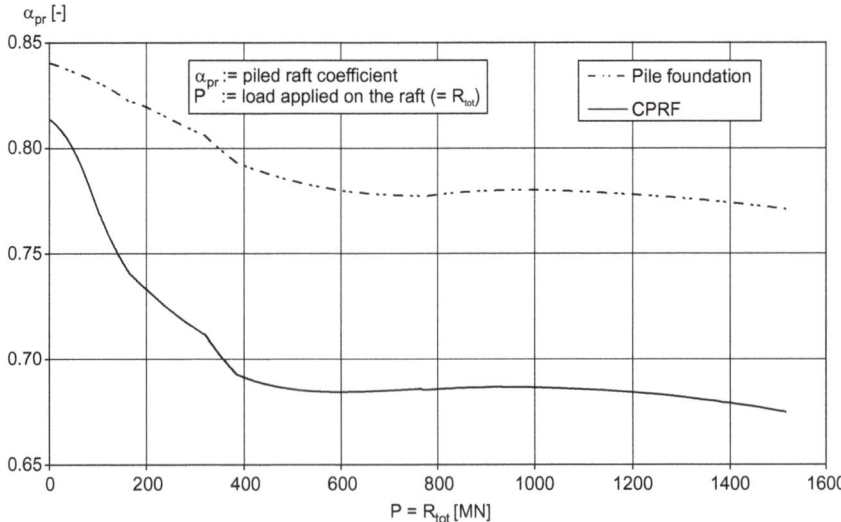

Figure 5.10 Variation of the piled raft coefficient with the load applied on the foundation.

For the pile foundation and the CPRF, the load distribution along the pile shaft is plotted in Figure 5.11 for four different piles for the load case $P_k = G_k + Q_k$.

5.5.3 Proof of the external bearing capacity for the ULS

The ULS was investigated by increasing the load applied on the raft to $4 \cdot (G_k + Q_k)$. Taking the total resistance of the different foundations at $4 \cdot (G_k + Q_k)$, $R_{tot,k} = 1512$ MN, and the respective factors of safety from the German code DIN 1054 as $\gamma_G = 1.35$, $\gamma_Q = 1.50$ and $\gamma_{R,v} = 1.40$ (for the ultimate capacity of a shallow foundation) one gets

$$E_d = G_k \gamma_G + Q_k \gamma_Q = 375\,\text{MN} \cdot 1.35 + 58\,\text{MN} \cdot 1.50 = 594\,\text{MN}$$
$$< R_{tot,d} = \frac{1}{\gamma_{R,v}} R_{tot,k} = \frac{1}{1.40} 1512\,\text{MN} = 1080\,\text{MN}$$

where G_k includes the weight of a 3-m-thick raft.

Therefore the proof of the external bearing capacity for the ULS according to ISSMGE TC 212 (2013) is fulfilled for CPRF as well as for the raft foundation and for the pile foundation.

5.5.4 Proof of the external serviceability for the SLS

The proof of the external serviceability for the SLS is frequently carried out for the settlement-inducing load which in this example is taken as the load

222 Combined Pile-Raft Foundations

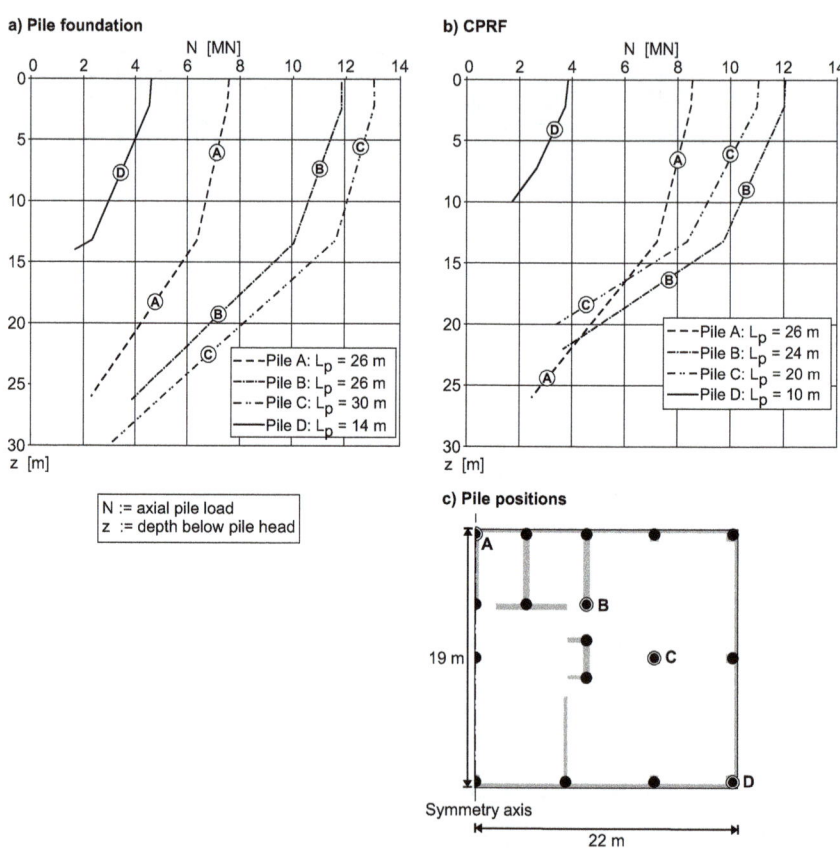

Figure 5.11 Pile load distribution for different pile positions ($P_k = G_k + Q_k$).

case $P_k = G_k + Q_k/3 - U_k$. The relevant SLS criteria have to be defined for each structure considering the specific constraints and limitations of the project. In this example, as defined in Section 5.4.1, there is one relevant SLS criterion to be met, namely for the angular distortion:

$$C_d = \alpha \leq 1/500$$

Taking the values documented in Table 5.5 one gets

Raft foundation: $\alpha_{max} = 1/174 > 1/500$
Pile foundation: $\alpha_{max} = 1/569 < 1/500$
CPRF: $\alpha_{max} = 1/605 < 1/500$

Therefore, the proof of the external serviceability for the SLS according to ISSMGE TC 212 (2013) is fulfilled for CPRF and for the pile foundation but not for the raft foundation.

5.6 STRUCTURAL ANALYSIS OF RAFT AND PILES

5.6.1 Structural analysis of raft and piles – model

The structural analysis was carried out with the program package SOFiSTiK Structural Desktop (SSD) (SOFiSTiK AG 2023a). The elastically supported raft was modelled with the 2D shell elements implemented in the program ASE (SOFiSTiK AG 2023b) where plate structural behaviour is based on Mindlin's plate theory as described in the implementations of Hughes and Tezduyar (1981), Tessler and Hughes (1983) and Crisfield (1984). For the piles the linear spring elements implemented in the program ASE (SOFiSTiK AG 2023b) were applied. The material parameters for steel and concrete considered in the structural analysis are summarised in Tables 5.6 and 5.7.

The equivalent pile spring stiffness and the distribution of Winkler's modulus shown in Figures 5.12–5.14 for the different configurations were derived from the 3D FEA for a load level of $P_k = G_{raft,k} + G_k + Q_k$ as is generally done in design practice. Again, as in the 3D FEA, in this simplified analysis only the raft has been considered, but no walls and columns of the

Table 5.6 Structural analysis: Material parameters for steel

Parameter		Unit	
Grade	-	-	B 500 B
Young's modulus	E	MPa	200000
Poisson's ratio	ν	—	0.3
Nominal elastic limit	$f_{y,k}$	MPa	500
Characteristic tensile strength	$f_{t,k}$	MPa	540
Density	ρ	kg/m³	7850[a]

[a] weight of the raft is not considered in the design

Table 5.7 Structural analysis: Material parameters for concrete

Parameter		Unit	
Grade	-	-	C35/45
Young's modulus	E	MPa	34000
Poisson's ratio	ν	—	0.2
Characteristic compression strength	$f_{c,k}$	MPa	35
Characteristic tensile strength	$f_{ct,k,0.05}$	MPa	2.2
Density	ρ	kg/m³	2400[a]

[a] Weight of the raft is not considered in the design

224 Combined Pile-Raft Foundations

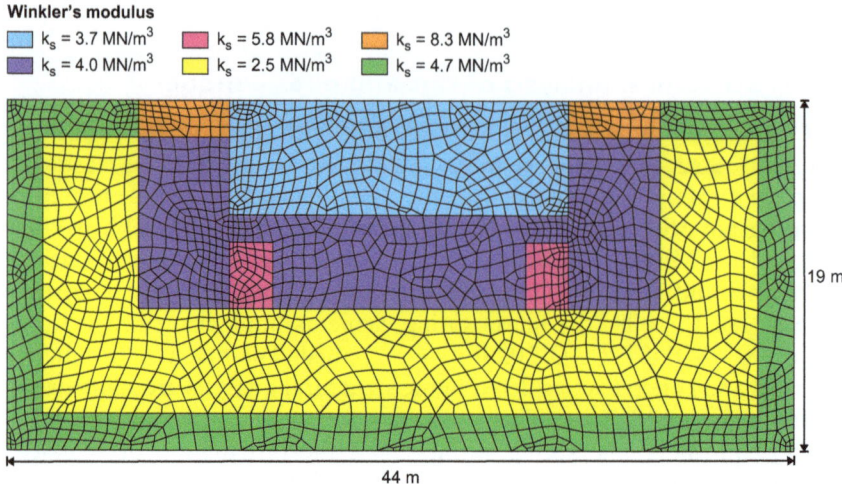

Figure 5.12 Raft foundation: Winkler's modulus in the structural analysis.

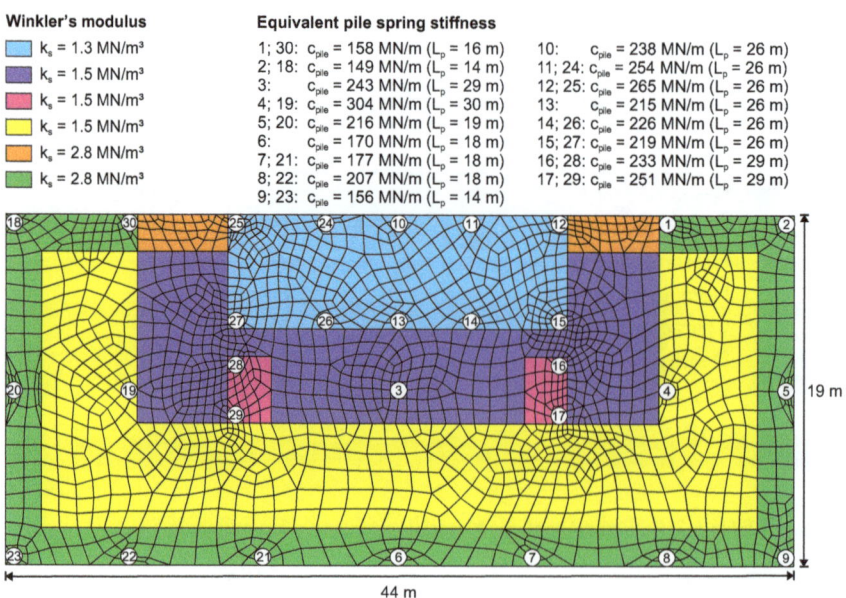

Figure 5.13 Pile foundation: Winkler's modulus and equivalent spring stiffness in the structural analysis.

Design example 225

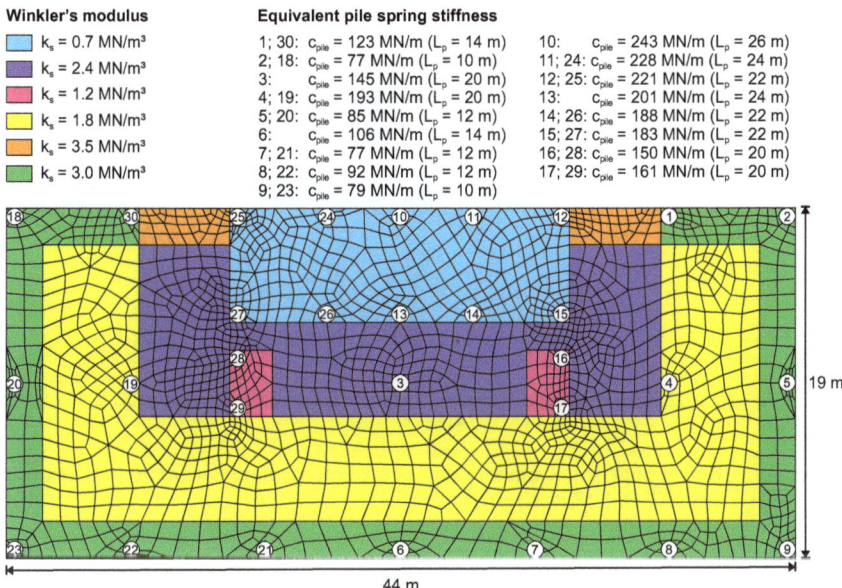

Figure 5.14 CPRF: Winkler's modulus and equivalent spring stiffness in the structural analysis.

superstructure/basement. The distribution of the Winkler's modulus was simplified as would be the case in a structural analysis.

5.6.2 Comparison of the structural analysis model and 3D FEA

To verify the structural analysis model, a comparison of the calculated deformations with the results of the 3D FEA for comparable load cases is useful. For the three different foundation configurations, Figures 5.15–5.17 compare the settlements achieved with structural analysis model with the results achieved with the 3D FEA for the load case $P_k = G_k + Q_k$; i.e. as in Section 5.5.2, settlements caused by the weight of the raft are not considered.

In general, a reasonable agreement was achieved. A better match could be obtained if the distribution of Winkler's modulus would be refined. However, it was decided to choose a relatively simple approximation (i.e. not too many areas with different moduli) as would be typical of the approach in a real project.

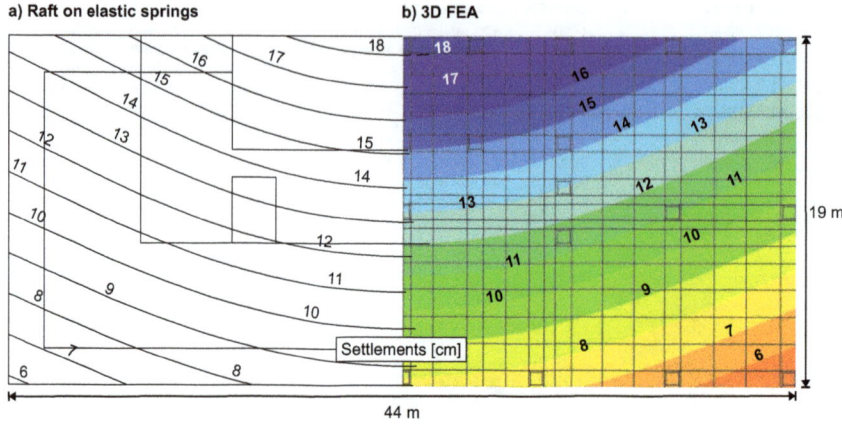

Figure 5.15 Raft foundation: Comparison of settlements between structural analysis and 3D FEA ($P_k = G_k + Q_k$).

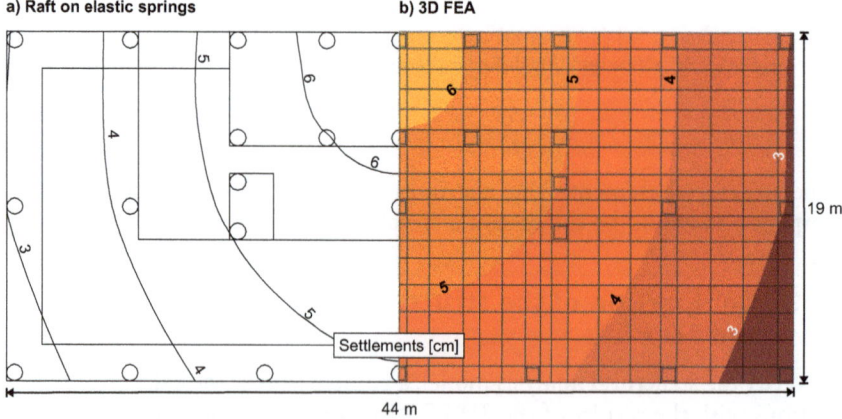

Figure 5.16 Pile foundation: Comparison of settlements between structural analysis and 3D FEA ($P_k = G_k + Q_k$).

5.6.3 Structural design

An estimation of the required amount of reinforcement steel for the raft was carried out based on the German versions of Eurocodes EC 1990 "Basis of structural design" (DIN EN 1990:2021-10) and EC 1992 "Design of concrete structures" (DIN EN 1992-1-1:2021-10) using the program BEMESS (SOFiSTiK AG 2023c). As a conservative approach, no combination coefficients ψ_i were applied to reduce the live loads, i.e.:

$$E_{d,ULS} = G_k \gamma_G + Q_k \gamma_Q \tag{5.2}$$

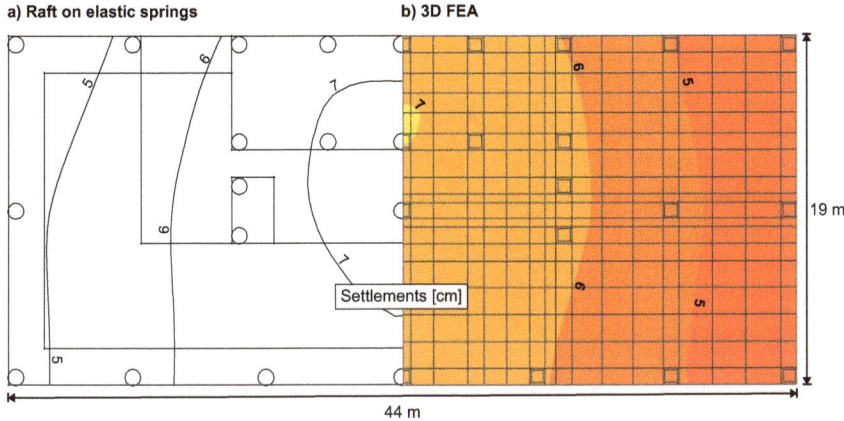

Figure 5.17 CPRF: Comparison of settlements between structural analysis and 3D FEA ($P_k = G_k + Q_k$).

$$E_{d,SLS} = G_k + Q_k \qquad (5.3)$$

where $E_{d,ULS}$ = design value for the effect of actions in the ULS; $E_{d,SLS}$ = design value for the effect of actions in the SLS; G_k = characteristic permanent load; Q_k = characteristic live (variable) load and γ_G, γ_Q = partial factors of safety.

Assuming that the raft transfers its own weight directly to the soil, the weight of the raft was not considered in the design. For the control of the crack width in the SLS crack width of $w_k = 0.2$ mm (soil side) and $w_k = 0.4$ mm (air side) were considered.

For the estimation of the required amount of reinforcement steel for the bored piles additionally the German version of the European code on bored piles (DIN EN 1537:2015-10) was applied considering geometrical imperfections of $e = 0.1$ m (load eccentricity at the pile head) and $i = 0.02$ m/m (pile inclination). The estimation of reinforcement steel was carried out using the program DC-Pfahl (DC-Software Doster & Christmann GmbH 2023) considering for each pile the pile load distribution derived from the 3D FEA for load case $P_k = G_k + Q_k$ as exemplified in Figure 5.11 for piles A, B, C and D.

Tables 5.8 and 5.10 summarise the design raft bending moments derived from the structural analysis which are the base for the determination of the required reinforcement. The costs for the different foundation configurations are shown in Table 5.11, where the costs of the piles and raft are scaled to the total cost of the CPRF and a ratio $p_{steel}/p_{concrete} = 15.3$ for the costs of 1 ton of reinforcement steel and 1 ton of concrete is assumed. It has to be mentioned that for the raft foundation additional costs would have to be considered due to the deeper excavation pit as was discussed for the case history of the WestendDuo (Section 4.3.2).

228 Combined Pile-Raft Foundations

Table 5.8 Partial factors of safety and reduction factors considered in the simplified design

Parameter		
Partial factor of safety for permanent loads	γ_G	1.35
Partial factor of safety for live loads	γ_Q	1.50
Partial factor for the material resistance of steel	γ_s	1.15
Partial factor for the material resistance of concrete	γ_c	1.50
Reduction factor to consider long term effects on the concrete	α_{cc}	0.85

Table 5.9 Design values for raft bending moments: SLS

			Raft foundation		Pile foundation		CPRF	
Parameter			Min	Max	Min	Max	Min	Max
Bending moment	m_{xx}	kNm/m	−1931	19223	−4004	4677	−4860	4719
Bending moment	m_{yy}	kNm/m	−2538	5132	−3070	2678	−3012	2660

Table 5.10 Design values for raft bending moments: ULS

			Raft foundation		Pile foundation		CPRF	
Parameter			Min	Max	Min	Max	Min	Max
Bending moment	m_{xx}	kNm/m	−2913	27230	−5677	6525	−6621	6801
Bending moment	m_{yy}	kNm/m	−4106	8417	−4260	4817	−4196	6206

Table 5.11 Comparison of relative costs for the different configurations

	Raft		Piles		
Configuration	Steel	Concrete	Steel	Concrete	Total
Raft foundation	0.437	1.013	-	-	1.450
Pile foundation	0.206	0.540	0.156	0.175	1.077
CPRF	0.208	0.540	0.117	0.135	1.000

REFERENCES

Brown, P.T., Gibson, R.E. (1979). Surface settlement of a finite elastic layer whose modulus increases linearly with depth. *International Journal for Numerical and Analytical Methods in Geomechanics*, 3, 37–47.
Crisfield, M.A. (1984). A quadratic Mindlin element using shear constraints. *Computers & Structures*, 18, 833–852.
DC-Software Doster & Christmann GmbH (2023). DC-Pfahl für Windows. Benutzerhandbuch, Version 8.4.
DIN EN 1536:2015-10 (2015). Execution of special geotechnical work – Bored piles; German version EN 1536:2010+A1:2015.
DIN EN 1990:2021-10 (2021). Eurocode: Basis of structural design; German version EN 1990:2002 + A1:2005 + A1:2005/AC:2010.
DIN EN 1992-1-1:2011-01 (2011). Eurocode 2: Design of concrete structures – Part 1-1: General rules and rules for buildings; German version EN 1992-1-1:2004 + AC:2010.
DIN EN 1997-1:2014-03 (2014). Eurocode 7: Geotechnical design – Part 1: General rules. German version EN 1997-1:2004 + AC:2009 + A1:2013.
DIN 1054:2021-04 (2021). Subsoil – Verification of the safety of earthworks and foundations – Supplementary rules to DIN EN 1997-1.
Davis, E.H. (1968). Theories of plasticity and failure of soil masses. In *Soil mechanics: selected topics*, ed. Lee, I.K., 341–354, New York: Elsevier.
Hughes, T.J.R., Tezduyar, T.E. (1981). Finite elements based upon Mindlin plate theory with particular reference to the four-node bilinear isoparametric element. *Journal of Applied Mechanics*, 48, 3, 587–596.
ISSMGE Technical Committee TC 212 (2013). ISSMGE Combined Pile-Raft Foundation Guideline. *Report of the ISSMGE Technical Committee TC 212 – Deep Foundations*, ed. Katzenbach, R., Choudhury, D., Technische Universität Darmstadt.
Mašín, D. (2019). Modelling of soil behaviour with hypoplasticity – Another approach to soil constitutive modelling. Springer Series in Geomechanics and Geoengineering, Springer Nature Switzerland AG. https://doi.org/10.1007/978-3-030-03976-9
Niemunis, A., Herle, I. (1997). Hypoplastic model for cohesionless soils with elastic strain range. *Mechanics of Cohesive-frictional Materials*, 2, 4, 279–299.
Tessler, A., Hughes, T.J.R. (1983). An improved treatment of transverse shear in the Mindlin-type four-node quadrilateral element. *Computer Methods in Applied Mechanics and Engineering*, 39, 311–335.
Tochnog Professional Company (2022). TOCHNOG PROFESSIONAL - User's Manual. Tochnog Professional Company, https://www.tochnogprofessional.nl, Last access on January 20th 2020.
Randolph, M.F. (2021). PIGLET: Analysis and design of pile groups. Users' Manual, Version 6-2, Perth.
SOFiSTiK AG (2023a). SOFiSTiK Structural Desktop (SSD). Service Pack 2023-1 Build 257.

SOFiSTiK AG (2023b). ASE - General Static Analysis of FE Structures. ASE Manual, Service Pack 2023-1 Build 41.

SOFiSTiK AG (2023c). BEMESS - Design of Plates and Shells. BEMESS Manual, Service Pack 2023-1 Build 41.

von Wolffersdorff, P.-A. (1996). A hypoplastic relation for granular materials with a predefined limit state surface. *Mechanics of Cohesive-frictional Materials*, 1, 3, 251–271.

Appendix

APPENDIX A: NOTATION

Unless stated otherwise, all parameters are characteristic values; i.e. no factors of safety have been applied. Units are listed according to the International System of Units (Système International SI), but alternative units may be adopted according to local practice.

Symbol	Name	Unit
a_{eq}	equivalent circular raft radius	m²
a_{gr}	pile group-raft area ratio	-
A_b	pile base cross section area	m²
A_g	plan area of the pile group as a block	m²
A_r	raft area	m²
A_s	pile shaft area	m²
c_{pile}	equivalent spring stiffness for the pile	MN/m
d_{eq}	diameter of the equivalent pier	m
d_p	pile diameter	m
e	distance between pile axes, i.e. pile spacing	m
E	Young's modulus	MPa
E_d	design value of actions	MN
G	dead load	MN
k_s	modulus of subgrade reaction (Winkler's modulus)	MN/m³
K_0	coefficient of earth pressure at rest	-
K_{rs}	raft-soil stiffness ratio for rectangular rafts	-
L_p	pile length	m
n_p	number of piles in the piled raft/pile group	-
N	axial pile load	MN

(Continued)

Symbol	Name	Unit
P_{eff}	effective applied load caused by the superstructure and the raft under consideration of the uplift	MN
P_s	load caused by the superstructure	MN
q_b	pile base resistance per unit area	kN/m²
q_s	pile shaft resistance per unit area / shaft friction	kN/m²
Q	live load	MN
R	pile resistance (equivalent to N at the pile head)	MN
R_b	pile base resistance	MN
$R_{b,ult}$	ultimate pile base capacity	MN
R_d	design value of resistances	MN
R_{raft}	resistance of the raft	MN
R_s	pile shaft resistance	MN
$R_{s,ult}$	ultimate pile shaft capacity	MN
R_{tot}	total resistance of the foundation	MN
R_{ult}	ultimate capacity of a pile	MN
s	settlement	cm
s_g	limit settlement of the pile head to establish the ultimate pile capacity (= $0.1 d_p$)	cm
s_u	undrained shear strength	kPa
t_r	thickness of the raft	m
u	pore pressure	kPa
U	uplift	MN
V_{ult}	ultimate capacity of a vertically loaded raft foundation	MN
α_{pr}	piled raft coefficient	-
φ	angle of friction	°
γ	unit weight	kN/m³
γ_G	partial factors of safety for dead loads	-
γ_Q	partial factors of safety for live loads	-
γ_R	partial factors of safety for resistances	-
η	global factor of safety	-
ν	Poisson's ratio	-
ΣR_{pile}	sum of all pile resistances of a pile group/piled raft	MN

APPENDIX B: SUMMARY OF CASE HISTORIES ON PILED RAFTS

Appendix B.1: General remarks

Only structures founded on piled rafts (for definition see chapter 1.2) have been in included in this summary where settlement measurements have been documented. Obviously, this summary is far from being complete.

The authors would be grateful for readers to contribute additional case histories on piled rafts to be included in future editions of this book.

Tables B.1 and B.2 explain the table headers and abbreviations used in the following summary of case histories on piled rafts.

Table B.1 **Table headers**

Table header	Explanation
year	year of completion
soil	dominating soil/rock layers dominating the foundation behaviour
H	height of the building above ground level
P	load of the structure as indicated in the literature which may include dead and live loads (*[1] indicates effective applied load caused by the superstructure and the raft under consideration of the uplift)
A_r	area of the raft
$t_{r,max}$	maximum raft thickness
$z_{r,max}$	maximum depth of the raft below ground level
n_p	number of piles
L_p	pile length
d_p	pile diameter
R_{max}	bandwidth of maximum measured pile resistances (axial pile load at the pile head) for the instrumented piles after the building is finished
s_{max}	maximum measured settlement
t	time of settlement measurement after completion of construction of building (*[2] indicates completion of shell only)

Table B.2 **Abbreviations**

Abbreviation	Explanation
b	pile base
CPRF	foundation designed according to DGGT/DIBt (2001) or ISSMGE Technical Committee TC 212 (2013), respectively (foundations not indicated as such have not been necessarily designed according to other design approaches)
N/A	information not available
bo	boulder
cl	clay
fi	artificial fill
gd	glacial drift
gr	gravel
ll	loess loam

(Continued)

Table B.2 (Continued) Abbreviations

Abbreviation	Explanation
ma	marl
mo	molasse
os	organic soil
pd	pyroclastic deposit
sa	sand
si	silt
casi	calcisiltite
clst	claystone
lst	limestone
sst	sandstone
sist	siltstone
ro	rock
wro	weathered rock
oc	overconsolidated

Concerning the maximum measured settlement s_{max} it has to be pointed out that for many case histories the reference date of the initial measurement is not known. Therefore, it is possible that in a number of case histories the initial settlements due to the weight of the raft or even the basement might not be included.

In a number of case histories, especially for the examples from Japan, the load of the structure has been calculated from an average contact pressure at the base of the raft and the area of the raft.

Appendix B.2: Regional summaries

Table B.3 London, UK

Building	Year [-]	Soil [-]	H [m]	P [MN]	A_r [m²]	$t_{c,max}$ [m]	$\tilde{z}_{c,ma}$ [m]	n_p [-]	L_p [m]	d_p [m]	R_{max} [MN]	s_{max} [cm]	t [a]
Shell Centre Williams (1957), Green and Cocksedge (1975)	1962	oc cl	107	588	1344	0.9	14.6	49	16÷21	1.4÷2.1/ 3.4÷4.6[b]	N/A	5.9	11
Cambridge Road Morton and Au (1975)	1967	oc cl	N/A	106	496	0.7	N/A	116	15.3	0.6	N/A	3.1	6
Commercial Union Building Green and Cocksedge (1975)	1968	oc cl	118	513	1386	3.0	23.0	12	34	2.0/4.7[b]	N/A	4.0	4
Hyde Park Cavalry Barracks Hooper (1973)	1970	oc cl	90	228	625	1.52	8.8	51	24.8	0.9/2.4[b]	2.9÷3.7	2.1	3
Guy's Hospital - Communications Tower Mould (1975)	1971	oc cl	135	366	856	3.05	8.5	N/A	20	0.8	N/A	2.0	2[*2]
Guy's Hospital - User Tower Mould (1975)	1971	oc cl	122	681	1595	3.05	8.5	N/A	20	0.8	N/A	2.5	2[*2]
Barbican Tower I Butler (1975)	1972	oc cl	N/A	394	680	1.8	1.8	N/A	21.7	N/A	N/A	3.4	2
Barbican Tower III Butler (1975)	1972	oc cl	N/A	351	900	1.8	1.8	N/A	17.1	N/A	N/A	3.3	2
Dashwood-House Hight and Green (1976), Green and Hight (1976)	1975	oc cl	68.4	286	1102	1.5	8,1	462	15.0	0.485	0.2÷0.7	3.3	0
Stonebridge Park Cooke et al. (1981)	1975	oc cl	41.6	156	831	0.9	0.9	351	13.0	0.5	0.3÷0.7	1.7	4
National Westminster Bank Tower Hooper (1979)	1978	oc cl	185	1442	2290	2.0 4.5	14.5	375	26.5	1.2/ 2.1[b]	N/A	4.0	0
Queen Elizabeth II Conference Centre Burland and Kalra (1986), Kalra and Willows (1986), Price and Wardle (1986)	1986	oc cl	34.5	N/A	2260	2.0	13.7	8 21	16.0 4.4	1.8 4.0	6.4	2.0	0

Table B.4 Frankfurt am Main, Germany

Building	Year [-]	Soil [-]	H [m]	P [MN]	A_r [m²]	$t_{r,max}$ [m]	$z_{t,max}$ [m]	n_p [-]	L_p [m]	d_p [m]	R_{max} [MN]	s_{max} [cm]	t [a]
Torhaus (Sections 1.3 and 2.2.5)	1985	oc cl	130	2×200[*1]	2×429	2.5	3.0	2×42	20.0	0.9	1.7÷6.9	14.0	2.00
Messeturm (Section 2.3.3)	1990	oc cl	256	1570[*1]	3457	6.0	14.0	64	26.9÷34.9	1.3	5.8÷20.1	14.4	8.00
American Express (Ripper and El Mossallamy 1999, Reul 2000)	1993	oc cl	75	723[*1]	3575	2.0	14.0	35	20.0	0.9	3.0÷5.3	5.5	1.00[*2]
Westend 1 (Section 3.2.2)	1993	oc cl	208	950[*1]	2940	4.7	14.5	40	30.0	1.3	9.2÷14.9	12.0	2.50[*2]
Congress Center (Barth and Reul 1997, Reul 2000)	1996	oc cl	52	1440[*1]	10200	2.7	14.2	141	12.5÷34.5	1.3	2.4÷5.9	5.8	0.00
Japan Center (Lutz et al. 1996, Ripper and El Mossallamy 1999)	1996	oc cl	115	630[*1]	1920	3.5	15.8	25	22.0	1.3	7.9÷13.8	6.5	0.50[*2]
Commerzbank II (Katzenbach et al. 1994, Holzhäuser 1998)	1997	oc cl/ lst	259	1346[*1]	2690	4.5	9.0	111	37.6÷45.6	1.8/1.5	5.9÷21.3	2.1	0.00
Forum – Kastor (Lutz et al. 1996)	1997	oc cl	95	750[*1]	2830	3.0	13.5	26	20.0÷30.0	1.3	5.0÷12.6	5.5	0.00[*2]
Forum – Pollux (Lutz et al. 1996)	1997	oc cl	130	760[*1]	1920	3.0	13.5	22	30.0	1.3	7.4÷11.7	7.0	0.00[*2]
Eurotheum[CPRF] (Katzenbach et al. 1998, Katzenbach et al. 2000, Moormann 2002)	1999	oc cl	110	425[*1]	1893	2.5	13.0	25	25.0÷30.0	1.5	1.8÷6.1	2.9	1.00[*2]
MAIN TOWER[CPRF] (Katzenbach et al. 1998a, Katzenbach et al. 2000, Moormann 2002)	1999	oc cl	200	1470[*1]	3800	3.8	21.0	112	30.0	1.5	1.4÷8.0	2.5	0.00
Neue Messehalle 3[CPRF] (Section 4.3.4)	2001	oc cl	45	2×319[*1] 2×59[h]	2×2824	1.4	4.1	2×14	15.0	1.5	N/A	4.4 0.8[u]	2.00

Building	Year	Type											
Main Forum^{CPRF}	2003	oc cl	80	849*1	7579	2.8	13.4	48	18.5÷20.0	1.2	3.3÷6.7	3.2	0.00*2
Westhafen Tower^{CPRF} (Section 2.3.3)	2003	oc cl	112	490*1	3731	4.2	21.4	32	12.5÷15.0	1.5	0.7÷10.9	3.5	0.00*2
Skyper (Richter and Lutz 2010)	2004	oc cl	154	810*1	1900	3.5	13.4	46	31.0÷35.0	1.5	N/A	5.5	4.00
WestendDuo^{CPRF} (Section 4.3.1)	2006	oc cl	96	615*1	4100	1.8	15.0	26	20.0÷25.0	1.2	7.4	4.7	0.00
Park Tower^{CPRF} (Sectio 4.3.2)	2007	oc cl	110	186*1,x1	1000	2.7	10.8	16	33.5	1.4	1.3÷8.6	2.7	9.00
Messehalle 11^{CPRF}	2009	sa/ oc cl/ wro	36	1672*1	25010	0.8	7.1	200	12.0÷17.5	1.2	2.5÷3.5	1.6	0.00*2
Portalhaus Messe^{CPRF}	2009	sa/ oc cl/ wro	33	345*1	3336	0.8	4.6	52	15.0÷17.5	0.9÷1.2	1.2	1.0	0.00*2
Leo^{CPRF} (Reul et al. 2024)	2013	oc cl	67	866*1	5570	1.8	10.8	40^{x2} 115^{x3}	15.0÷20.0 7.0÷9.5^{x3}	1.2 0.9÷1.2^{x3}	2.1	1.6	0.25*2
European Central Bank^{CPRF}	2014	sa/lst/oc cl	185	2170*1	6800	2.6	8.7	97	20.0÷37.0	1.2	N/A	4.3^{x4}	6.00
Taunusturm^{CPRF}	2014	oc cl	170	1369*1	5400	3.0	13.2	46 12	20.0÷30.0 15.0	1.5 0.9	10.2÷13.9	5.7	0.00
Praedium^{CPRF}	2017	sa/ oc cl	66	756	4572	1.2	7.0	51	15.0÷20.0	0.9	1.7÷2.9	3.5	0.00
Messehalle 12^{CPRF}	2018	sa/oc cl	31	2414	26400	0.8	7.0	381	8.0÷41.0	1.2÷1.7	0.5÷3.4	2.8	1.00
Omniturm^{CPRF} (Section 4.3.3)	2019	oc cl	190	1170*1	2600	2.9	16.0	41 5	34.4÷27.4 12.0÷20.0	1.5 1.2	12.9÷14.2	6.6	0.50
Cascada^{CPRF}	2021	sa/ oc cl	66	332*1	2850	1.2	10.5	29	16–21	1.5	1.3÷6.0	1.8	1.00
Messehalle 5^{CPRF}	2023	oc cl	25.5	97*1	739^{x5}	0.8	1.0	16	150	1.2	0.7÷2.4	1.9	1.00
Tower One^{CPRF}	2023	oc cl	189	1374*1	5000	3.5	12.8	68	30.0÷45.0	1.5	N/A	8.4	1.25*2

h: horizontal load; u: horizontal displacement; *1: load related to the demolition of the SGZ Bank (Section 4.3.2); *2: additionally, some sections of the perimeter secant pile wall contribute in the load transfer; x3: jet grouting columns; x4: initial settlement due to the weight of the raft and the basement not included; x5: area of the building founded as a CPRF.

238 Appendix

Table B.5 Germany except Frankfurt am Main

Building	Year [-]	Soil [-]	H [m]	P [MN]	A_r [m²]	$t_{r,max}$ [m]	$z_{r,max}$ [m]	n_p [-]	L_p [m]	d_p [m]	R_{max} [MN]	s_{max} [cm]	t [a]
Bahn Tower, Berlin[CPRF] (Section 4.3.6)	1998	sa/gd	103	655[*1]	2600	2.5	9.0	44	15.0÷25.0	1.5	15.0	2.8	0.00
Beisheim Center BC-1, Berlin[CPRF] (Vrettos and Borchert 2011)	2004	sa/gd	70	N/A	1000	2.0	12.0	N/A	8.0÷10.0	1.2	N/A	2.6	0.00
Treptowers, Berlin[CPRF] (Section 4.3.7)	1998	sa	121	632[*1]	1376	3.0	8.0	54	12.5÷16.0	0.9	6.7÷7.0	7.3	0.00
Post Tower, Bonn[UCPR] (Löffler et al. 2001, Sommer et al. 2004)	2002	sa / clst	163	1100	7350	3.5	20.0	60	15.0	1.5	6.0	3.2	1.00
Power plant Boxberg, Brandenburg[CPRF] (Placzek and Jentzsch 1997, Jentzsch et al. 2001)	1999	sa	160	6160	15747	5.0	12.0	N/A	7.0÷8.5	0.5	N/A	5.5	1.00
Weser Tower, Bremen[CPRF] (Section 4.3.8)	2010	sa	82	269[*1]	836	1.8	7.0	31	16.0÷22.0	0.9	2.3÷3.9	3.3	0.00
Hochschule Darmstadt[CPRF] (Reul and Krajewski 2010)	2012	sa	65.5	37[*1X2]	115[X1]	1.0	7.4	18	10.0÷13.0	0.9	0.6÷1.3	0.7	0.00
Kesselhaus Darmstadt[CPRF] (Pitteloud et al. 2004)	2002	sa/cl/ sa	50.0	31[*1]	289	1.8	1.8	12	8.5	1.2	0.7÷0.8	1.0	0.50
Zentrale Arcor, Eschborn (Thaher et al. 2002)	2002	sa/cl	N/A	920	10000	2.0	11.5	57	10.0	1.3	0.6÷2.5	3.0	0.00[*2]
Deutsche Messe AG, Hannover (Katzenbach et al. 2000)	1998	gd/sa/gr/ cl/clst	81.4	N/A	952	N/A	6.6	32	17.5	1.2	N/A	1.4	0.00[*2]
Haus der Wirtschaft, Offenbach[CPRF] (Section 4.3.5)	1999	oc cl	68	605[*1]	5120	2.0	8.5	47 6	25.0 37.5÷41.0	1.2	1.4÷3.1	2.5	0.00
Hegau Tower, Singen[CPRF] (Section 4.3.9)	2008	gr/gd	68	350	3400	2.0	8.3	11 17	7.5÷9.0 10.0÷11.0	1.2 0.9	1.7÷3.9	1.7	0.00[*2]
R+V Versicherung, Wiesbaden[CPRF]	2010	lll/ oc cl	21.3	412[*1]	5085	0.8	4.5	168	13.0÷17.5	0.9	0.9÷1.9	0.5[X3]	0.00

[UCPR]: piles and raft are uncoupled by means of a crushable foam layer at the pile head; [X1]: piles and raft are uncoupled by means of crushable foam; [X2]: only extension area of the building; [X3]: at the time of the initial measurement the raft, the one-storey basement and parts of the ground floor level had already been constructed

Table B.6 Europe except London and Germany

Building	Year [-]	Soil [-]	H [m]	P [MN]	A_r [m^2]	$t_{r,max}$ [m]	$z_{r,na}$ [m]	n_p [-]	L_p [m]	d_p [m]	R_{max} [MN]	s_{max} [cm]	t [a]
Ghent Grain Terminal, Belgium (Goosens and Van Impe 1991)	1978	sa/cl	75	N/A	2919	1.2	1.2	697	13.4	0.52/ 0.80[b]	N/A	20.0	10.00
Pier 4, Neuville-sur-Oise, France (Van Impe and De Clerq 1995)	N/A	si,sa/sa	N/A	12	47	1.5	2.0	5	6.7	1.00	N/A	1.7	0.00
New Law Court, Napoli, Italy (Caputo et al. 1991)	1989	fi/sa/os/pd	110	1500	6954	1.0	1.0	241	42.0	1.5÷2.2	N/A	6.0	1.00
New Mail Center, Napoli, Italy (Caputo 1991)	1990	fi/sa/os/pd	N/A	300	1625	N/A	N/A	136	35.0	1.5	N/A	1.5	0.00
Pier 7, Garigliano Bridge, Italy (Mandolini et al. 2005)	1995	cl/sa/gr	N/A	117	201	4.0	N/A	144	48	0.406/ 0.356[x1]	0.9÷1.1	5.2	9.50
Tank 12 Port of Napoli, Italy	2003	fi/sa	15	23[x2]	123	N/A	N/A	13	11.3	0.6	N/A	2.7[x3]	0.00
Tank 14 Port of Napoli, Italy (De Sanctis and Russo 2008)	2003	fi/sa	15	22[x2]	87	N/A	N/A	13	11.3	0.6	N/A	1.6[x3]	0.00
Appartment Building, Rotterdam, Netherlands (Hooper 1979)	1978	sa	67	289	1175	1.5	1.5	222	17	0.45/ 0.71[b]	1.1	3.0	0.75
Östra Nordstaden, Gothenburg, Sweden (Hansbo et al. 1973)	1972	cl	N/A	N/A	N/A	6.0	1.4	N/A	18÷20	0.125	N/A	5.5	0.00
Residential Building Olskroken, Gothenburg, Sweden (Hansbo 1993)	1982	cl	N/A	54	900	N/A	0.4	104	18	0.3	0.3	3.6	N/A
Roche Building 1, Basel Switzerland (Pitteloud and Meier 2019)	2015	ma/mo	178	2100	2670	2.5	19.6	153	12÷24	1.2	2.6÷5.9	2.2	0.25

[x1]: lower half of the pile; [x2]: maximum load during the reported observational period; [x3]: maximum average settlement during the reported observational period.

240 Appendix

Table B.7 Japan

Building	Year [-]	Soil [-]	H [m]	P [MN]	A_r [m^2]	$t_{r,max}$ [m]	$z_{r,max}$ [m]	n_p [-]	L_p [m]	d_p [m]	R_{max} [MN]	s_{max} [cm]	t [a]
Coal silo (Kakurai et al. 1987)	N/A	sa/si	11.9	8	106	0.6	1.2	5	22.7	0.4	3.4[1]	3.6	0.75
5-storey building, Urawa City (Yamashita et al. 1993, 1994)	N/A	oc cl / sa	17.1	48	552	N/A	2.4	20	15.8	0.7÷0.8	0.2÷2.0	2.0	0.00
4-storey building, Urawa City (Yamashita and Kakurai 1991 after O'Neill et al. 1996)	1987	oc cl, si/sa	14.1	30	559	0.3[x2]	2.1	16	15.1	0.5÷0.6	0.5÷2.1	1.1	0.00
21-storey building complex, Niigata (Majima and Nagao 2000)	1993	sa/si	125.0	N/A	3300	3.7	19.0	157	20.0÷25.0	1.0÷1.8	1.2÷2.1	2.2	0.75
7-storey office building, Minamisuna Tokyo[x3] (Yamashita and Yamada 2009, Yamashita et al. 2011b, 2016[e])	2004	sa/si	29.4	378	3849	0.3	2.2	70	30.0	0.6÷0.9	5.1÷5.7	3.1	4.00
Amuplaza building, Kagoshima (Sonoda et al. 2009, Kitiyodom Pastsakorn et al. 2009)	2004	sa/si	45.0	1062	9000	0.6/ 1.2[x4]	6.5	160	20.0÷25.0	1.5÷2.0	N/A	1.3	0.00
11-storey office building, Aichi (Yamashita et al. 2011a)	2005	sa/ sa, gr	60.8	582	3216	0.8	3.6	40	26.9÷27.5	1.1÷1.5/ 1.4÷1.8[b]	12.9	1.0	−0.25
13-storey hospital, Osaka (Yamashita et al. 2011a)	2005	sa/ sa, gr	51.3	345	2409	0.6	6.4	17[x5]	19.0	0.8 to 1.0	5.5÷7.5	2.6	3.75
19-storey residential building, Kagoshima[x6] (Yamashita et al. 2011a)	2006	sa/si/ cl/pd	75.8	173	673	0.6	3.2	28	62.8	1.2÷1.3/ 1.8÷2.2[b]	5.5÷6.0	2.2	1.75
4-storey parking garage, Urayasu[x7] (Yamashita et al. 2019[e])	2006	fi/sa/ cl/si	N/A	660	14678	0.5/ 1.2[x8]	2.4	152	35.0÷62.0	0.5÷1.0	3.2÷3.9	10.3	11.75

Appendix 241

Building (Reference)	Year	Soil										
Hadron experimental hall, Ibaraki (Yamashita et al. 2011a, 2012b[e], 2014[e])	2007	sa/si	19.0	1164	N/A	13.4	371	22.0÷25.5	0.6÷0.8	1.0÷1.9	2.0	6.75
12-storey residential building, Tokyo[x9] (Yamashita et al. 2011b, 2012a[e], 2013a[e], 2018[e])	2008	sa/cl/si/sa,gr	38.7	199	1.5	4.8	16	45.0	0.8÷12	8.3÷15.3	1.8	2.50
47-storey residential building, Aichi (Yamashita et al. 2010, 2011a)	2009	sa/si/sa,gr	161.9	981	0.5	4.3	36	50.0	1.5÷1.9/ 1.8÷2.2[b]	24.3÷24.9	2.4	1.50
12-storey office building, Tokyo[x10] (Yamashita et al. 2013a[e], 2013b[e], 2014[e])	2011	fi/cl/si/sa/sa,gr	55.7	2244	0.6	7.2	180	42.5÷46.1	0.6÷1.2	4.7÷11.2	2.4	0.00
Abeno Harukas, Osaka (Hirakawa et al. 2016)	2014	sa/gr	300	3166	4.5	30.5	68	32.0	2.3÷2.5/ 3.4÷4.2[b]	54	3.3	0.00

[e]: includes measurements during an earthquake event; [x1]: average pile resistance; [x2]: the raft consists of a mat ($t = 0.3$ m) and a grid of beams ($t = 1.8$ m); [x3]: the foundation includes a grid of 12-m-deep cement mixing walls as a protection against liquefaction; [x4]: the raft consists of a mat ($t = 0.6$ m) and a grid of beams ($t = 1.2$ m); [x5]: perimeter diaphragm wall contributes in the load transfer; [x6]: the foundation includes a grid of 20-m-deep cement mixing walls as a protection against liquefaction; [x7]: the foundation includes a grid of 13-m to 15-m-deep cement mixing walls as a protection against liquefaction; [x8]: the raft consists of a mat ($t = 0.5$ m) and pile caps ($t = 1.2$ m); [x9]: the foundation includes a grid of 16-m-deep cement mixing walls as a protection against liquefaction; [x10]: the foundation includes a grid of 20-m-deep cement mixing walls as a protection against liquefaction.

Table B.8 Australia and Asia except Japan

Building	Year [-]	Soil [-]	H [m]	P [MN]	A_r [m²]	$t_{r,max}$ [m]	$z_{t,max}$ [m]	n_p [-]	L_p [m]	d_p [m]	R_{max} [MN]	s_{max} [cm]	t [a]
QV1 Building, Perth, Australia (Smith and Randolph 1990, Randolph and Clancy 1994)	1991	sa/cl/sist	163	1144	1603[x1]	2.5	6.0	280	20.0	0.8	N/A	4.0	0.00
6-storey building B2, Shanghai, China (Tang et al. 2014)	1991	cl/sa	16.8	61	420[x2]	N/A	1.2	278	16.0	0.2[x3]	0.1÷0.2	5.9	3.75
6-storey building B3, Shanghai, China (Tang et al. 2014)	1991	cl/sa	16.8	61	420[x2]	N/A	1.2	278	16.0	0.2[x3]	0.1÷0.2	6.1	3.75
6-storey building B1, Shanghai, China (Tang et al. 2014)	1999	cl/si	16.8	63	434[x2]	N/A	1.2	262	16.0	0.2[x3]	0.1	2.3	11.00
Jin Mao Tower, Shanghai, China (Korista et al. 1997, Sarkisian et al. 2006)	1998	cl	421	2033	3844	4.0	19.0	429	65.0	0.914	N/A	5.5	N/A
30-storey residential building, Xiamen, China (Zhou et al. 2016)	2004	cl, bo	94	1008	3370	1.6	9.4	65[x4]	10.0	0.9	1.3÷1.8	5.3	1.75
Shanghai World Financial Center, Shanghai, China (Wang et al. 2006)	2008	cl/si/sa/sa, gr	492	N/A	4900	4.5	19.0	985	41.7÷60.7	0.7	1.0÷5.0	13.0	0.50
Shanghai Tower, Shanghai, China (Su et al. 2014, Tang and Zhao 2014, Zhao and Liu 2017)	2014	cl/sa	632	8910	8945	6.0	31.2	955	52.0÷56.0	1.0	N/A	6.0	−1.25
Palace Regency, Chennai, India (Balakumar et al. 2021)	N/A	si/sa	36	N/A	800	0.6	3.0	93	14.0	0.6	N/A	1.4	1.25
Ammonia tank, India (Balakumar and Anirudhan 2011)	N/A	cl/sa	N/A	153	855	0.4	N/A	437	10.0	0.45	N/A	6.5	0.00

Appendix 243

Project	Year	Soil											
Chimney 1–2 Hadera power station, Israel	1979	sa/cl	N/A	166	796	3.0	3.0	48	55.0	1.5	N/A	1.0	16.00
Chimney 3–4 Hadera power station, Israel (Wiseman 1995 after O'Neill et al. 1996)	1979	sa/cl	N/A	166	796	3.0	3.0	48	55.0	1.5	N/A	1.5	16.00
10-storey building, Republic of Korea (Roh et al. 2019)	N/A	fl/wro/ro	41.4	8.6	9^{x5}	1.1	1.1	5	23.0	0.508	0.7÷2.3	1.1	0.25
Tower 1^{x6}, Petronas Twin Towers, Kuala Lumpur, Malaysia (Baker et al. 1994, Baker et al. 1998)	1997	sa,si, cl/wro	451.9	2680	2867	4.5	25.0	104	35.0÷105.0	$1.2\times2.8/$ 0.8×2.8^{x7}	6.4÷60.4	3.5	0.00
Port of Singapore Authority (PSA), Singapore (Leung and Radhakrishnan 1985, Leung et al. 1988)	1986	wro	183	N/A	2569	2.0	7.0	202	6.8÷12.8	1.15÷1.50	3.9÷12.9	0.4	0.75
Jumeirah Emirates Towers Hotel, Dubai, UAE	2000	sa/ wro/ ro	309	N/A	1375	1.5	6.0	92	40.0÷45.0	1.2÷1.5	N/A	1.0	−1.00
Emirates Office Tower, Dubai, UAE (Poulos and Davids 2005)	2000	sa/ wro/ ro	355	N/A	1375	1.5	6.0	102	40.0÷45.0	1.2÷1.5	N/A	N/A	N/A
Burj Khalifa, Dubai, UAE (Poulos and Bunce 2008, Badelow and Poulos 2016, Poulos 2017)	2010	sst/ casi	828	N/A	3523^{x8} 1.0^{x9}	3.7^{x8} 1.0^{x9}	7.6^{x8} 4.9^{x9}	196^{x8} 730^{x9}	47.5^{x8} $30.0\div35^{x9}$	1.5^{x8} 0.9^{x9}	N/A	4.3	−2.00

^{x1}: five separate rafts; ^{x2}: the raft consists of strip foundations and a mat of unknown thickness; ^{x3}: edge length of the square pile; ^{x4}: each pile has a so called deformation adjustor at the pile head; ^{x5}: only instrumented piled raft; the foundation comprises three separated piled rafts with similar dimensions; ^{x6}: Tower 1 and attached bustle; ^{x7}: cross section of barrettes; ^{x8}: tower; ^{x9}: podium.

Table B.9 America

Building	Year [-]	Soil [-]	H [m]	P [MN]	A_r [m²]	$t_{r,max}$ [m]	$z_{r,max}$ [m]	n_p [-]	L_p [m]	d_p [m]	R_{max} [MN]	s_{max} [cm]	t [a]
La Azteca, Mexico City, Mexico (Zeevaert 1957)	1955	cl	N/A	78	662	2.5[x1]	6.0	83	16.0	0.4	N/A	21.0	0.50
Bridge, Mexico City, Mexico (Mendoza and Romo 1998, Mendoza et al. 2000[e])	1996	cl	N/A	32	327	2.7[x2]	3.0	77	27.4	0.5[x3]	0.5÷0.8	24.5	0.50
19 storey building, USA (Koerner and Partos 1974)	N/A	sa/si	N/A	95	748[x4]	N/A	N/A	132	7.6	0.41/ 0.76[b]	N/A	8.4	0.75
The Pyramid, Memphis, USA (Reese et al. 1993, O'Neill et al. 1996)	N/A	cl/ oc cl	N/A	25[x5]	165[x6]	N/A	N/A	27 [x7]	30.9	0.406	N/A	0.2	0.00

[e]: includes measurements during an earthquake event; [x1]: the raft consists of a mat of unknown thickness and a grid of beams (t = 2.5 m); [x2]: the raft consists of a mat (t = 0.25 m) and a grid of beams (t = 2.7 m); [x3]: edge length of the square pile; [x4]: ground area of the building; the foundation comprises 4 pile caps with 7 piles each and 52 pile caps with 2 piles each; [x5]: Pyramidal building with plan dimensions of 140×140 m; corners supported by rectangular piled rafts of 5.5 m×30 m; similar rafts placed at other locations around the perimeter of the structure; [x6]: vertical load including weight of the raft; horizontal load H = 1.3 MN; moment load M = 328 MNm; [x7]: all but six piles were battered.

REFERENCES

Badelow, F., Poulos, H.G. (2016). Geotechnical foundation design for some of the world's tallest buildings. *Proceedings of the 15th Asian Regional Conference on Soil Mechanics and Geotechnical Engineering*, 96–108.

Baker, C. N., Azam, T., Joseph, L. S. (1994). Settlement analysis for 450 m tall KLCC Towers. *Proceedings of the Conference on Vertical and Horizontal Deformations of Foundations and Embankments*, Texas, ASCE Geotechnical Special Publication No. 40, Vol. 2, 1650–1671.

Baker, C. N., Drumright, E., Joseph, L. S., Azam, T. (1998). Foundation design and performance of the world's tallest building. *Proceedings of the 4thConference on Case Histories in Geotechnical Engineering*, 175–187.

Balakumar, V., Anirudhan, I.V. (2011). Piled raft behaviour – Model studies and field performance. *Proceedings of Indian Geotechnical Conference*, Kochi, Paper No. N-322, 947–950.

Balakumar, V., Huang, M.J., Oh, E., Jayasiri, N. S., Hwang, R., Balasubramaniam, A. S. (2021). Piled raft on sandy soil - An observational study. *Geotechnical Engineering Journal of the SEAGS & AGSSEA*, 52, No. 3, 61–65.

Barth, U., Reul, O. (1997). CongressCenter Messe Frankfurt - Kombinierte Pfahl-Plattengründung zur Beherrschung der großen Lastexzentrizitäten. Vorträge des 4 Darmstädter Geotechnik-Kolloquiums Mitteilungen des Institutes und der Versuchsanstalt für Geotechnik der Technischen Hochschule Darmstadt, Heft 37, 117–129.

Burland, J. B., Kalra, J. C. (1986). Queen Elizabeth II Conference Centre: geotechnical aspects. *Proceedings of the ICE*, Part 1, No. 80, 1479–1503.

Butler, F. G. (1975). Heavily over-consolidated clays - Review paper: Session III. *Proceedings of the Conference Settlements of Structures*, Cambridge, 531–578, London: Pentech.

Caputo, V. (1991). Equivalent elastic analysis of settlement for piled foundations. *Proceedings of the 10th ECSMFE*, Vol. 4, 1346–1348, Rotterdam, Balkema.

Caputo, V. Mandolini, A., Viggiani, C. (1991). Settlement of a piled foundation in pyroclastic soil. *Proceedings of the 10th ECSMFE*, Vol. 1, 353–358, Rotterdam, Balkema.

Cooke, R. W., Bryden-Smith, D. W., Gooch, M. N., Sillett, D. F. (1981). Some observations of the foundation loading and settlement of a multi-storey building on a piled raft foundation in London Clay. *Proceedings of the ICE*, No. 70, Part 1, 433–460.

Deutsche Gesellschaft für Geotechnik (DGGT)/Deutsches Institut für Bautechnik (DIBt) (2001). Richtlinie für den Entwurf, die Bemessung und den Bau von Kombinierten Pfahl-Plattengründungen (KPP) – KPP-Richtlinie.

De Sanctis, L., Russo, G. (2008). Analysis and Performance of Piled Rafts Designed Using Innovative Criteria. *Journal of Geotechnical and Geoenvironmental Engineering*, 134, 8, 1118–1128.

Goosens, D., Van Impe, W.F. (1991). Long-term settlements of a pile group foundation in sand overlying a clayey layer. *Proceedings of the 10th ECSMFE*, 1, 425–428, Rotterdam: Balkema.

Green, P. A., Hight, D. W. (1976). The instrumentation of Dashwood House London. *CIRIA Report*, No. 78.

Green, P. A., Cocksedge, J. E. (1975). The settlement behaviour of three tall buildings in London. *Proceedings of the Conference Settlement of Structures*, Cambridge, 159–168, London: Pentech.

Hansbo, S. (1993). Interaction problems related to the installations of pile groups. *Proceedings of the Conference Deep Foundations on Bored and Auger Piles*, Ghent, 59–66, Rotterdam: Balkema.

Hansbo, S., Hofmann, E., Mosesson, J. (1973). Östra Nordstaden, Gothenburg. Experiences concerning a difficult problem and its unorthodox solution. *Proceedings of the 8th ICSMFE*, Moscow, Vol. 2.2, 105–110.

Hight, D. W., Green, P. A. (1976). The performance of a piled-raft foundation for a tall building in London. *Proceedings of the 4th ECSMFE*, Vol. 1.2, 467–472.

Hirakawa, K. Hamada, J., Yamashita, K. (2016). Settlement behavior of piled raft foundation supporting a 300 m tall building in Japan constructed by top-down method. *Proceedings of the 15th Asian Regional Conference on Soil Mechanics and Geotechnical Engineering*, 166–169.

Holzhäuser, J. (1998). Experimentelle und numerische Untersuchungen zum Tragverhalten von Pfahlgründungen im Fels. *Mitteilungen des Institutes und der Versuchsanstalt für Geotechnik der TU Darmstadt*, Heft 42.

Hooper, J. A. (1973). Observations on the behaviour of a piled-raft foundation on London Clay. *Proceedings of the ICE*, Vol. 55, Part 2, 855–877.

Hooper, J. A. (1979). Review of behaviour of piled raft foundations. *CIRIA Report*, No. 83.

ISSMGE Technical Committee TC 212 (2013). ISSMGE Combined Pile-Raft Foundation Guideline. Report of the ISSMGE Technical Committee TC 212 – Deep Foundations, ed. Katzenbach, R., Choudhury D, Technische Universität Darmstadt.

Jentzsch, E., Schulte, K. Placzek, D. (2001). A contribution to the analysis and the design concept of piled raft foundations. *Proceedings of the 15th ICSMFE*, Istanbul, 985–989.

Kakurai, M., Yamashita, K., Tomono, M. (1987). Settlement Behavior Piled Raft Foundation on Soft Ground. *Proceedings of the 8th Asian Conference SMFE*, Kyoto, 373–376.

Kalra, J. C., Willows, K. R. (1986). Queen Elizabeth II Conference Centre: design and construction. *Proceedings of the ICE*, Part 1, 80, 1451–1477.

Katzenbach, R., Arslan, U., Gutwald, J. (1994). A numerical study on pile foundation on the 300 m high Commerzbank Tower in Frankfurt am Main. *Proceedings of the 3rd European Conference on Numerical Methods in Geomechanics*, Manchester, 271–277, Rotterdam: Balkema.

Katzenbach, R., Arslan, U., Moormann, C. (1998). Design and safety concept for piled raft foundations. *Proceedings of the Conference Deep Foundations on Bored and Auger Piles*, Ghent, 439–448, Rotterdam: Balkema.

Katzenbach, R., Arslan, U., Moormann, C. (2000). Piled raft foundation projects in Germany. In *Design Applications of Raft Foundations*, ed. Hemsley, J.A., 323–391. London: Thomas Telford.

Pastsakorn, K., Matsumoto, T., Sonoda, R. (2009). A post-analysis of a large piled raft foundation constructed using reverse construction method. eds: Van Impe and Haegeman, *Proceedings of the Deep Foundations on Bored and Auger Piles*, BAP V, 127 – 133.

Koerner, R.M., Partos, A. (1974). Settlement of Building on Pile Foundation in Sand. *ASCE Journal of the Geotechnical Engineering Division*, 100, 3, 265–278.

Korista, D. S., Sarkisian, M. P., Abdelrazaq, A. K. (1997). Design and construction of China's tallest building. Eds: Viswanath, H. R., Tolloczko, J. J. A., Clarke, J. N., *Proceedings of the Conference Multi-purpose High rise Towers and Tall Buildings*, 289–304, Taylor & Francis Ltd.

Leung, C. F., Radhakrishnan, R. (1985). The behaviour of pile-raft foundation in weak rock. *Proceedings of the 11th ICSMFE*, San Francisco, Vol. 3, 1429–1432.

Leung, C. F., Radhakrishnan, R., Wong Y. K. (1988). Observations of an instrumented pile-raft foundation in weak rock. *Proceedings of the ICE*, Part 1, No. 84, 693–711.

Löffler, M., Reinke, H.G., Meyer, N. (2001). Hochhausgründung in Bonn durch eine modifizierte Pfahl-Plattengründung, Pfahl-Symposium 2001, Mitteilungen des Instituts für Grundbau und Bodenmechanik der TU Braunschweig, Heft 65, 69–82.

Lutz, B., Wittmann, P., El Mossallamy, Y., Katzenbach, R. (1996). Die Anwendung von Pfahl-Plattengründungen - Entwurfspraxis, Dimensionierung und Erfahrungen mit Gründungen in überkonsolidierten Tonen auf der Grundlage von Messungen. *Vorträge der Baugrundtagung 1996 in Berlin*, 153–164, Essen: DGGT,

Majima, M., Nagao, T. (2000). Behaviour of piled raft foundation for tall building in Japan. In *Design Applications of Raft Foundations*, ed. Hemsley, J.A., 393–410. London: Thomas Telford.

Mandolini, A., Russo, G., Viggiani, C. (2005). Pile foundations: Experimental investigations, analysis and design. *Proceedings of the 16th ICSMGE*, Osaka, 177–213.

Mendoza, M.J., Romo, M.P. (1998). Performance of a friction pile-box foundation in Mexico City. *Soils and Foundations*, 38, 4, 239–249.

Mendoza, M.J., Romo, M.P., Orozco, M., Dominguez, L. (2000). Static and seismic behavior of a friction pile-box foundation in Mexico City clay. *Soils and Foundations*, 40, 4, 143–154.

Moormann, C. (2002). Trag- und Verformungsverhalten tiefer Baugruben in bindigen Böden unter besonderer Berücksichtigung der Baugrund-Tragwerk- und Baugrund-Grundwasser-Interaktion. *Mitteilungen des Institutes und der Versuchsanstalt für Geotechnik der TU Darmstadt*, Heft 59.

Morton, K., Au, E. (1975). Settlement observations on eight structures in London. *Proceedings of the Conference Settlement of Structures*, Cambridge, 183–203, London: Pentech.

Mould, G. (1975). Guy's Hospital Tower Block, London - settlement records. *Proceedings of the Conference Settlement of Structures*, Cambridge, 204–211, London: Pentech.

O'Neill, M. W., Caputo, V., De Cock, F., Hartikainen, J., Mets, M. (1996). Case histories of pile supported rafts. *Report for ISSMFE TC18*, University of Houston, Texas.

Pitteloud, L., Meier, J. (2019). High-frequency monitoring results of a piled raft foundation under wind loading. *International Journal of Geotechnical and Geological Engineering*, 13, 3, 90–102.

Pitteloud, L., Frößl, B., Gündling, N. (2004). Tragverhalten einer Kombinierten Pfahl-Plattengründung im sandigen Baugrund. Vorträge des 4. Kolloquiums Bauen in Boden und Fels; Technische Akademie Esslingen.

Placzek, D., Jentzsch, E. (1997). Pile-raft-foundation under exceptional vertical loads - Bearing behaviour and settlements. *Proceedings of the 14th ICSMFE*, Hamburg, Vol. 2, 1115–1118, Rotterdam: Balkema.

Poulos, H.G. (2017). *Tall Building Foundation Design*. CRC Press, London.

Poulos, H.G., Bunce, G. (2008). Foundation design for the Burj Dubai – The world's tallest building. *Proceedings of the 6th Conf, Case Histories in Geotechnical Engineering*, Paper 1.47.

Poulos, H. G., Davids, A. J. (2005). Foundation design for the Emirates Twin Towers, Dubai. *Canadian Geotechnical Journal*, 42, 716–730.

Price, G., Wardle, I. F. (1986). Queen Elizabeth II Conference Centre: monitoring of load sharing between piles and raft. *Proceedings of the ICE*, Part 1, No. 80, 1505–1518.

Randolph, M. F., Clancy, P. (1994). Design and performance of a piled raft foundation. *Proceedings of the Conference on Vertical and Horizontal Deformations of Foundations and Embankments*, Texas, ASCE Geotechnical Special Publication No. 40, 314–324.

Reese, L.C., Wang, S.T., Reuss, R. (1993). Test of auger piles for design of pile-supported rafts. Deep Foundations on Bored an Auger Piles II, Ed: Van Impe, W.F., 343–347, Rotterdam: Balkema.

Reul, O. (2000). In-situ-Messungen und numerische Studien zum Tragverhalten der Kombinierten Pfahl-Plattengründung. *Mitteilungen des Institutes und der Versuchsanstalt für Geotechnik der Technischen Universität Darmstadt*, Heft 53.

Reul, O., Krajewski, W. (2010). Geotechnische Aspekte bei Umbau und Modernisierung von Hochhäusern. Vorträge des 7. Kolloquiums Bauen in Boden und Fels; Technische Akademie Esslingen, 73–81.

Reul, O., Ganal, A., Kissel, W., Ramm, H. (2024). Untersuchungen zum zeitabhängigen Tragverhalten von Gründungen. Bautechnik, 101, Heft 9.

Richter, T., Lutz, B. (2010). Berechnung einer Kombinierten Pfahl-Plattengründung am Beispiel des Hochhauses "Skyper" in Frankfurt/Main. Bautechnik 87, Heft 4, 204–2011.

Ripper, P., El Mossallamy, Y. (1999). Entwicklungen der Hochhausgründungen in Frankfurt. *Hochhäuser - Darmstädter Statik-Seminar 1999*, Bericht Nr. 16 - Technische Universität Darmstadt, Institut für Statik.

Roh, Y., Kim, G., Kim, I., Kim, J., Jeong, S., Junhwan, L. (2019). Lessons learned from the field monitoring of instrumented piled raft bearing in rock. *ASCE Journal of Geotechnical and Geoenvironmental Engineering*, 145, 8. https://doi.org/10.1061/(ASCE)GT.1943-5606.0002078

Sarkisian, M. P., Mathias, N. J., Long, E., Mazeika, A., Gordon, J., Chakar, J. P.(2006). Jin Mao Tower's influence on China's new innovative tall buildings. *Shanghai International Seminar of Design and Construction Technologies of Super High-Rise Buildings*.

Smith, D.M.A., Randolph, M.F. 1990. Piled raft foundations - a case history. *Proceedings of the Conference on Deep Found*. Practice, Singapore, 237–245.

Sommer, M., Amann, P., Brun, B., Löffler, M. (2004). Die Entkoppelte Pfahl-Platten-Gründung – Ein alternatives Konzept zur Verbesserung der Gebrauchstauglichkeit hochbelasteter Gründungen. Beiträge zum 19. Christian Veder Kolloquium, Mitteilungen der Gruppe Geotechnik der Technischen Universität Graz, Heft 21, 153–173.

Sonoda, R., Matsumoto, T., Pastsakorn, K., Moritaka, H., Ono, T. (2009). Case study of a piled raft foundation constructed using a reverse construction method and its post-analysis. *Canadian Geotechnical Journal*, 46, 2, 142–159.

Su, J., Xia, Y., Xu, Y., Zhao, X., Zhang, Q. (2014). Settlement monitoring of a super-tall building using the Kalman filtering technique and forward construction stage analysis. *Advances in Structural Engineering*, 17, 6, 881–893.

Tang, Y., Zhao, X. (2014). 121-story Shanghai Center Tower foundation re-analysis using a compensated pile foundation theory. *The Structural Design of Tall and Special Buildings*, 23, 854–879.
Tang, Y. J., Pei, J., Zhao, X. H. (2014). Design and measurement of piled-raft foundations. *Proceedings of the ICE, Geotechnical Engineering*, 167, GE5, 461–475.
Thaher, M., Konrad, K., Norweg, T., Eickenberg, E. (2002). Die Gründungsausführung des Neubaus Zentrale Arcor in Eschborn. Vorträge der Baugrundtagung 2002 in Mainz, 217–224, Essen: DGGT.
Van Impe, W.F., De Clerq, L. (1995). A piled raft interaction model. *Geotechnica*, 73, 1–23.
Vrettos, C., Borchert, K.M. (2011). Combined foundation of a high-rise building complex on sand: Analysis and observation. *Soils and Foundations*, 51, 2, 343–350.
Wang, W.D., Wu, J.B., Li, Q. (2016). Design and performance of the piled raft foundation for Shanghai World Financial Center. *Proceedings of the 15th Asian Regional Conference on Soil Mechanics and Geotechnical Engineering*, 162–165.
Williams, G.M.J. (1957). Design of the Foundations of the Shell Building, London. *Proceedings of the 4th ICSMFE*, London, Vol. I, 457–461.
Wiseman, G. (1995). Pile supported raft foundation for a 250 m high chimney. Case History Report, Soil Engineering Ltd, Haifa, Israel.
Yamashita, K. Kakurai, M. (1991). Settlement behavior of the raft foundation with friction piles. Takenaka Technical Research Report No. 46.
Yamashita, K., Yamada, J. (2009) Settlement and load sharing of a piled raft with ground improvement on soft ground. *Proceedings of the 17th International Conference on Soil Mechanics and Geotechnical Engineering*, 1236–1239.
Yamashita, K., Kakurai, M., Yamada, T., Kuwabara, F. (1993). Settlement behavior of a five- story building on a piled raft foundation. *Proceedings of the Symposium Deep Foundation on Bored and Auger Piles*, Ghent, 351–356, Rotterdam: Balkema.
Yamashita, K., Kakurai, M., Yamada, T. (1994). Investigation of a piled raft foundation on stiff clay. *Proceedings of the 13th ICSMFE*, Vol. 2, 543–546, Rotterdam: Balkema.
Yamashita, K., Hamada, J., Soga, Y. (2010). Settlement and load sharing of piled raft of a 162 m high residential tower. *Proceedings of the GeoShanghai 2010*, ASCE Geotechnical Special Publication No. 205, 26–33.
Yamashita, K., Yamada, T., Hamada, J. (2011a). Investigation of settlement and load sharing on piled rafts by monitoring full-scale structures. *Soils and Foundations*, 51, 3, 513–532.
Yamashita, K., Hamada, J., Yamada, T. (2011b). Field Measurements on Piled Rafts with Grid-Form Deep Mixing Walls on Soft Ground. *Geotechnical Engineering Journal of the SEAGS & AGSSEA*, 42, 2.
Yamashita, K., Hamada, J., Onimaru, S., Higashino, M. (2012a). Seismic behavior of piled raft with ground improvement supporting a base-isolated building on soft ground in Tokyo. *Soils and Foundations*, 52, 5, 1000–1015.
Yamashita, K., Hashiba, T., Ito, H. (2012b). Settlement and load sharing behaviour of a piled raft subjected to strong seismic motion. *Proceedings of the 11th Australia-New Zealand Conference on Geomechanics*, 1514–1519.
Yamashita, K., Hamada, J., Wakai, Shuichi, Takinawa, T. (2013a). Long-term monitoring of piled raft foundations with grid-form deep mixing walls on soft ground. Takenaka Technical Research Report, No. 69, 1–16.

Yamashita, K., Wakai, S., Hamada, J. (2013b). Large-scale piled raft with grid-form deep mixing walls on soft ground. *Proceedings of the 18th International Conference on Soil Mechanics and Geotechnical Engineering*, 2637–2640.

Yamashita, K., Hamada, J., Wakai, Shuichi, Takinawa, T. (2014). Settlement and load sharing behaviour of piled raft foundations based on long-term monitoring. Takenaka Technical Research Report, No. 70, 29–40.

Yamashita, K., Hamada, J., Tanikawa, T. (2016). Static and seismic performance of a friction piled raft combined with grid-form deep mixing walls in soft ground. *Soils and Foundations*, 56, 3, 559–573.

Yamashita, K., Shigeno, Y., Hamada, J., Chang, D.-W. (2018). Seismic response analysis of piled raft with grid-form deep mixing walls under strong earthquakes with performance-based design concerns. *Soils and Foundations*, 58, 65–84.

Yamashita, K., Tanikawa, T., Uchida, A. (2019). Long-term behaviour of piled raft with DMW grid on reclaimed land. *Geotechnical Engineering Journal of the SEAGS & AGSSEA*, 50, 3.

Zeevaert, L. (1957). Compensated Friction-pile Foundation to Reduce the Settlement of Buildings on the Highly Compressible Volcanic Clay of Mexico City. *Proceedings of the 4th ICSMFE*, 3, 81–86.

Zhao, X., Liu, S. (2017). Foundation differential settlement included time-dependent elevation control for super tall structures. *International Journal of High-Rise Buildings*, 6, 1, 83–89, https://doi.org/10.21022/IJHRB.2017.6.1.83

Zhou, F., Lin, C., Zhang, F., Lin, S. Z., Wang, X. D. (2016). Design and field monitoring of piled raft foundations with deformation adjustors. *ASCE Journal of Performance of Constructed Facilities*, 30, 6.

Index

Pages in *italics* refer to figures, pages in **bold** refer to tables.

Abeno Harukas building, Osaka, 164, **241**
Accumulation, 85, 87
Actions, 81, 85, 86, 90, 101, 102, 104, 107, 108, 114, 211, 227, **231**
A-frames, 172–177
American Express Building, Frankfurt, 145, **236**
Amuplaza building, Kagoshima, 164, **240**
Anchoring point, 134, *135*, 176, 180
Angle of friction, **144, 210, 217, 232**
Angular distortion, 6, 104–106, 108, *149*, 150, 155–157, 159, **160**, *169*, 182, 192, 210, 211, 214, 219, 220, 222
Apartment Building, Rotterdam, **239**
Atrium building, Frankfurt, *see* Park Tower
AVISA model parameters
    accumulation rate factor, **74**
    anisotropic coefficient, **74**
    compression index, **74**
    critical state slope, **74,** *75*
    ISA hardening parameter, **74**
    ISA yield surface radius, **74**
    loading surface factor, **74**
    maximum void ratio, **74**
    maximum ISA exponent, **74**
    minimum ISA exponent, **74**
    Poisson's ratio, **74**
    stiffness factor, **74**
    swelling index, **74**
Axial spacing, 142; *see also* pile spacing

Back analysis, 11, 14, 25, 68, 69, 75, 87, 90, 92, 120, 123, *129*, 188, 195

Bahn Tower, Berlin, 185–189, **238**
Barbican Tower, London, **235**
Basement, 7, 17, 104, 109, 113, 122, 143, 147–149, 151, 152, *154*, *156*, 158, 159, 164, 166, *167*, 169, 178, *181*, 182, 192, 197, 202, 203, 208, 209, 225, 234, **237, 238**
Beisheim Center, Berlin, *186*, **238**
Bending moment, 5, 14, 17, 49–51, 57–59, 62, *88*, 90–92, 117, 118, 137, 208, 210, 212, 227, **228**
    normalised bending moment per unit length, 24, 49, 50
    normalised maximum positive bending moment per unit length, *see* normalised bending moment per unit length
Block
    deformation, 10, 37, 161
    failure, 54
Borehole, 89, 126, 127, 134, *135*, *167*, 168, 176, 188, 196
    extensometer, *see* extensometer
    log, 69, 76, 144, 160
    test, 127, 128
Boundary conditions, 13, 67, 85, 91, 114, 167
Burj Khalifa, Dubai, **243**

Cambridge Road, London, **235**
Cantilever beam, 154
Cap model parameters
    intersection of the conical yield surface with the t-axis, **26, 117**
    Poisson's ratio, **26, 117**
    shape parameter of the cap, **26, 117**

shape parameter of the cone, 26, **117**
shape parameter of the transition surface between cone and cap, 26, **117**
slope of the conical yield surface in the p-t plane, 26, **117**
Young's modulus, 26, **117**
Capacity, 1, 4, 38, 41–43, 84, 126, 136, **232**
  bearing, 15, 54, 104, 108, 111, 115, 120, 125, 141, 214
  external bearing, 107, 221
  internal bearing, 107, 108, 111, 120
  lateral, 174
  ultimate, 4, 10, 13, 16, 17, 24, 27, 38, 51, 54, 55, 101, 117, 123, 219, **232**
  structural, 69
  yield, 51
Cascada, Frankfurt, **237**
Casting the raft, 182; *see also* Concreting the raft
Cavity, *155*
Characteristic value, 4, 103, 107, 108, 211, 231
Clay, 7, 24, 62, 74–76, 81, 82, **89**, 106, 144, 163, 168, 179, 183, 196, 197, 199, 202, 233
  Frankfurt Clay, *see* Frankfurt Formation
  Kaolin, 14, 61, **89**
  London Clay, 7, **8**, 13, 15, 65, 141–144
  overconsolidated, 4, 8, 63, 68
  soft, 6, 62, **82**, 88, **89**, 91
  stiff, 39, 62, 141
Coefficient of earth pressure at rest, 26, 53, **65**, 75, 76, 117, **231**
Coefficient for
  average settlement, 5, 44, *45*, 57, *58*
  differential settlement, 5, 46, *48*, 57, *58*
  maximum settlement, 5, 44, *46*, 47, *48*
  maximum positive bending moment, 5, *57*, *58*
Cohesive soil, 13, 16, 38, 39, 61, 120, 122
Combination coefficient, 226
Commercial Union Building, London, **235**
Commerzbank Tower, Frankfurt, 15, *165*, 166, 185, **236**

Compression index, **74**
Concrete, 1, 2, 6, 51, 64, 70, 87, **89**, 108, 128, 130–136, 149, 157, 158, 161, 183, 192, 201, 202, 223, 226, 227, **228**
  reinforced, 9, 10, 120, 147, 152, 164, 171, 178, 185, 188, 195, 199
  sample, 161
  strength, 157, **161**
Concreting the raft, 146, 168, 171, 181, 192
Cone/piezocone penetration test (CPT/CPTU), 126, 127
Congress Center, Frankfurt, **236**
Connection
  hinge, 82, 83, 86, 90, 91
  rigid, 82, 83, 86, 90, 91, 157
Consolidation, 61–63, 65, 69, 71, **73**, 75, 113, 122, 135, 151, 175, 203
Constitutive model
  AVISA visco-hypoplastic, 63, 72, 73, 75, 80, 144, 156, 163; *see also* AVISA model parameters
  cap, 11, 24, 44, 64, 68, 117, 123, 174, 195; *see also* cap model parameters
  elasto-plastic, 11, 15, 23, 24, 44, 51, 62, 63, 68, 87, 90, 117, 120, 123, 158, 174, 195
  elasto-viscoplastic, 61, 62
  hardening soil (HS), 87, 158
  hardening soil (HS) small strain, 90, 91
  hypoplastic, 73, 120, 195, 217
  linear-elastic, 72, 117, 120, 123, 217
  Masin's visco-hypoplastic, 81
  modified Cam Clay (MCC), 62
  Tresca, 55, 56
  visco-hypoplastic, 69
  Wolffersdorff hypoplasticity, 215, 217; *see also* Wolffersdorff model with intergranular strain extension
Constraints, 5, 14, 49, 57, 59, **60**, 82, 85, 90, 103, 137, 149, 164, 210, 214, 222
Construction, 1, 7, 10, 15, 27, **29**, 69–72, 73, 84, 87, 102–103, 105, 110, 119–120, *121*, 122–123, 128, 130, 135–136, 142–144, 146–149, 151–152, *153*, 155, 156, 157, 158, 161, 163, 164,

Index

166, 167–168, 170–171, 173, 175, 178–181, 183, 186–187, 192, 194, 196, 202–203, 218, 233
Construction process, *see* construction
Contact, 1, 2, 25, 31, 49, 71, 91, 107, 122, 185, 215
   perfectly rough, 24, 64, 71, 215
Contact pressure, 7, 49, 50, *130*, 133, 166, 170, 182, 184, 188, 189, 203, 234
   cell, 7, 8, 10, *11*, 69, 70, 122, *123*, 130, 133, 143, *145*, 146, 150, *160*, 161, *169*, 170, *179*, 180, *186*, 188, 189, 192, 196, *197*, *201*, 203
   transducer, *see* contact pressure cell
Contact surface, 115, 120
Contact stress, *see* Contact pressure
Contact normal stress, *see* Contact pressure
Core sample, 158
Costs, 17, 49, 118, 137, 149, 150, 208, 227, **228**
Crack width, 108, 227
Creep, 61–63, 69, 72, **73**, 75, 81, 120, 135, 151

Damping, 92
Dashwood House, London, 142, **235**
Deep mixing walls (DMW), **89**, 90
Deep excavation, 17, 129, *130*, 148, 164
Design concepts
   combined pile-raft foundation (CPRF), 1, 6, 12, 15, 22, 71, 101, 104, 105, 107, 109, 110, 125, 126, 128, 129, 135, 137, 141, 146, 147, **149**, **150**, 152, 158–160, 167, 168, 174, 175, 180, 187, 191, *195*, 196, 208, 214–*216*, 218–222, 225, 227, **228**, **233**, **236–238**
   compensated piled raft foundation, 17
   conventional design, 16, 208, 211, **219**
   "creep pile" design, 7, 10, 16, 38, 122
   Japanese design approach, 16
   modified conventional design, 11, 16, 69, 146
   pile-enhanced raft, 17
   raft-enhanced pile groups, 17
Design
   example, 208

   optimisation, 57
   parameters for structural engineering, 5
   process, 2, 6, 10, 15, 44, 57, 59–61, 69, 101, 104, 109–111, 122, 129, 148, 149, 159, 166, 180, 188, 191, 210
   value, 17, 103, 104, 107, 108, 176, 211, 227, **228**, **231**, **232**
Deutsche Bank, Frankfurt, 164–167
Deutsche Messe AG, Hannover, **238**
Diameter of the equivalent pier, 22, **231**
Displacement
   horizontal, 83, 90, 114, 175–178, 180, **237**
   vertical, 87, 114, 168, 175, 176, 188
Drained
   analysis, 34, 63
   conditions, 27, 38, 54, 55
Drainage path, 76
Dresdner Bank, Frankfurt, 9
Drilled borehole, 126, 127
Discretisation, *see* mesh discretisation
Dynamic probing (DP), 126

Earth pressure, 109, 133, 167, 177, 178
   cell, *173*, *176*
   passive, 81, 91
Earthquake, 12, 84, **89**, 90, 92, 102, 109, **241**, **244**; *see also* loading
EC7, 85, 102, 103, 104, 107, 127, 211
Element
   continuum, 71, 115, 215
   interface, 71, 115, 119, 120
   plate, 115
Embedded pile, 115, 119
Emirates Office Tower, Dubai, **243**
Equivalent circular raft radius, 23, **231**
Equivalent spring stiffness for the pile, 40, *109*, *110*, 224, 225, **231**
Eurocode, 101–103, 107, 108, 127, 226
European Central Bank, Frankfurt, **237**
Eurotheum, Frankfurt, **236**
Excavation
   level, 71, 159
   pit, 102, 120, 122, 130, 134, 148, 150, 159, 164, 166–168, 180, 202, 227
Extensometer, 10, *11*, 69, *70*, 79, 122, *123*, 130, 134, 135, 142, 160–*162*, *173*, 176, *177*, *179*, 180, 181, 183

254 Index

Factor of safety, 16, 17
  global, 101, **232**
  overall, 101–103; *see also* global factor of safety
  partial, 102, 103, **228**
Failure, 51, 101–104, 108, 117
  load, 17, 56, 125
  mechanism, 56, 57
  surface, 54
Fibre optic measurements, 132
Finite Difference Method (FDM), 111, 114
Finite Element Analysis (FEA), 49, **82, 86, 89**, 125
  axis-symmetric, 15
  coupled, 61–63, 67, 68, 71, 72, 122, 156, 163, 168
  coupled displacement-pore pressure, *see* coupled
  two-dimensional (2D) FEA, 61
  three-dimensional (3D) FEA, 15, 22, 62, 71, 87, 90–92, 124, 148, 158, 168–170, 174, 188, 191, 195, 196, 202, 208, 211, 212, 214, 215, 217–220, 223
Finite Element Limit Analysis (FELA), 56, 57, 125, 219
Finite Element Method (FEM), 14, 88, 111, 113
Four, Frankfurt, 164, *165*, *170*, 171
Foundation base, 54, 166, 180, 183, 187
Frankfurt Formation, 7, 8, 10, *11*, 24–26, 44, 48, 69–71, 74–76, 122, 123, 141, 143, 144, 148, *153*, *154*, 156, 158, *160*, 164, 166, *167*, 171, *173*, 174, *176*, 180, 183
Frankfurt Limestone, 10, 24, 71, 143, 144, 146, 148, 158, *160*, 166, 168
Freestanding pile group (FPG), 1, 2, 4, 13, 25, 31–39, 54, 55, 56, 57, 81–83, 85–87, 89, 91, 92, 111, 211

Garden Tower, Frankfurt, *165*, 166
Geodetical, *130*, 136
Geodetic measurement, 129, 135, *162*, 170, 180, 191, 196, 203
Geophysical techniques, 127
Ghent Grain Terminal, **239**
Glacial
  deposits, 186
  drift, *186*, 187, 191, **233**

Groundwater, 27, 70, 124, 144, 148, 154, 168, 180, 200, 208
  drawdown, 61, 62, 71, 77, 78, *121*, 122, 144, 148, 151, *152*; *see also* groundwater lowering
  level, 10, 69–73, 76–78, 120, 122, 133, 134, 148, 151, 158, 163, 166, 174, 182, 185, 187, 188, 191, 196, 200, 202, 208
  lowering, 70, *163*, 168, 171
  rise, 70, 122, 163
Group settlement ratio, 111, 112
Guy's Hospital, London, **235**

Hadron experimental hall, Ibaraki, **241**
Halfspace, 49; *see also* Winkler halfspace
Heave, **102**, 104, 156, 168, 182
Hegau Tower, Singen, 199–203, **238**
High-rise building, 6–11, 16, 54, 90, 109, 110, 128, 135, 142, 145, 146, 152, 154–156, 158, 164, 166–168, 178, 180, 182, 185, 187, 188, 190–192, 199, 202
Hochschule Darmstadt, 152, **238**
Hogging, 46, 51, 59, **60**, *105*, **106**
Hybrid mudmat, 83
Hyde Park Cavalry Barracks, London, 7, 8, 12, 14, 142, **235**
Hydraulic pressure pad, 176
HyPR, 124, 125

Inclinometer, 122, *123*, 130, 134, *173*, 176, *177*, 180
Inertial forces, 84
Instrumentation, 136, 142
Integral bridge, 84
Interaction-pile
  pile-pile, *12*, 31
  pile-raft, *12*, 31, 37, 38
Investigation depth, 127
ISSMGE Combined Pile-Raft Foundation Guideline, 1, 15, 22, 105, 141, 146

Japan, 12, 16, 90, 146, 164, 234, **240, 241**
Japan Center, Frankfurt, 30, 142, *165*, 166, **236**
Jin Mao Tower, Shanghai, **242**
Jumeirah Emirates Towers Hotel, Dubai, **243**

Kastor, Frankfurt, 146, **236**
Kesselhaus, Darmstadt, **238**
KPP-Richtlinie, 1, 12, 105–107, 141, 146, 148; *see also* ISSMGE Combined Pile-Raft Foundation Guideline

La Azteca building, Mexico City, 6, 7, 14, 17, **244**
Laboratory test, 74, 92, 126, 127, 144
Lauenburg strata, 196, *197*
Leo, Frankfurt, **237**
Limestone band, 69, 74, 75, 144, 148, 164, 166
Limit state design, 101, 102, 109, 169
Liquefaction, 90, 91, **241**
Load
  caused by the superstructure, 24, 208, **232, 233**
  dead, 109, 147, 164, 178, 196, 199, 203, 208, **218, 219,** 231, **232;** *see also* structural weight
  effective applied load, 24, 41–42, *43*, 232, **233**
  live, *70*, 84, 109, 120, 122, 147, 164, 178, 196, 199, 203, 208, 211, *213*, **218, 219,** 226–227, **228,** **231–233**
  permanent, 211, *213*, 227, **228**
Load
  level, 4, 15, 17, 38–42, 44, 49–51, 62–65, 67, 117, 118, 137, 220, 223
  transfer, 2, 6, 10, 14, 16, 34, 37–39, 48, 51, 62, 86, 87, 92, 124, 128, 133, 141–143, 149, 151, 166, 180, 185, 201, *202*, 211, 214, **237,** **241**
Load cell
  pile base, 3, 130, *131*, 142
  pile head, 2, 7, *8*, 69, 130, *131*, 142, 143, 146, 150, 169, 180, 188, 191, 196
Loading/load
  combined, 101, 102
  core-edge, 25, 41–44, *45*, 49–51, 57–59, 137
  cyclic, 84–87, 92
  dynamic, 84, 85, 88, **89,** 91, 92
  earthquake, 90, 91
  eccentric, 180

horizontal, 81, 86–88, 90–92, 157, 158, 172–174, **237, 244;** *see also* lateral
impact, 84, 85
lateral, 12, 17, 81, **82,** 85–87, **89,** 101, 102, 136; *see also* horizontal
periodic, 84, 85
seismic, 17, **89,** 90
uniform, 23, 25, **29,** 41, 42, 44, *47*, 49–51, 137, *209*, *213*, *216*
vertical, 12, 14, 22, **29,** 51, 55, 56, 81, 82, 85–**89,** 92, 105, 137, 172, 173, **244**
wind, 90, 109
Loading cycles, 14, 61, 87
Load transfer layer (LTL), 2, 62, 201, 202
Long term bearing behaviour, 61–63, 68, 120

Main Forum, Frankfurt, **237**
Main Tower, Frankfurt, **236**
Marriott Hotel, Frankfurt, 8
Measurement device, 3, 10, 12, 69, *70*, 122, 130, 136, 146, 150, 151, 160, 169, *173*, 176, 177, 201
Measurement
  geodetic, 129, 135, *162*, 170, 180, 191, 196, 203
  geotechnical, 7, 10, 12, 69, *130*, 135, 143
Mesh discretisation, 115
Mesh refinement
  h-refinement, 116–118
  p-refinement, 116–118
Messehalle, Frankfurt, **236, 237**
Messeturm, Frankfurt, 11, 12, 14, 30, 34, 61, 68–81, 145, 151, *172*, **236**
Mindlin
  plate theory, 223
  solution, 14
Mixed-in-place (MIP) walls, 90
Model
  one-phase, 67, 68
  two-phase, 67, 68
Model test
  centrifuge, 14, 47, **82,** 86, 87, **89,** 91, 92
  large-scale, 13, 39, 54, 81
  shaking table, **89,** 91, 92
  small-scale, 13, 34, 37, 62, **82,** 83, 85–87, **89,** 91, 92, 112

Modulus of subgrade reaction, *see* subgrade reaction modulus
Monitoring, 90, 101, 103, 129, 130, 142, 163
Monolithic, 84, 146, 152, 157, 178, 180
Moraine sediments, 199–202
Multipoint borehole extensometer, *see* extensometer

National Westminster Bank Tower, London, 142, **235**
Neue Messehalle 3, Frankfurt, 81, 146, 171–178, **236**
New Law Court, Napoli, **239**
New Mail Center, Napoli, **239**
Non-cohesive soil, 38, 208, 209
Non-destructive testing technique, 158
Non-reinforced concrete columns, 201, 202
Number of piles in the pile group, 65, **231**

Östra Nordstaden, Gothenburg, 6, **239**
Omniturm, Frankfurt, 146, 163–171, **237**
Opernturm, Frankfurt, 163
Optimum design, *see* Design optimization
Overall stiffness of a piled raft, 137
  normalised, 23
Overconsolidation ratio (OCR), 61–63, 75, 76, 81

Palace Regency, Chennai, **242**
Parametric study, 14, 22, 24, 25, 27–28, 29, 39, 48, 55, 57, 61, 88, 92, 168, 174
Park Tower, Frankfurt, 75, 80, 132, 152–163, **237**
Partial factors of safety, 4, 101–104, 107, 211, 227, **228**, **232**
Permeability, 63, **65**, 75, 76, 80, 133, *134*
  system, 76, 79, 80
Petronas Twin Towers, Kuala Lumpur, **243**
Pier 4, Neuville-sur-Oise, **239**
PIGLET, 149, 217
Pile base, *3*, 7, 52, 67, 107, *127*, 130, 131, 142, 146, *162*, 168, *169*, 171, 176, 201
  area, 115, 211, **231**
  resistance, 10, 13, 16, 34, 37, 169–171

resistance per unit area, 119, 135, **232**; *see also* unit base resistance
  unit base resistance, **210**, 211, 214
Pile-box foundation, 90
Pile cap, 1, 4, 16, 141, 211, 217, **241**, **244**
Pile foundation, 7, 15, 16, 85, 106, *113*, 166, 180, 208, 211, 212, 214, 218–222, 224, 226, **228**
Pile group-raft area ratio, 23, 43, 44, **231**
Pile integrity test, 128
Pile length, 13, 14, *28*, 30, 44, 46, 47, 56, 59–62, 108, 112, 137, 148, **149**, 168, 210, 212, 214, **231**, **233**
  active, 92
  relative, 23, 44, *45*
  total, 44–49, 57–61, 137, 149
Pile load, 2, 7, 31, 33, 34, 69, 79, 90, 102, 110, 122, 125, *130*, 133, 161, 191, 194, 202, 227
  axial, 2, 3, 4, 17, 31, 33, 131, 132, 222, **231**, **233**
Pile load test, 107, 122, 135, 136, 187, 188, 209, **210**, 214
  dynamic (high strain), 136
  static axial, 136
  static bi-directional (Osterberg cell), 136
  static lateral, 136
  Statnamic, 136
Pile position
  centre, 10, 31, 33–38, *40*, 53, 65, 66, 68
  corner, 10, 31, 33, 37, *40*, 65, 66, 68, 142
  edge, 10, *35*, *36*, *53*, 66–68, 142
Pile resistance, 2, *3*, 4, *5*, 10, 11, 15, 17, 31, 32, 37–41, 51, *52*, 63, 65, 70, 71, 76–78, 90, 117, 123–125, 142, *145*, 146, 151, 152, 161, *162*, 169–171, 180, 183–185, 188, 189, 192–194, 196–199, 203, 211, 214, **232**, **233**, **241**
  normalised, 65–68
Pile shaft, 3, 17, 31, 33, 34, 54, 79, 107, 115, 119, 130, 132, 191, 192, 194, 221, **232**
  area, 4, 115, 132, 211, **231**
  friction, 3, 4, 10, 31, 33, 34, 37, 38, 61, 120, 131, 132, 135, 185, 187, 191, 194, 209–211, 214, **232**

resistance, 10, 13, 16, 37, 55, 62, 125, 142, *187*, 211, **232**
  resistance per unit area, **232**; *see also* shaft friction
Pile skin friction, *see* pile shaft friction
Pile spacing, 7, 13, 15, *28*, 31, 38, 39, 40, 54, 56, 62, 112, 137, **231**
Pile type, **82, 86, 89**, 128
  bored pile, 2, 7, *9*, 10, 13, 61, 69, 71, **82, 86, 89**, 122, 132, 143, 148, 166, 167, 174, 180, 187, 201, 215, 227
  displacement pile, **82, 86, 89**, 128, 136
Piled raft, 1–2, 4–7, 10–15, 17, 22–92, 105–106, 108, 110–118, 120, 122–126, *129*, 136–137, 141–143, 145, 146, 148–150, 160, 164, 166, 169, 208, 211, 214–215, 219–222, 232, 233, **243, 244**
Piled raft coefficient, 4, 5, 11, 13, 15, 17, 38, 39, 44, 46–48, 62–68, 76–78, 87, 91, 105, 106, 117, 118, 120, 122–125, 142, 143, 149, **160, 169**, 215, 219–221, **232**
Plan area of the pile group as a block, 22, **231**
Plastic deformation, 42
Poisson's ratio, 23, **65**, 73, 125, 218, **223, 232**
Pollux, Frankfurt, 146, *172*, **236**
Pore pressure, 61–63, 67, 68, 69, 71, 72, 85, 91, 122, *130*, 133, 151, 152, 156, 161, 163, 170, 171, 180, 182, 184, 185, 196, 203, **232**
  excess, 67, 68, 80
Pore pressure cell, 69, 70, 122, *123*, 130, 133, 134, 146, 150, *160*, 161, 169–171, *179*, 185, 196, *197, 201*
Pore pressure transducer, *see* pore pressure cell
Pore water pressure, *see* pore pressure
Portalhaus Messe, Frankfurt, **237**
Port of Napoli, **239**
Port of Singapore Authority, **243**
Post Tower, Bonn, **238**
Power
  cooling, 148
  heating, 148
Power plant, Boxberg, **238**

Praedium, Frankfurt, **237**
Pressuremeter test (PMT), 126
Primary support, 167, 168
Principle of effective stress, 185
Proof, 107, 108, 174, 214, 221, 222
  selected pile length, 212
Pyramid, Memphis, **244**

Queen Elizabeth II Conference Centre, London, 143, 163, **235**
QV1 building, Perth, 14, 16, **242**

Raft
  area, 23, 25, 27, 218, **231**
  foundation, 1, 2, 4, 5, 7, *9*, 14, 23–25, 27, 31, *32*, 41, 42, 49–51, 55–57, 59, 62, 63, 80–82, 85–87, **89**, 91, 92, 108, 117, 137, 148–150, 159, 160, 166, 195, 208, 211, 218–222, 224, 226–228, **232**
  resistance, 31; *see also* resistance of the raft
  thickness, 8, 41, 57, 60, 148, 168, 202, 211, 214, **233**
Raft-soil stiffness ratio, 23, 28, 30, 31, 41–43, 49, 137, **231**
Railway tunnel, 178–182
Ratio of the effective applied load, 24
Reinforcement, 51, 130–132, *136*, 150, 157, 158, 188, 212, 226, 227
Relief well, *167*, 168
Reloading, 14, 61, 62, 119
Resonance frequency, 92
Residential Building Olskroken, Gothenburg, **239**
Resistance
  lateral, 81
  of the raft, 5, 31, *32*, 52, 108, **232**; *see also* raft resistance
Resistance-settlement curve, 3, 5, 31, 37, 39, 107, *187*
Retaining wall, 148, 167, 168, 180
Rigid inclusions (RI), 2, 201
Roche Building 1, Basel, **239**
Rupel clay, 179, 183
R+V Versicherung, Wiesbaden, **238**

Sagging, 51, *105*, **106**, 149
Sand, *8*, 10, 13, 34, 38, 39, 69, 81, 82, 85–87, **89**, 90–92, **106**, 112, *133*, 144, 148, 152, *154*, 158, 164, 173, 174, 176, 179, 199, 200, 209, 210, 214–217

band, 69, 74–76, 144, 148
Berlin sand, 186–195
layer, 24, 144, 166, 179
Weser sand, 196, 197
Seismic profile, 92
Serviceability, 5, 6, 9, 54, 150, 159, 182, 192, 220
  external, 108, 120, 221, 222
  internal, 108, 111
  limit state (SLS), 14, 104, 108, 110, 169
Settlement
  average, 5, 9, 23, 24, 39, 42, 44, 59, 60, 111, 112, 117, *118*, 137, **169**, 219, **239**
  differential, 5, 6, 9, 10, 24, 37, 47, 48, 54, 57, 59, 61, 104–106, 108, 117, *118*, 137, 156, 169
  limit settlement of pile head, **232**
  maximum, 5, 7, 8, 15, 16, 41, 47, 49, 142, 143, 146, 149, 151, **160**, 161, 163, 169, *170*, 181, 182, 192
  normalised differential, 24, 42–44
Settlement profile, 79, 161, *162*, 183
SGZ-Bank, *see* Park Tower, Frankfurt
Shaft grouting, 122, 191
Shanghai Tower, **242**
Shanghai World Financial Center, **242**
Shear modulus, 92
  initial equivalent, 24
  profile, 217
Shear strength, 16, 85, 103, 115, 119
  undrained, 56, **179**, 180, **232**
Sheet pile wall, 202
Shell (of the building), 70, 125, 146, 155, *170*, 171, 181–185, 188, 189, *193*, 197, *198*, 203, **233**
Shell elements, 25, 117, 223
Shell Centre, London, **235**
Shoring system, 202
Single pile, 1, 3, 14, 15, 17, 25, 37, 38, 41–43, 81, **82**, 85–87, **89**, 91, 101, 102, 107, 111, 112
  resistance, 31, 33
  shaft friction, 34
  shaft resistance, 13, 33
  stiffness, 40
  ultimate capacity, 4, 24
Site investigation, 76, 126, 127, 166, 196
Skyper, Frankfurt, **237**
Sleeve, *179*, 180, 187

Sleeve excluding shaft friction, *see* sleeve
Slip-form construction method, 154
Softening, 85
Sony Center, Berlin, 185
Spring stiffness, *see* equivalent spring stiffness for the pile
Standard penetration test (SPT), 126
Steady state conditions, 67
Step-by-step analysis, 27, **29**, 72, 73, *121*, 123, 218
Stonebridge Park, London, 12, 16, 44, **235**
Strain gauge, 69, *88*, 130–132, 146, 160, 161, 169, 171, 173, 176, 192, 202
Stress, 101, 128, 168, 212
  deviatoric, 52–54
  effective, 34, 38, 91, 126
  equivalent pressure, *see* mean stress
  gradient, 115
  horizontal, 75, 176
  in situ stress state, **29**, 73, *121*, 218
  level, 13, 34, 81, 214, 217
  mean, 34, *36*, 37, 51–54
  principal, 34
  shear, 91, 137
  state, 13, 38, 51, 52, 54, 55, 120
  vertical, 64, 117
Strip foundation, 61, 152, 154–156, **243**
Structural
  analysis, 110, 159, 212, 220, 223–227
  behaviour, 84, 223
  capacity, 69
  component, 115, 120; *see also* structural element
  damage, **106**
  design, 157, 169; *see also* structural analysis
  element, 104, 108, 111, 208
  engineer, *see* structural engineering
  engineering, 109, 110, 111, 218
  load, 146, 168
  material, 103
  model, *110*, 158
  weight, 120, 122; *see also* dead load
Subgrade reaction modulus, 4, 40, 110, **231**; *see also* Winkler's modulus
Sum of all (mobilised) pile resistances, 4, 5, 40, 65, **232**
Superstructure, 4, 9, 15, 17, 24, 27, 71, 72, 84, 92, 107, 108, 111, *121*,

147, 157, 164, 168, 169, 171, 178, 190, 192, 195, 196, 199, 202, 208, 218, 225, **232**, **233**
Swelling index, 75
Swelling pressure, 183

Tall building, 25, 28, 30, **106**
Taunusturm, Frankfurt, 146, *165*, 237
Tide-dependent water level, 196
Tilting, 9, 10, 154, 156
Time-dependent behaviour, 61, 62, 68, 107
Top-down method, 143, 164, 166
Torhaus Messe, Frankfurt, 10, 11, 12, 16, **30**, 31, 44, 48, 145, *172*, **236**
Total
  building load, 7, 69, 170, 188
  resistance of the foundation, 5, 40, 107, **232**
Tower One, Frankfurt, *172*, 237
Treptowers, Berlin, **30**, 34, 190–195, 197, **238**
Truss forces, *174*, 177, 178

Ultimate capacity, 10, 16, 17, 27, 51, 54, 55, 101, 117, 123, 219, 221, **232**
Ultimate limit state (ULS), 13, 16, 103, 104, 107–110, 115, 125, 126, 214, 221, 227, **228**
Utilization of the deviatoric stress, 51, 52
Undercutting, 9, 154, *155*
Undrained, 55, 63
Uniaxial compressive strength, 160, **161**
Unit weight, **26**, **65**, **73**, **74**, **117**, 179, 210, **217**, **218**, 232
Unloading, 14, 61, 119, 183
Uplift, 27, **29**, 70, 71, **102**, 104, *121*, 148, 151, 168, 169, 171, 187, 188, 218, **219**, **232**, **233**

Victoria Street Redevelopment, London, 113, 143
Viscosity index, 63, 79, 80, 81

Weser Tower, Bremen, **30**, 195–199, **238**
Westend 1, Frankfurt, 11, **30**, 122–125, 145, **236**
WestendDuo, Frankfurt, 71, 147–152, **227**, **237**
Westhafen Tower, Frankfurt, 71, 78, **237**
Wind turbine, 87, 88
Winkler
  halfspace, 4, *45*
  modulus, 109–111, 208, 212, 220, 223–225, **231**; *see also* subgrade reaction modulus
  springs, 109, 195
Wished-in-place, 72, 122, 136, 218
Wolffersdorff model with intergranular strain extension model parameters
  Critical angle of friction, **217**
  Exponent, **217**
  Granular hardness, **217**
  Initial void ratio, **217**
  Maximum void ratio, **217**
  Parameters of the intergranular strain concept, **217**
  Void ratio at a state of maximum density, **217**
  Void ratio at critical state, **217**
Woolwich and Reading Beds, 141

Young's modulus, 23, 24, 49, 69, 70, 73, 79, 106, **125**, 131, 202, 218, **231**
  concrete, 160, **161**, 188, 223
  equivalent, 23, 24, 202
  steel, 223

Zentrale Arcor, Eschborn, **238**